Fundamentals of

METAL FORMING

Fundamentals of METAL FORMING

Robert H. Wagoner
The Ohio State University

Jean-Loup Chenot
École Nationale Superieure des Mines de Paris

John Wiley & Sons, Inc.

New York Chichester Brisbane Toronto Singapore

ACQUISITIONS EDITOR Cliff Robichaud
MARKETING MANAGER Debra Riegert
PRODUCTION EDITOR Ken Santor
MANUFACTURING COORDINATOR Dorothy Sinclair
ILLUSTRATION COORDINATOR Rosa Bryant

This book was set in 10/12 Palatino by John Wiley & Sons and printed and bound by Hamilton Printing Company. The cover was printed by New England Book Components, Inc.

Recognizing the importance of preserving what has been written, it is a policy of John Wiley & Sons, Inc. to have books of enduring value published in the United States printed on acid-free paper, and we exert our best efforts to that end.

The paper on this book was manufactured by a mill whose forest management programs include sustained yield harvesting of its timberlands. Sustained yield harvesting principles ensure that the number of trees cut each year does not exceed the amount of new growth.

***Library of Congress Cataloging in Publication Data*:**

Wagoner, R. H. (Robert H.)
Fundamentals of metal forming / Robert H. Wagoner, Jean-Loup Chenot.
p. cm.
Includes index.
ISBN 0-471-57004-4 (cloth: alk. paper)
1. Metals—Mechanical properties. 2. Metal-work. I. Chenot, J.
L. II. Title
TA460.W318 1996
671—dc20

Printed in the United States of America

10 9 8 7 6 5 4 3 2 1 95-47898

CIP

Preface

We intend this book as preparation for senior undergraduate or first-year graduate students in the fundamental knowledge (mechanics, materials, basic numerical methods) needed to understand the analysis of metal-forming operations. It spans the considerable gap between traditional materials approaches (centering on material properties and structure) and purely mechanical ones (centering on force balances, and numerical methods). While seeking simplicity and comprehension, we do not avoid complex issues or mathematics (such as tensors) needed as preparation for further study leading to high-level modeling capability. In view of the progress in numerical methods and computation, these techniques are now so widespread as to be essential for many applications in the mechanics of materials.

The breadth of the intended audience (four distinct groups depending on educational discipline and rank) requires that various parts of the book be emphasized for a given course. For more than ten years, the first author has taught a senior undergraduate course in metal forming to metallurgical engineering students, in which certain parts of this book are used thoroughly while other parts are omitted. For example, such a course may consist of the following:

Senior-level course on Metal Forming
(For materials or mechanical engineering students)

Chapter 1	Complete
Chapter 2-4	Very briefly, little tensor manipulation
Chapter 7	von Mises and Hill's normal quadratic yield
Chapter 9	Complete
Chapter 10	Complete

The first author has also used the material in this book to teach a first-year graduate course to materials and mechanical engineering students on the subjects of elasticity and plasticity, where the emphasis has been different:

Graduate course on Elasticity and Plasticity
(For materials or mechanical engineering students)

Chapter 1	Second part, analysis of tensile test
Chapter 2-4	Covered moderately, very brief on large-strain measures
Chapter 5	Very brief, to introduce work methods (more for ME)
Chapter 6	Complete
Chapter 7	Complete
Chapter 8	Complete (less for ME)

Colleagues have used early versions of the material in this book to teach graduate courses on the subject of continuum mechanics to engineering mechanics students, with good results.

Graduate course in Continuum Mechanics

(For engineering mechanics or mechanical engineering students)

Chapter 2	Complete, brief review
Chapter 3	Complete
Chapter 4	Complete
Chapter 5	Complete
Chapter 6	Continuum aspects, omit crystal symmetry
Chapter 7	von Mises isotropic plasticity

The second author has taught the techniques and material in this book in a variety of settings, including: undergraduate courses in several French engineering schools, one-week training courses for engineers in industry, and graduate courses in the École des Mines and French universities.

The background of the students and the disciplines in which these courses were offered included continuum mechanics, plasticity and viscoplasticity, and general numerical methods and finite elements.

We are convinced that students learn by doing, not by reading, and thus have included numerous problems in each chapter, ranging in difficulty from using equations and derivations presented in the text (*Proficiency Problems*) to ones requiring creative and original thought (*Depth Problems*). Solutions for all of these problems are available in a Solutions Manual for this volume. We have added a third class of problems, *Numerical Problems*, requiring students to write and use their own computer programs. We believe that this approach, while frequently slow and painful, encourages greater understanding and mastery of the material.

This book is limited to a presentation of the fundamental mechanics, materials, and basic numerical concepts used in metal forming analysis. An advanced treatment that starts from this basis and develops the finite element method and illustrates its use in actual analysis, is nearing completion. Therefore, concepts particular to finite element analysis and complex forming application have been deferred to that volume.

Robert H.Wagoner, *Columbus, Ohio*
Jean-Loup Chenot, *Sophia-Antipolis, France*

February 1996

To our wives.

Acknowledgements

This book was made possible by the selfless contributions of close friends upon whom we imposed to read, write, edit, and criticize. Robyn K. Wagoner, without an iota of technical training, typed much of this manuscript and made nearly all of the many changes required in figures, equations, and text. Robert H. Wagoner, Sr., also without technical training, read and edited every chapter several times, finding all of the inevitable misreferenced equations and figures, improper English, and many subtle errors that occur when technical people from two countries collaborate. Donna Smith copied, packaged, and mailed innumerable versions of manuscript and proofs. Beatrice Chenot and Robyn Wagoner gave up weekends with their husbands over a period of years in order to make this book possible.

Chapter 8 was contributed by Peter M. Anderson, of The Ohio State University. Many of the original figures were drafted by Weili Wang, wife of Dajun Zhou, who provided criticism for the technical content, especially in the area of numerical methods. Sriram Sadagopan contributed text and figures for several of the friction tests mentioned in Chapter 9. J. K. Lee tested and evaluated much of the material in this book by teaching continuum mechanics to graduate students using it. He also solved many of the problems in Chapters 2 through 7. Yeong-Sung Suh solved in parallel many of the problems in the first four chapters.

Without these many services, provided without complaint or compensation, this book would not exist, or it would have been even later, by years, than it is now.

Robert H.Wagoner, *Columbus, Ohio*
Jean-Loup Chenot, *Sophia-Antipolis, France*

February 1996

Contents

A note about notation

There are nearly as many kinds of notations for the various symbols representing mathematical quantities as there are authors in the field. We have attempted to follow a pattern that is intuitively consistent without striving for complete mathematical precision in each and every usage. In fact, no set of symbols, no matter how carefully defined, can replace understanding of the physical concepts and the underlying meaning of the equations and operations. We shall not attempt to cover every exception or eventuality in this note, but rather hope to provide a guide to our concept of the notation. In the first few chapters, we will provide examples of alternate notation while in later ones we will use the most convenient form.

We will not refer to curvilinear coordinates. Thus, all indices refer to physical axes or components and are placed in the subscript position. Also, vectors and tensors refer only to physically real quantities, rather than the generalized usage commonly found in the finite element community. Here are examples of the notation guidelines we follow.

SCALARS (a, A, α, t, T, a_1, a_{12}...)

Scalar quantities are physical quantities represented by a simple number, upper case or lower case and in any style (*italics*, Latin, Greek, etc.).

VECTORS (**a**, **b**, **c**...)

Physical vectors (as opposed to generalized 1-D arrays of numbers) will be represented by lower case bold letters*, with several subsidiary techniques for emphasizing the components rather than the entire vector quantity:

$$\mathbf{a} = a_1\hat{e}_1 + a_2\hat{e}_2 + a_3\hat{e}_3 = |\mathbf{a}|\hat{g} \leftrightarrow a_1, a_2, a_3 \leftrightarrow \begin{bmatrix} a_1 \\ a_2 \\ a_3 \end{bmatrix} = [a_i] = [a] \leftrightarrow a_i$$

The vector **a** is shown as the sum of three scalar components multiplied by three base vectors ($\hat{\mathbf{e}}$). Since base vectors are the same as any other vector except for their intended use, any lower-case bold letter can represent them. We use $\hat{\mathbf{e}}$ specifically to emphasize the use of the vectors to represent other vectors (i.e. as base vectors). (In some cases, we will use $\hat{\mathbf{x}}_1, \hat{\mathbf{x}}_2, \hat{\mathbf{x}}_3$ to emphasize that the base vectors form a Cartesian set.) The superposed carat emphasizes that the vector is of unit length. The third representation emphasizes that the vector **a** can be decomposed into a magnitude $|\mathbf{a}|$ (or sometimes simply "a") and a direction represented by a unit vector $\hat{\mathbf{a}}$.

* We will on occasion use capital letters for vectors or their components to indicate some special quality or connection with another vector. An example is the expression of a given vector's components in an alternate coordinate system, or the same material vector at different times.

The symbol "$\leftrightarrow$" means that we will treat the other notations as equivalent even though they are not precisely equal. The right-hand side therefore emphasizes the components of **a** while ignoring the base vectors, which presumably have been defined elsewhere and are unchanged throughout the problem of interest. The brackets indicate an emphasis on all components, either in matrix or indicial form. (Note that the vector nature of **a** is not shown on the bracketed form, since the components need not apply to a physical vector.) The second "$\leftrightarrow$" indicates that a_i can be used to indicate a single component of **a** (the ith one), or in the general sense, all of the components (i.e. for i=1,2,3 for 3-D space).

We will occasionally use a similar notation to denote numerical "vectors" such as those encountered in F.E.M. and other applications. In these cases, the range on the subscripts will be determined by the problem itself, rather than being limited to 3, as for usual 3-D space.

A few other vector notations are useful:

$\times$ indicates the vector cross product
. indicates the vector or tensor dot product

TENSORS OF RANK GREATER THAN 1 (**A**, **B**, **C**...)

Our tensor notation follows precisely the vector usage except we shall attempt to use upper case letters to represent tensors of rank greater than one. (Exceptions will apply to tensors which have almost ubiquitous usage, such as σ for the stress tensor or ε for the infinitesimal strain tensor.) Therefore, our tensor notation closely follows our vector notation:

$$\mathbf{A} \leftrightarrow \begin{bmatrix} A_{11} A_{12} A_{13} \\ A_{21} A_{22} A_{23} \\ A_{31} A_{32} A_{33} \end{bmatrix} = \left[A_{ij}\right] = [A] \leftrightarrow A_{ij}$$

MATRICES ([A], [a], [a],...)

Our matrix notation follows the bracketed forms introduced for physical vectors and tensors, without regard to whether the components represent physical quantities:

$$\begin{bmatrix} A_{11} A_{12} A_{13} \\ A_{21} A_{22} A_{23} \\ A_{31} A_{32} A_{33} \end{bmatrix} = \left[A_{ij}\right] = [A] \leftrightarrow A_{ij}$$

As before, the right-most form can be interpreted as a single component (the $i^{th}j^{th}$ one) or as all of the components ($i = 1,2,3; j = 1,2,3$ for general 3-D space). The bracketed forms emphasize all of the components, in the order shown by the first matrix.

We will tend to use lower case letters for column matrices, which correspond closely to vectors, but this may not always be possible while preserving clarity. Several other common usages for matrices:

$$[A]^T \leftrightarrow \text{transpose of}[A]$$

$$[A]^{-1} \leftrightarrow \text{inverse of}[A]$$

$$|A| \leftrightarrow \text{determinant of}[A]$$

$$[A][B][C] \leftrightarrow \text{matrix multiplication}$$

$$[I] \leftrightarrow \text{identity matrix}$$

Each form can also include subscript indices to emphasize the components.

INDICIAL FORM (Aij, $B_{ij,}$ C_{ij}...)

The indicial forms of vectors, tensors, and matrices have been introduced above by the loose equivalence with other notations. Thus, for example, the multiplication of matrices [A] and [B] can be represented by $A_{ij}B_{jk} = C_{ik}$, which can be interpreted for any one component (the $i^{th}k^{th}$ one), or for all components (where i = 1,2,3; k = 1,2,3 for 3-D space, for example). However, the subscript j is a dummy index because we follow Einstein's summation convention where any repeated index is summed over to obtain the result. [i and k, conversely, are free indices because we are free to choose their values].

A few symbols are useful for representing matrix and tensor operations:

δ_{ij} = Kronecker delta, with the property that $\delta_{ij} = \begin{Bmatrix} 0 \text{ if } i \neq j \\ 1 \text{ if } i \neq j \end{Bmatrix}$

ε_{ijk} = Permutation operator,

with the property that $\varepsilon_{ijk} = \begin{Bmatrix} 0 & \text{if} & i & = j; i = k; j = k \\ 1 & \text{if} & ijk & = 1,2,3; 2,3,1; \text{or } 3,1,2 \\ -1 & \text{if} & ijk & = 3,2,1; 1,2,3; \text{or } 2,1,3 \end{Bmatrix}$

FUNCTIONS AND OPERATORS (L,F,K...)

We will denote functions and operators with script symbols without regard to the domain or range of the operation. These spaces will be clarified when the operator is first introduced. For example, $\mathbf{x} = \mathbf{L}(\mathbf{A})$ indicates that **L** is an operator which operates on a tensor and results in a vector. 2nd ranked tensors are often defined, for example, as linear vector operators [i.e. a linear operator that operates on a vector and has a vector result].

CHAPTER

1

The Tensile Test and Basic Material Behavior

Fundamental and practical studies of metal mechanical behavior usually originate with the **uniaxial tension** test. Apparently simple and one-dimensional, a great deal of hidden information may be obtained by careful observation and measurements of the tensile test. On the other hand, the underlying physical complexity means that interpretation must be quite careful (along with the procedure followed to conduct the test) if meaningful results are to be realized.

In this chapter, we present a range of concepts at the simplest, most intuitive level. Many of these concepts will be expounded and presented in more mathematical form in subsequent chapters. This approach has the advantage of introducing basic mathematical and numerical ideas without the formalisms that are a source of confusion to many students when first encountered. However, the downside is that some of the introductions and treatments in this chapter are not precise and cannot be extended to more general situations without additional information.

Structurally, this chapter is divided into two parts: the first introduces the tensile test and the standard measures of material response. In the second section, we introduce all of the techniques necessary for analyzing a nonuniform tensile test. This analysis is not of great research interest because this 1-D form omits important aspects of the physical problem that can now be simulated using more powerful numerical methods. By omitting multi-axial complexity, however, we lose little of qualitative importance, and are able to illustrate the use of several numerical methods that will be needed in subsequent chapters, especially Chapter 10.

1.1 TENSILE-TEST GEOMETRY

Standard tensile test analysis is based on an ideal view of the physical problem. A long, thin rod is subjected to an extension (usually at a constant extension rate) and the corresponding load is measured.[1] The basic assumptions are that the loading is purely axial and the deformation takes place uniformly, both along the length of the specimen and throughout the cross-section. Under these conditions, it is sufficient to measure just two macroscopic quantities for much of the desired information: extension and load.

Two kinds of tensile specimens are used for standard tests: a round bar for bulk material (plates, beams, etc.), and a flat specimen for sheet products. Each is subject to ASTM specifications and has a nominal **gage length** of 2 inches. The gage length refers to the distance between ends of an extension gage put on the specimen to measure extension between these points. The **deforming length** is the length of the specimen that undergoes plastic deformation during the test. This length may change but should always be significantly longer than the gage length in order to ensure that the stress state is uniaxial and deformation is quasi-uniform over the gage length. Figure 1.1 shows the general geometry.

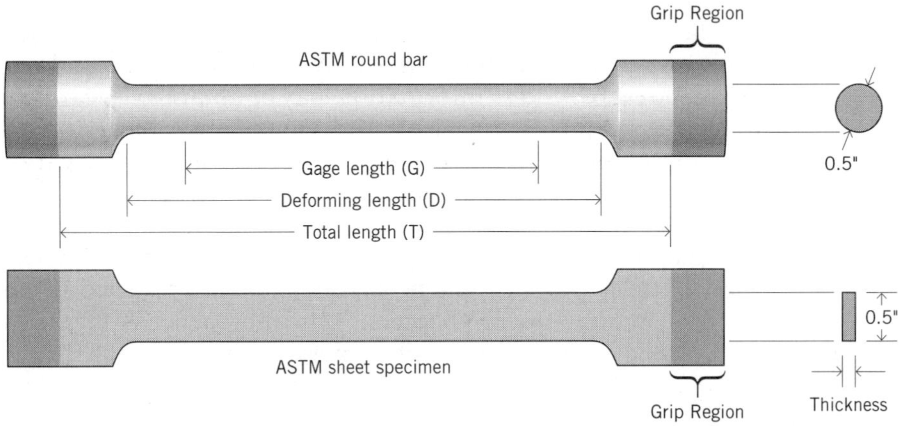

Figure 1.1 Standard tensile specimen shapes.

1.2 MEASURED VARIABLES

A standard tensile test is carried out by moving one end of the specimen (via a machine *crosshead*) at a constant speed, v, while holding the other end fixed. The primary variables recorded are **load** (P) and **extension** (Δl). Note that the extension could be obtained by multiplying v times t (time). This is done in some cases, but usually the "lash" (looseness) in the system necessitates use of an extension gage for accuracy. A typical load-extension curve appears in Figure 1.2.

---

[1] Variations on the basic method include imposing a certain load and measuring the extension or extension rate, or imposing jumps in rates or loads. These tests will not be considered in detail here, although they are very useful for materials research and for high-temperature deformation investigations.

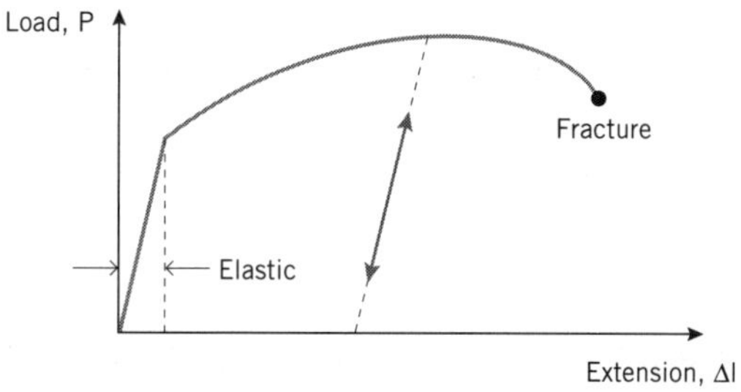

Figure 1.2 Typical response of a metal under uniaxial tension.

Note that the load-extension variables depend on specimen size. If, for example, the specimen were twice as large in each direction, the load would be four times as great, and the extension would be twice as great. Since we want to measure **material properties**, we normalize the measured variables to account for specimen size. The simplest way to do this is to normalize to the original specimen geometry.

1.3 ENGINEERING VARIABLES (NORMALIZED TO ORIGINAL SPECIMEN SIZE)

The variables may be defined as follows:

$\sigma_e =$ $P/A_o =$ **engineering stress**,

$e =$ $\Delta l/l_o =$ **engineering strain** (sometimes called **elongation**)
(Note: l_o is usually equated to the gage length, L, Fig. 1.1.)

where the following notation is used:

$A_o =$ initial cross-sectional area

$l_o =$ G = initial gage length

$\Delta l =$ $l - l_o =$ change in gage length, extension

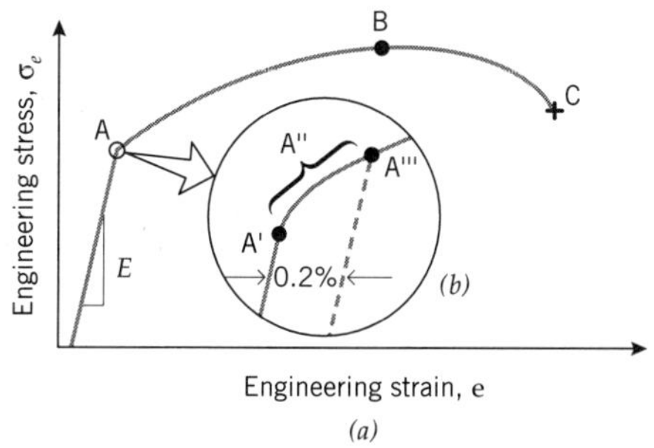

Figure 1.3 Points of special interest in a tensile test.

Engineering stress has units of force per area, and engineering strain is dimensionless (mm/mm, for example). Engineering strain is often presented as a percentage by multiplying by 100.

Other basic engineering quantities may be derived from the other ones as follows. The point at $A(e_y, \sigma_y)$ is of great interest and it may be used to define

E= **Young's modulus** = σ/e in the elastic region

σ_y= **yield point, yield stress, yield strength, elastic limit,** or **proportional limit.**

When finer resolution is available (Figure 1.3*b*), the apparent point *A* is seen as part of a smooth curve. This smoothness leads to ambiguity in defining *A*, so several other points are used:

$\sigma_{A'}$ = **proportional limit,** or **stress** at which the stress-strain curve ceases to be linear. (This is rather subjective and depends on the resolution and magnification of the curve.)

$\sigma_{A''}$ = **elastic limit** or **yield stress,** the minimum stress required to produce a permanent, plastic deformation. It can be found only by repeated loadings and unloadings. $\sigma_{A''}$ may be less than or greater than $\sigma_{A'}$ and $\sigma_{A'''}$.

$\sigma_{A'''}$ = **0.2% offset strength,** often called the **yield strength,** is obtained by drawing a line parallel to the elastic line but displaced by $\Delta e = 0.002$ (0.2%) and noting the intersection with the stress-strain curve.

The point $B(e_u, \sigma_{uts})$, at the maximum load sustainable by the specimen, defines

e_u = **uniform elongation** (elongation before necking begins)

σ_{uts} = **ultimate tensile strength**

The point C (e_t, σ_t) defines the limit of the tensile ductibilty or formability.

e_t = **total elongation**

e_{pu} = $(e_t - e_u)$ = **post-uniform elongation**

The **engineering-strain rate**, $\dot{e}$, is defined as de/dt, and is the rate at which strain increases. This quantity can be obtained simply by noting that all the strain takes place in the deforming length, D, so that the crosshead speed, v, is the same as the extension rate of D (See Fig. 1.1). That is,

$$\dot{e} = \frac{de}{dt} = \frac{dD/D_o}{dt} = \frac{dT/D_o}{dt} = \frac{v}{D_o} = \frac{\text{crosshead speed}}{\text{deforming length}} \tag{1.1}$$

The third equality is correct because the region outside of D is rigid; that is it does not deform, so that the velocity of all points outside of D is the same, and dD = dT (T = total length, Figure 1.1).

1.4 TRUE VARIABLES (NORMALIZED TO CURRENT CONFIGURATION)

Assume the original tensile test shown in Figure 1.4*a* is stopped at point X, and the specimen is unloaded to point Y. If the tensile test is then restarted, the dashed line will be followed approximately, and the specimen will behave as if no interruption occurred. If, instead, we remove the specimen and hand it to a new person to test, as in Figure 1.4*b*, the result will be quite different. The second person will measure the cross-sectional area and find a new number, A'_o, because the previous deformation reduced the width and thickness while increasing the length.

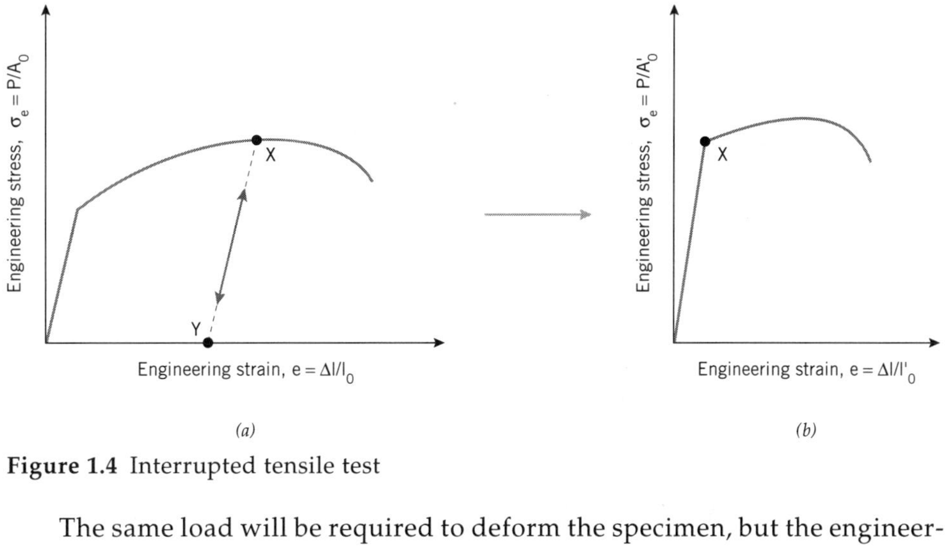

Figure 1.4 Interrupted tensile test

The same load will be required to deform the specimen, but the engineering stress will be different: $\sigma'_e = P/A'_o$. Obviously, if yield stress is to have a real material meaning, $\sigma_y \equiv \sigma'_y$, independent of who tests it. Similarly, a small extension at point X will produce different measured engineering strains for the same reason:

$$e_a = \frac{\Delta l}{l_o}, \quad e_b = \frac{\Delta l}{l'_o} \tag{1.2}$$

To take care of this problem, we introduce real or **"true" strain**, an increment of which refers to an infinitesimal extension per unit of *current* length. We limit ourselves to a small extension to insure that the current length is constant and well known. By assuming that the incremental strain over the current gage length is uniform, we can write mathematically that the true strain increment is $d\varepsilon = dl/l$ (not dl/l_o). We can express the total true strain as a simple integral:

$$\varepsilon = \int_{\varepsilon_t=o}^{\varepsilon_t} d\varepsilon = \int_{l_o}^{l} \frac{dl}{l} \Rightarrow \varepsilon = \ln\frac{l}{l_o} \tag{1.3}$$

Similarly, the real or true stress refers to the load divided by the *current* cross-sectional area:

$$\sigma_t = \frac{P}{A} \quad (\text{not } \frac{P}{A_o}) \tag{1.4}$$

Exactly analogous to the definition of engineering-strain rate, the **true-strain rate** is defined as $d\varepsilon/dt$. As in Section 1.3, this rate is simply related to the crosshead speed:

$$\dot{\varepsilon} = \frac{d\varepsilon}{dt} = \frac{dD/D}{dt} = \frac{dT/D}{dt} = \frac{v}{D} = \frac{\text{crosshead speed}}{\text{current deforming length}} \tag{1.5}$$

1.5 RELATIONSHIP AMONG TRUE AND ENGINEERING VARIABLES

Definitions	Engineering	True	
Strain	$e = \dfrac{l_f - l_o}{l_o}$	$\varepsilon = \ln l_f/l_o$	(1.6)
Stress	$\sigma_e = P/A_o$	$\sigma_t = P/A$	(1.7)

Equation 1.6 yields the relationship between the two strain representations with only simple manipulation:

$$e = \exp(\varepsilon) - 1 \qquad \varepsilon = \ln(1 + e) \tag{1.8}$$

Equation 1.7 cannot be solved simultaneously until a relationship between the original and current cross-sectional area (A_o, A) is known. A material assumption is required, namely that plastic deformation produces no net change in volume. This condition is called **plastic incompressibility** and is a very accurate assumption for metals and for most other liquids and dense solids.

Consider a small right parallelepiped of material in a tensile specimen, as shown in Figure 1.5.

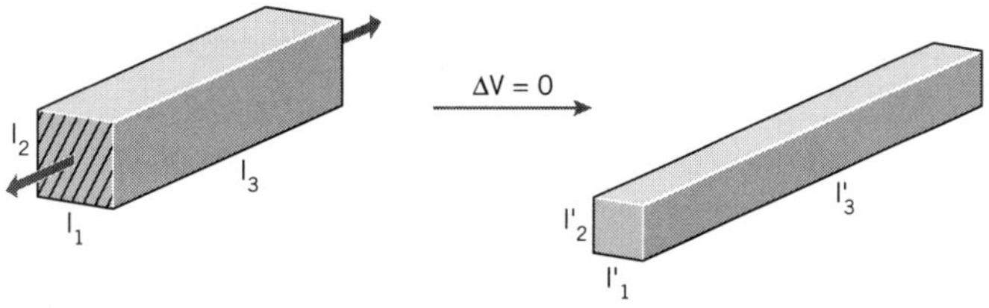

Figure 1.5 Dimensional changes of a volume element.

$$\Delta V = 0 = l_1' l_2' l_3' - l_1 l_2 l_3, \text{ so } \frac{l_1' l_2' l_3'}{l_1 l_2 l_3} = 1$$

For each direction we can substitute true or engineering strains for the corresponding dimensional changes using Equations 1.6 and 1.8:

$$\frac{l_i'}{l_i} = \exp(\varepsilon_i) = 1 + e_i, \text{ such that} \tag{1.9}$$

$$\exp(\varepsilon_1)\ \exp(\varepsilon_2)\ \exp(\varepsilon_3) = 1, \tag{1.10a}$$

or, by taking logarithms of both sides,

$$\varepsilon_1 + \varepsilon_2 + \varepsilon_3 = 0 \tag{1.10b}$$

$$(1 + e_1)(1 + e_2)(1 + e_3) = 1 \tag{1.10c}$$

Consider now the cross-sectional area:

$$l_1 l_2 = A_o \quad l_1' l_2' = A$$
$$\Delta V = 0 = A l_3' - A_o l_3 \ (l_3 = \text{axial length})$$

Solving this, we obtain a relationship for the current cross-sectional area in terms of the original one and the strain:

$$\frac{A_o}{A} = \frac{l_3'}{l_3} = \exp(\varepsilon) \text{ or } A = A_o \exp(-\varepsilon) \tag{1.11}$$

We can now solve for the relationship between σ_e and σ by substituting Equation 1.11 into Equation 1.7:

$$\sigma_e = \frac{P}{A_o} \qquad \sigma = \frac{P}{A_o}\exp(\varepsilon),$$

from which we can obtain the following relationships in terms of the strains:

$$\begin{aligned} \sigma_e &= \sigma \exp(-\varepsilon) \\ \sigma &= \sigma_e \exp(\varepsilon) \end{aligned} \qquad (1.12)$$

or the equivalent relationships in terms of engineering strains:

$$\begin{aligned} \sigma_e &= \frac{\sigma}{(1+e)} \\ \sigma &= \sigma_e(1+e) \end{aligned} \qquad (1.13)$$

1.6 ANALYSIS OF WORK HARDENING

J. H. Hollomon[2] discovered in 1945 that many engineering alloys, particularly ferrous alloys, obey a simple true stress-strain relationship in the plastic regime. His equation states that:

$$\sigma = k\,\varepsilon^n \text{ (Hollomon equation)} \qquad (1.14)$$

where k and n are constants known as the **strength coefficient** and **work-hardening rate**, or **work-hardening exponent**, respectively.

The constants k and n are determined from the true stress-strain curve by taking logarithms of both sides of Equation 1.13:

$$\begin{aligned} (\ln \sigma) &= n(\ln \varepsilon) + \ln k \\ (Y) &= A(X) + B \end{aligned} \qquad (1.15)$$

Note that Equation 1.15 is the equation of a line whose slope is n and whose y-intercept is (ln k). By taking logarithms of experimental true stress-strain pairs and plotting as a straight line, k and n can be obtained. If the logarithmic plot is not linear, it shows that the material does not truly obey Equation 1.14, and n is

---

[2] J. H. Hollomon, *A.I.M.E. Trans.* 162 (1945): 268 O. Hoffman and G. Sachs, *Introduction to the Theory of Plasticity for Engineers* (New York, McGraw-Hill, 1953),P. 45.

not a constant. In this case, the work-hardening rate has meaning only as a function of strain and is defined as

$$n(\varepsilon) = \frac{d(\ln \sigma)}{d(\ln \varepsilon)}$$

It is often useful to think of n at a grain strain, even though Equation 1.14 is not obeyed.

1.7 NECKING, UNIFORM ELONGATION

Necking is the **localization of strain** that occurs near the end of a tensile test. Once it has begun, none of the equations we have developed is applicable because the strains and stresses are no longer uniform over our length of measurement, the gage length. That is, the current deforming length D is less than our gage length, G. It is therefore necessary to know the limit of **uniform elongation**. This limit is also a measure of **formability,** because it approximates the end of the tensile test for most metals at room temperature.

The most widely used estimate of the uniform elongation is attributed to Considere[3]. The onset of necking occurs at the point of maximum load (or engineering stress) according to this view. Note that the necking phenomenon is a result of the competition between work hardening (σ increases with increasing ε[4]) and the reduction of cross-sectional area because of continuing extension. At the start of a tensile test, the strain hardening dominates the geometric softening and the load increases. Eventually, the reduction of cross-section dominates and the load decreases. These effects are just balanced when dP=0.

The maximum load point is readily available on load-elongation curves or engineering stress-strain curves, but also may be found purely by knowing the true work-hardening law. For a Hollomon material, we can use Equations 1.14 and 1.11 to find the load:

$$P = k\,\varepsilon^n A_o \exp(-\varepsilon), \qquad \text{or} \qquad \ln P = \ln k + n \ln \varepsilon + \ln A_o - \varepsilon.$$

We can now set the slope to zero to obtain the desired expressions:

$$0 = \frac{d(\ln P)}{d\varepsilon} = \frac{n}{\varepsilon} - 1 \Rightarrow \quad n = \varepsilon \text{ at uniform elongation} \qquad (1.16)$$

This surprising result, known as the **Considere criterion**, says that the onset of necking and the end of uniform elongation occur when the true work-hardening rate (expressed as $d \ln \sigma / d \ln \varepsilon$) exactly equals the true strain. Equation 1.16 is valid even if the Hollomon equation (Equation 1.14) is not obeyed,

---

[3] A. Considere: *Ann. Ponts Chaussees* 9 (1885): 574-775.

[4] Nearly all metals exhibit work hardening or strain hardening, up to very large strains. Other materials or some metals under special circumstances can strain soften, but we ignore these cases here.

but n must be interpreted as the quantity (d ln σ/d ln ε), a quantity that may vary with strain. In fact, the general form of Considere's criterion states that

$$\frac{d\sigma}{d\varepsilon} = \sigma \quad \text{or} \quad \frac{d(\ln \sigma)}{d\,(\ln \varepsilon)} = \varepsilon \tag{1.16a}$$

at the onset of instability (at the limit of uniform elongation).

Note that dP = 0 is not in general the proper condition for plastic instability, although this condition yields an identical result for rate-sensitive materials (See Exercise 1.1). For this reason, and because of the simplicity of the derivation based on dp = 0, most standard texts present the derivation as shown above.

1.8 STRAIN-RATE SENSITIVITY

Suppose that we perform two tensile tests on identical specimens but at two different crosshead velocities, v_1 and v_2, where $v_2 > v_1$. Many materials are more difficult to deform at higher rates, so that the superimposed tensile tests will look like the one in Figure 1.6.

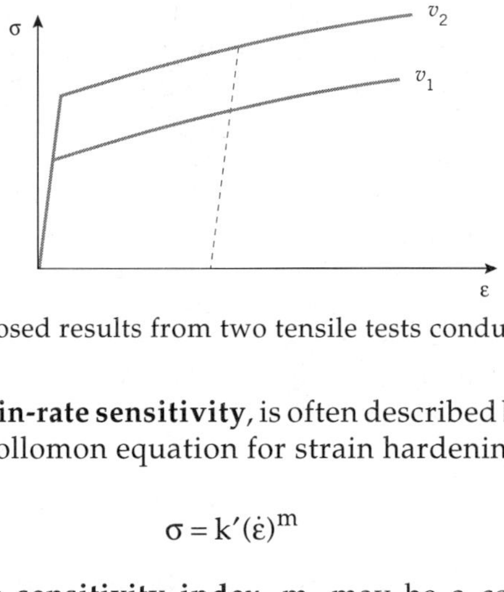

Figure 1.6 Superimposed results from two tensile tests conducted at different rates.

This effect, **strain-rate sensitivity**, is often described by a power law exactly analogous to the Hollomon equation for strain hardening:

$$\sigma = k'(\dot{\varepsilon})^m \tag{1.17}$$

The **strain-rate sensitivity index**, m, may be a constant as implied by Equation 1.17, or may depend on strain rate (and strain). In a general treatment, *m* may be defined as $d\ln\sigma/d\ln\dot{\varepsilon}$.

In order to analyze tensile data to obtain m values, we examine Equation 1.17 for two true strain rates (corresponding to two crosshead velocities):

$$\sigma_2 = \sigma\,(\text{at } v_2) = k'\,(\dot{\varepsilon}_2)^m, \quad \sigma_1 = \sigma\,(\text{at } v_1) = k'\,(\dot{\varepsilon}_1)^m$$

which can be arranged to yield

$$\frac{\sigma_2}{\sigma_1} = \left(\frac{\dot{\varepsilon}_2}{\dot{\varepsilon}_1}\right)^m \Rightarrow \ln\frac{\sigma_2}{\sigma_1} = m\ln\frac{\dot{\varepsilon}_2}{\dot{\varepsilon}_1} \Rightarrow m = \frac{\ln(\sigma_2/\sigma_1)}{\ln(\dot{\varepsilon}_2/\dot{\varepsilon}_1)} \tag{1.18}$$

Equation 1.18 shows how to obtain m from two tensile tests conducted at different extension rates. Since only the ratio of the true strain rates is required, the ratio of known crosshead velocities will suffice: $v_2/v_1 = \dot{\varepsilon}_2/\dot{\varepsilon}_1$, provided $D_1 = D_2$. The same is true for the stresses:

$$\frac{\sigma_2}{\sigma_1} = \frac{\sigma_{e2}}{\sigma_{e1}} = \frac{P_2}{P_1}$$

The strain-rate-sensitivity index may therefore be obtained very simply, without reference to stresses or strains:

$$m = \frac{\ln(\sigma_2/\sigma_1)}{\ln(\dot{\varepsilon}_2/\dot{\varepsilon}_1)} = \frac{\ln(\sigma_{e2}/\sigma_{e1})}{\ln(\dot{\varepsilon}_2/\dot{\varepsilon}_1)} = \frac{\ln(P_2/P_1)}{\ln(v_2/v_1)} \tag{1.19}$$

1.9 PHYSICAL SIGNIFICANCE OF n AND m

As a simple rule, the work-hardening rate affects the stress-strain curve primarily up to the uniform strain, and the strain-rate-sensitivity index affects behavior primarily in the *post-uniform* or necking region. Increasing n and m increases the total strain to failure and therefore increases the formability of the material.

Increasing n from 0.3 to 0.4, for example, increases the uniform true strain by 0.1. A similar increase occurs in the post-uniform region when m is increased by a much smaller amount, say from 0.010 to 0.030.

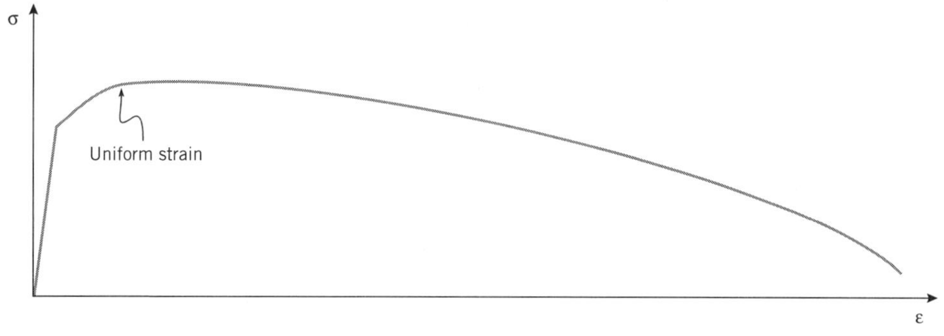

Figure 1.7 Tensile test of a superplastic material.

A special class of materials, called **superplastic** alloys, have very high m values, in the range of 0.2 to 0.4. The post-uniform elongation in these materials is enormous, up to *thousands* of percent. These alloys generally have low n and are not very strong, but are used for forming very complex shapes that would normally need to be constructed of many parts and then joined. A typical superplastic engineering stress-strain would look like the one shown in Figure 1.7.

Exercise 1.1 **Derive the plastic instability point (limit of uniform elongation) for a material with power-law hardening and strain-rate sensitivity:**

$$\sigma = k\,\varepsilon^n\,\dot{\varepsilon}^m$$

Consider two separate criteria: Consider (dP = 0) and an alternative approach introduced by Hart. In the Hart approach, one introduces a small variation in cross sectional area in a small region along the specimen length and observes whether this perturbation shrinks or grows. Which of the criteria is more realistic physically?

For either criterion, we need to relate specimen length, area, strain and their corresponding strain rates:

$$\varepsilon = \ln\frac{l}{l_o} = -\ln\frac{A}{A_o}, \text{ or } d\varepsilon = \frac{dl}{l} = \frac{-dA}{A} \tag{1.1-1}$$

$$\dot{\varepsilon} = \frac{d\varepsilon}{dt} = \frac{\dot{l}}{l} = -\frac{\dot{A}}{A} \tag{1.1-2}$$

$$\ddot{\varepsilon} = \frac{\ddot{l}}{l} - \left(\frac{\dot{l}}{l}\right)^2 = -\frac{\ddot{A}}{A} + \left(\frac{\dot{A}}{A}\right)^2 \tag{1.1-3}$$

but, for a constant velocity tensile test, $\ddot{l} = 0$, so $\ddot{\varepsilon} = \left(\frac{\dot{l}}{l}\right)^2 = \left(\frac{\dot{A}}{A}\right)^2 = -\dot{\varepsilon}^2$. Then, we write in a general way:

$$\sigma = \sigma\,(\varepsilon, \dot{\varepsilon}) \tag{1.1-4}$$

$$\dot{\sigma} = \left(\frac{\partial\sigma}{\partial\varepsilon}\right)_{\dot{\varepsilon}} \dot{\varepsilon} + \left(\frac{\partial\sigma}{\partial\dot{\varepsilon}}\right)_{\varepsilon} \ddot{\varepsilon} = \left(\frac{\partial\sigma}{\partial\varepsilon}\right)_{\dot{\varepsilon}} \dot{\varepsilon} - \left(\frac{\partial\sigma}{\partial\dot{\varepsilon}}\right)_{\varepsilon} (\dot{\varepsilon})^2 \tag{1.1-5}$$

where the quantities in parentheses are material properties at an instant in time, whether the function in Eq. 1.1-4 is known explicitly or not.

For the Considere Criterion, we find the point at which $\dot{P} = 0$:

$$0 = \dot{P} = \sigma\dot{A} + A\dot{\sigma} \Rightarrow \frac{\dot{\sigma}}{\sigma} = \frac{-\dot{A}}{A}, \text{ or } \frac{\dot{\sigma}}{\sigma} = \dot{\varepsilon} \tag{1.1-6}$$

Eq. 1.1-6 is the same as Eq. 1.16a, except that the differentials have been replaced by time derivatives (with no change of meaning). For a rate-sensitive material, we then substitute Eq. 1.1-5 for $\dot{\sigma}$:

$$\frac{\dot{\sigma}}{\sigma} = \left(\frac{\partial \ln \sigma}{\partial \varepsilon}\right)_{\dot{\varepsilon}} \dot{\varepsilon} - \left(\frac{\partial \ln \sigma}{\partial \dot{\varepsilon}}\right)_{\varepsilon} \dot{\varepsilon}^2 = \dot{\varepsilon} \tag{1.1-7}$$

where we have used the standard relationship $\frac{dX}{X} = d\left(\ln X\right)$. Rearranging yields the Considere condition (i.e. where $\dot{P} = 0$) for a strain-rate sensitive material:

$$\left(\frac{\partial \ln \sigma}{\partial \varepsilon}\right)_{\dot{\varepsilon}} - \left(\frac{\partial \ln \sigma}{\partial \dot{\varepsilon}}\right)_{\varepsilon} = 1 \quad \text{(Considere Criterion)} \tag{1.1-8}$$

For the particular case of a power-law material (for strain and strain-rate hardening), $n = \left(\frac{\partial \ln \sigma}{\partial \ln \varepsilon}\right)_{\dot{\varepsilon}}$, and $m = \left(\frac{\partial \ln \sigma}{\partial \ln \dot{\varepsilon}}\right)_{\varepsilon}$ so

$$\frac{n}{\varepsilon} - m = 1, \quad \text{or} \quad \varepsilon_{\text{Considere}} = \frac{n}{1+m} \tag{1.1-9}$$

For the second criterion, we imagine that in a small region along the specimen length, a fluctuation in cross-section is introduced, ΔA , with corresponding fluctuations of $\Delta \dot{A}$ and Δl. In order to maintain equilibrium, the load transmitted throughout the specimen length is equal in the uniform area and small perturbed region:

$$P = \sigma A = (\sigma + \Delta\sigma)(A + \Delta A), \text{ or} \tag{1.1-10}$$

$$0 = \sigma \Delta\sigma + A \Delta\sigma + \Delta\sigma \Delta A. \tag{1.1-11}$$

For an infinitesimally small variation (i.e. $\Delta A \rightarrow dA$), the third term is second-order such that we obtain:

$$\sigma dA + A d\sigma = 0, \quad \text{or} \quad \frac{d\sigma}{\sigma} = -\frac{dA}{A} = d\varepsilon \tag{1.1-12}$$

We can rewrite $d\sigma$ in terms of de and $d\dot{\varepsilon}$:

$$d\sigma = \left(\frac{\partial\sigma}{\partial\varepsilon}\right)_{\dot{\varepsilon}} d\varepsilon + \left(\frac{\partial\sigma}{\partial\dot{\varepsilon}}\right)_{\varepsilon} d\dot{\varepsilon} \tag{1.1-13}$$

where de and $d\dot{\varepsilon}$ refer to the difference in strain increment and strain rate increment between the section with area fluctuation and the remainder of the specimen.

The corresponding differences in ε and $\dot{\varepsilon}$ may be found directly:

$$d\varepsilon = \frac{dl}{l} = -\frac{dA}{A} \tag{1.1-14}$$

$$\dot{\varepsilon} = \frac{\dot{l}}{l} = -\frac{\dot{A}}{A}, \quad \text{so} \quad d\dot{\varepsilon} = \frac{d\dot{l}}{l} - \frac{\dot{l}}{l^2} dl = -\frac{d\dot{A}}{A} + \frac{\dot{A}}{A^2} dA \tag{1.1-15}$$

With these relationships, we can rewrite the stress difference:

$$d\sigma = -\left(\frac{\partial\sigma}{\partial\varepsilon}\right)_{\dot{\varepsilon}} \frac{dA}{A} + \left(\frac{\partial\sigma}{\partial\dot{\varepsilon}}\right)_{\varepsilon} \left(-\frac{d\dot{A}}{A} + \frac{\dot{A}}{A^2} dA\right) \tag{1.1-16}$$

Eq. 1.1-16 may be substituted into Eq. 1.1-12 to obtain:

$$\frac{d\sigma}{\sigma} = \frac{dA}{A} = -\left(\frac{\partial \ln\sigma}{\partial\varepsilon}\right)_{\dot{\varepsilon}} \frac{dA}{A} + \left(\frac{\partial \ln\sigma}{\partial\dot{\varepsilon}}\right)_{\varepsilon} \left(-\frac{d\dot{A}}{A} + \frac{\dot{A}}{A^2} dA\right) \tag{1.1-17}$$

where the substitutions $\frac{1}{\sigma}\left(\frac{\partial\sigma}{\partial\varepsilon}\right)_{\dot{\varepsilon}} = \left(\frac{\partial \ln\sigma}{\partial\varepsilon}\right)_{\dot{\varepsilon}}$, and $\frac{1}{\sigma}\left(\frac{\partial\sigma}{\partial\dot{\varepsilon}}\right)_{\varepsilon} = \left(\frac{\partial \ln\sigma}{\partial\dot{\varepsilon}}\right)_{\varepsilon}$ have been made. Eq. 1.1-17 is then solved to obtain an expression for $\frac{d\dot{A}}{A}$:

$$\frac{d\dot{A}}{dA} = \frac{1 - \left(\frac{\partial \ln\sigma}{\partial\varepsilon}\right)_{\dot{\varepsilon}} - \left(\frac{\partial \ln\sigma}{\partial\dot{\varepsilon}}\right)_{\varepsilon} \dot{\varepsilon}}{\left(\frac{\partial \ln\sigma}{\partial \ln\dot{\varepsilon}}\right)_{\varepsilon}}, \tag{1.1-18}$$

where Eq. 1.1-2 has been used to obtain $\dot{\varepsilon}$ in place of $-\frac{\dot{A}}{A}$ on the right-hand side.

The criterion may be completed by physical reasoning. If a fluctuation of dA (negative) leads to a larger difference in strain rate ($d\dot{A}$ negative), then the plastic deformation is unstable to small fluctuations along the specimen length and necking proceeds. If, on the other hand, a negative dA produces a tendency to reduce the area difference ($d\dot{A}$ positive), then deformation is stable to such small fluctuations and any **incipient neck** will die out. The critical condition between these two regimes, or the **bifurcation point**, is reached when $d\dot{A} = 0$. Thus, the condition reached at the point when necking begins (plastic instability point) is

$$\frac{d\dot{A}}{dA} = 0 = 1 - \left(\frac{\partial \ln \sigma}{\partial \varepsilon}\right)_{\dot{\varepsilon}} - \left(\frac{\partial \ln \sigma}{\partial \ln \dot{\varepsilon}}\right)_{\varepsilon} \tag{1.1-19}$$

For the particular case of a power-law material (for strain and strain-rate hardening), $n = \left(\frac{\partial \ln \sigma}{\partial \ln \varepsilon}\right)_{\dot{\varepsilon}}$ = a constant, and $m = \left(\frac{\partial \ln \sigma}{\partial \ln \dot{\varepsilon}}\right)_{\varepsilon}$ = a constant, and Eq. 1.1-19 becomes:

$$0 = 1 - \frac{n}{\varepsilon} - m, \quad \text{or} \quad \varepsilon_u = \frac{n}{1-m} \tag{1.1-20}$$

(Onset of necking for a power-law, rate-sensitive material)

Compare Eq. 1.1-20 with Eq. 1.1-9, which is based on the simple but physically-irrelevant condition dP = 0. Eq. 1.1-20 shows that deformation is stabilized for a rate-sensitive material, in agreement with experimental evidence. Note, also, that Eq. 1.1-20 and Eq. 1.1-9 reduce to the standard result for a rate-insensitive material (for which m=0):

$$\varepsilon_u = n \tag{1.1-21}$$

(Onset of necking for a rate-insensitive power-law material)

As a final note, Eq. 1.1-20 is valid for any material for which the flow stress depends on strain and strain rate as long as the quantities n and m are interpreted as instantaneous values of the material hardening properties:

$$n_i = \left(\frac{\partial \ln \sigma}{\partial \ln \varepsilon}\right)_{\dot{\varepsilon}}, \text{and} \quad m_i = \left(\frac{\partial \ln \sigma}{\partial \ln \dot{\varepsilon}}\right)_{\varepsilon}$$

Also, the instability analysis and criterion presented are the same for other kinds of linear perturbations along the specimen length.

1.10 NUMERICAL TENSILE-TEST ANALYSIS

In the second part of this chapter, we introduce the basic concepts of numerical simulation by considering a nonuniform tensile test carried out for materials of known **constitutive equation**. (That is, we will assume that we know the precise relationship between stress and kinematic variables such as strain and strain rate.) We will use this knowledge of the material and of the specimen geometry to predict the engineering stress-strain curve (equivalent to the load-elongation curve), which represents the macroscopic behavior observed.

In order to illustrate the various numerical principles, we consider a nonuniform tensile bar made of a strain-hardening and strain-rate-sensitive material. The procedure for analyzing this problem in 1-D is outlined below.

Problem

Calculate the engineering stress-strain curve and total elongation for the tensile test of a strain-hardening, strain-rate-sensitive material.

Approach

1. Set up a time-discretized version of the 1-D differential equation for a segmented tensile specimen (i.e., a space-discretized one) with a taper ignoring biaxial effects and temperature changes.

2. Use time discretization to transform the time part of the governing differential equation into a series of regular equations.

3. Apply boundary conditions and solve the regular equations at each time, and move forward sequentially and re-solve for all times.

   We now proceed in sequence according to this approach. Derive the governing differential equation. Consider 1/2 of a tensile specimen as shown in Figure 1.8.

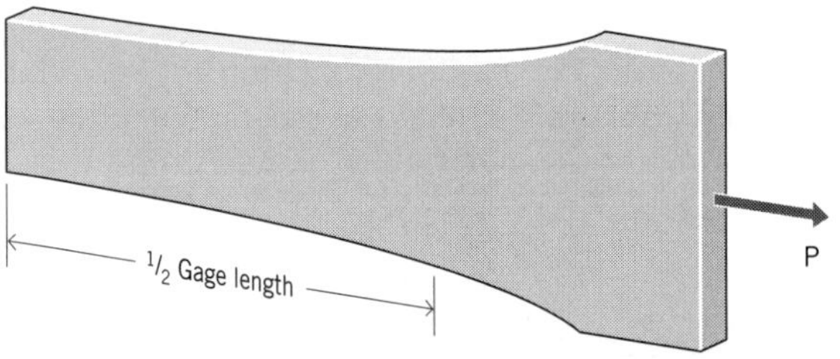

Figure 1.8 One half of tensile test specimen, reduced by symmetry.

Break the gage length into a finite number of elements (10 shown) to obtain a spatial discretization, as shown in Figure 1.9. The center is optionally labeled C or 1, where the C has a special significance because of boundary conditions.

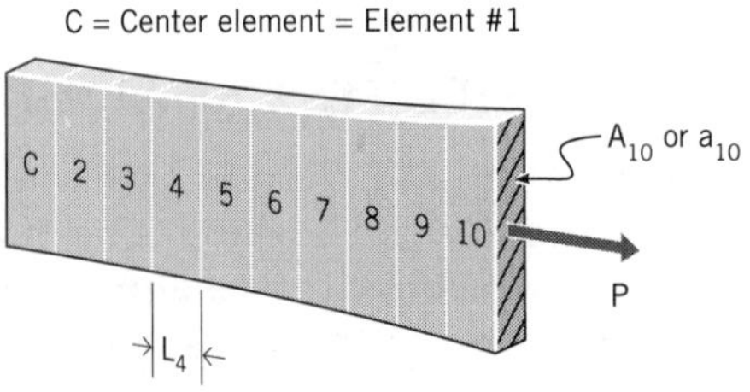

Figure 1.9 One half of tensile test specimen, as divided into elements for numerical analysis.

Each element has an original and current length and cross-sectional area as shown:

$$L_i,\ A_i = \text{original length, area} \qquad l_i,\ a_i, = \text{Current length, area} \tag{1.20}$$

We can note that by the definition of true strain (Equation 1.3) that

$$\ln \frac{l_i}{L_i} = \varepsilon_i \tag{1.21}$$

and, by $\Delta V_p = 0$ (constancy of plastic volume),

$$l_i\, a_i = L_i\, A_i \text{ or } \frac{l_i}{L_i} = \frac{A_i}{a_i} \tag{1.22}$$

Thus we obtain an expression for the current cross-sectional area of each element in terms of its original area and its strain.

$$a_i = A_i \exp(-\varepsilon_i) \tag{1.23}$$

Applying the equilibrium, just as in a chain or rope, the force on each element is equal to $F_c = F_i = F_j$, and so forth.

This discrete expression is equivalent to a continuous differential equation of the form $dF/dx = 0$. By the definition of true stress, however, $F = \sigma a$. Therefore, we can write the equilibrium equation in terms of stresses and strains:

$$\sigma_c\, a_c = \sigma_i\, a_i \quad \text{(for all elements, } i = 1, N) \tag{1.24}$$

where subscript c refers to the center or boundary element and i is any element; or, in terms of original element cross-sectional areas,

$$\sigma_c A_c \exp(-\varepsilon_c) = \sigma_i A_i \exp(-\varepsilon_i) \tag{1.25}$$

We assume that the flow stress of the material is some function of strain and strain rate—$\sigma = f(\varepsilon,\ \dot{\varepsilon},)$. Then the equilibrium condition becomes

$$\text{equilibrium equation} \quad A_c f(\varepsilon_c,\ \dot{\varepsilon}_c) \exp(-\varepsilon_c) = A_i f(\varepsilon_i, \dot{\varepsilon}_i) \exp(-\varepsilon_i) \tag{1.26}$$

For a particular hardening law—say, $\sigma = k\varepsilon^n \dot{\varepsilon}^m$,—we obtain an explicit expression of the equilibrium equation in terms of elemental strains and strain rates:

$$\text{equilibrium equation} \quad \dot{\varepsilon}_i = \left[\frac{A_c}{A_i} \exp(\varepsilon_i - \varepsilon_c) \left(\frac{\varepsilon_c}{\varepsilon_i} \right)^n \right]^{1/m} \dot{\varepsilon}_c \tag{1.27}$$

where Equation 1.27 represents N equations corresponding to the number of spatial elements of our spatial discretization. Note that the equilibrium equation relates the strain rate in any element to the strain rate in any other element depending on the total strain each element has undergone. We have chosen to watch the center element arbitrarily, since it will undergo the highest strains and strain rates as necking proceeds. Note also that the equation does not care what happens in other elements—it can be applied between elements i and j without considering k. This is an unusual simplification that arises in the 1-D case because all elements are connected in series.

The problem now is that we have one equation and four unknowns—$\dot{\varepsilon}_i$, $\dot{\varepsilon}_c$, ε_i, ε_c,—although two of these will be determined by boundary conditions. For example, we can apply fixed values of $\dot{\varepsilon}_c$ and ε_c and simply watch what happens elsewhere. This still leaves ε_i and $\dot{\varepsilon}_i$ to be related to each other.

Time Discretization

The real relationship between ε_i and $\dot{\varepsilon}_i$ is as follows:

$$\dot{\varepsilon}_i = \frac{d\varepsilon_i}{dt} \quad \text{or} \quad \varepsilon_i = \int_o^t \dot{\varepsilon}_i \, dt \tag{1.28}$$

To write the true **differential equation**, we would have to substitute these quantities in Equation 1.27.

For a numerical solution, an approximate relationship between ε and $\dot{\varepsilon}$ is required. Note that the equilibrium equation may be satisfied at a given time, but we do not know how to move forward. The usual approximation is $\Delta\varepsilon = \dot{\varepsilon}\Delta t$, or

$$\varepsilon^{(2)} = \varepsilon^{(1)} + \Delta\varepsilon = \varepsilon^{(1)} + \dot{\varepsilon}\,\Delta t \tag{1.29}$$

where superscript (j) means at the jth time increment . The problem is that $\dot{\varepsilon}$ is changing in a time interval, and we do not know which value to use: $\dot{\varepsilon}$ at the startor end of the time interval, or some intermediate value.

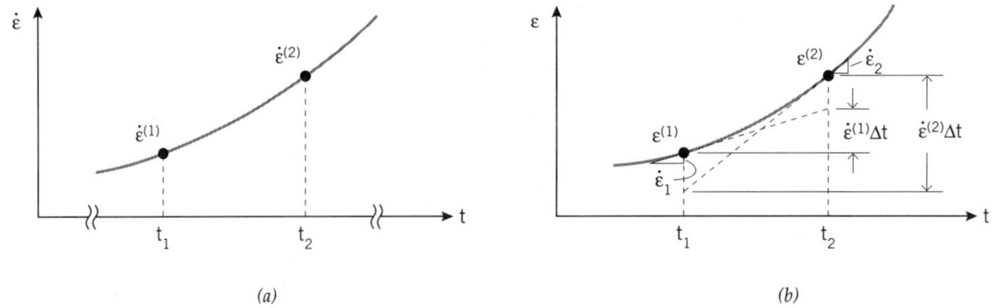

Figure 1.10 Variation of $\dot{\varepsilon}$ and ε with time and approximate relationships.

Figure 1.10 illustrates that there are alternative ways to choose $\varepsilon^{(2)}$, even if $\varepsilon^{(1)}$, $\dot{\varepsilon}^{(1)}$, and $\dot{\varepsilon}^{(2)}$ are known. If we really knew $\dot{\varepsilon}(t)$ and could integrate (Equation 1.28), we would have no need to solve the problem numerically. In fact, $\dot{\varepsilon}(t)$ is the solution we seek approximately.

The usual ways to **time discretize** are shown below. In a given application, any one approach may be better than another, but in all cases, as $\Delta t \to 0$, the result must reproduce the true differential relationship. That is, any method will give a good answer if $\Delta t \to 0$, but the methods will differ in accuracy for larger Δt. (Note: Usually any numerical result is tested by letting Δt, Δx, etc., get smaller until the result does not change more than a certain desired tolerance.)

The simplest discretization relies only on known quantities, and is thus **explicit**:

$$\Delta\varepsilon \cong \dot{\varepsilon}^{(1)}\Delta t, \qquad \varepsilon^{(2)} \cong \varepsilon^{(1)} + \dot{\varepsilon}^{(1)}\Delta t \quad \text{(explicit time discretization)} \tag{1.30}$$

where ε at the end of the step is immediately available from ε and $\dot{\varepsilon}$ at the beginning of the step. This method has a big advantage, as we will see, because $\varepsilon^{(2)}$ depends only on known values.

The **implicit** method is conceptually more consistent because the current strain depends on the current-strain rate (and previous strain):

$$\Delta\varepsilon \cong \dot{\varepsilon}^{(2)}\Delta t, \qquad \varepsilon^{(2)} \cong \varepsilon^{(1)} + \dot{\varepsilon}^{(2)}\Delta t \quad \text{(implicit time discretization)} \tag{1.31}$$

The equation at each time will be more difficult to solve because $\varepsilon^{(2)}$ and $\dot{\varepsilon}^{(2)}$ are both unknowns, although they are related by Equation 1.31.

A more general method can be constructed that incorporates aspects of both the explicit and implicit methods. Let $\Delta\varepsilon$ be a linear combination of $\dot{\varepsilon}^{(1)}\Delta t$ and $\dot{\varepsilon}^{(2)}\Delta t$, as follows:

$$\text{i.e.,} \qquad \varepsilon^{(2)} = \varepsilon^{(1)} + (1-\beta)\dot{\varepsilon}^{(1)}\Delta t + \beta\dot{\varepsilon}^{(2)}\Delta t \qquad \begin{pmatrix} \beta = 0, \text{ explicit} \\ \beta = 1, \text{ implicit} \\ \beta = \frac{1}{2}, \text{ semi-implicit} \end{pmatrix} \tag{1.32}$$

Note*: There is really no advantage, in terms of computational ease, of any particular β value except for $\beta = 0$, in which case all of the coefficients at a given time are known from the previous time. The real advantages become apparent whenever larger time steps can be used without reducing the accuracy of the solution. However, while the optimum value for β will depend on the nature of the solution, it would appear that the* ***semi-implicit scheme****, $\beta = \frac{1}{2}$ is likely to be better than either fully explicit or fully implicit.*

In order to solve the discrete equations (discrete in space and time) we have generated, we must simply substitute one of Equations 1.30, 1.31, or 1.32 (in our time discretizations) into Equation 1.27 to apply initial or boundary conditions, and then move forward in time, solving as we go.

There is one trivial choice that must be made; that is, whether to eliminate the current $\dot{\varepsilon}$ or ε from the resulting equations. Put another way, do we wish to treat current strain rates or strains as our primary variables? Once we have chosen an assumed relationship between the two (Equations 1.30, 1.31, and 1.32), there is no real distinction between the two approaches. For our illustration, we choose to watch $\dot{\varepsilon}$ at each time step and thus eliminate the current strain from consideration. After solving to find $\dot{\varepsilon}$ at each time step, we will find ε from the time discretization used.

Let's first approach the problem using the explicit time scheme. Equation 1.30 applies to each element i under consideration over the jth time interval:

$$\varepsilon_i^{(j)} = \varepsilon_i^{(j-1)} + \dot{\varepsilon}_i^{(j-1)} \Delta t^{(j)} \tag{1.33}$$

This applies to the first (i = 1) element, also called the c element. Equations 1.33 and 1.27, taken together, show how to proceed forward in time for each element, given the strains at the end of the last step and some condition expressing the current strain rate.

Let's summarize our equations for the explicit case in general form and then proceed to apply boundary conditions. Our discretizations give us the following form:

$$\dot{\varepsilon}_i^{(j)} = f\left[\varepsilon_i^{(j)}, \varepsilon_c^{(j)}\right] \dot{\varepsilon}_c^{(j)} \tag{1.34}$$

where $\varepsilon_i^{(j)}$ and $\varepsilon_c^{(j)}$ are known from the solution at the previous time step via Equation 1.33. The specific case we are considering can be written from Equations 1.30 and 1.27 directly:

$$\dot{\varepsilon}_i^{(j)} = \left[\frac{A_c}{A_i} \exp\left[\varepsilon_i^{(j)} - \varepsilon_c^{(j)}\right] \left[\frac{\varepsilon_c^{(j)}}{\varepsilon_i^{(j)}}\right]^n \right]^{1/m} \qquad \dot{\varepsilon}_c^{(j)} = f\left[\varepsilon_i^{(j)}, \varepsilon_c^{(j)}\right] \dot{\varepsilon}_c^{(j)} \tag{1.35}$$

$$\text{where:} \quad \varepsilon_i^{(j)} = \varepsilon_i^{(j-1)} + \dot{\varepsilon}_i^{(j-1)} \Delta t^{(j)} \tag{1.35a}$$

$$\varepsilon_c^{(j)} = \varepsilon_c^{(j-1)} + \dot{\varepsilon}_c^{(j-1)} \Delta t^{(j)} \tag{1.35b}$$

Equations 1.35 represent the entire explicit discretization.

Now, in order to solve Equations 1.35, we must start with boundary conditions. Let's choose the simplest ones, as follows:

$$\varepsilon_i = 0 \text{ at } t = 0 \tag{1.36}$$

$$\varepsilon_i = k \text{ (constant for all t).} \tag{1.37}$$

With the help of these boundary conditions, which correspond to a constant-true-strain-rate tensile test (*not* a constant crosshead speed, which is standard but more difficult to simulate), we can fill in a table to show how the calculation proceeds. Table 1.1 uses constant time steps of Δt.

Table 1.1 Explicit Numerical Solution of Equations 1.35

Step Number	Time	ε_i	ε_c	$\dot{\varepsilon}_c$	$\dot{\varepsilon}_i$
0	0	0 (bdy. cond.)	0 (bdy. cond.)	k (bdy. cond.)	$\left(\frac{A_c}{A_i}\right)^{\frac{1}{m}} k$
1	Δt	$\left(\frac{A_c}{A_i}\right)^{\frac{1}{m}} k\Delta t$	$k\Delta t$	k	$f\left[\varepsilon_i^{(1)}, \varepsilon_c^{(1)}\right]k$
2	$2\Delta t$	$\varepsilon_i^{(1)} + \dot{\varepsilon}_i^{(1)}\Delta t$	$2k\Delta t$	k	$f\left[\varepsilon_1^{(2)}, \varepsilon_c^{(2)}\right]k$
⋮	⋮	⋮	⋮	⋮	⋮
n	$n\Delta t$	$\varepsilon_i^{(n-1)} + \dot{\varepsilon}_i^{(n-1)}\Delta t$	$nk\Delta t$	k	$f\left[\varepsilon_1^{(n)}, \varepsilon_c^{(n)}\right]k$

Now that we have solved the explicit form of the problem, let's consider the implicit form. The only change appears in the assumed relationship which gives the current strains. For the implicit case, the governing equations equivalent to Equations 1.35 are

$$\dot{\varepsilon}_i^{(j)} = \left\{\left[\frac{A_c}{A_i}\exp\left[\varepsilon_i^{(j)} - \varepsilon_c^{(j)}\right]\left[\frac{\varepsilon_c^{(j)}}{\varepsilon_i^{(j)}}\right]^n\right]\right\}^{1/m} \dot{\varepsilon}_c^{(j)} = g\left[\varepsilon_i^{(j)}\right]\dot{\varepsilon}_c^{(j)} \tag{1.38}$$

$$\text{where } \varepsilon_i^{(j)} = \varepsilon_i^{(j-1)} + \dot{\varepsilon}_i^{(j)}\,\Delta t^{(j)} \tag{1.38a}$$

$$\varepsilon_c^{(j)} = \varepsilon_c^{(j-1)} + \dot{\varepsilon}_c^{(j)} \Delta t^{(j)}. \tag{1.38b}$$

Equations 1.38 cannot be solved algebraically at each time step because $\dot{\varepsilon}_i^{(j)}$ appears on both sides of the equality in **transcendental form**. Therefore, at each time, we will need to solve the equation using a **trial-and-error** or **iterative** method. While trial-and-error and direct substitution methods can be used, they have little generality and are mainly curiosities applicable to very small problems today. Instead of spending time on these (but see Problem 27), we prefer to introduce a very powerful and general method, the **Newton** or **Newton-Raphson method**[5].

Consider a general function, f, of a single variable, x, for which we seek the x value that satisfies

$$f(x) = 0 \tag{1.39}$$

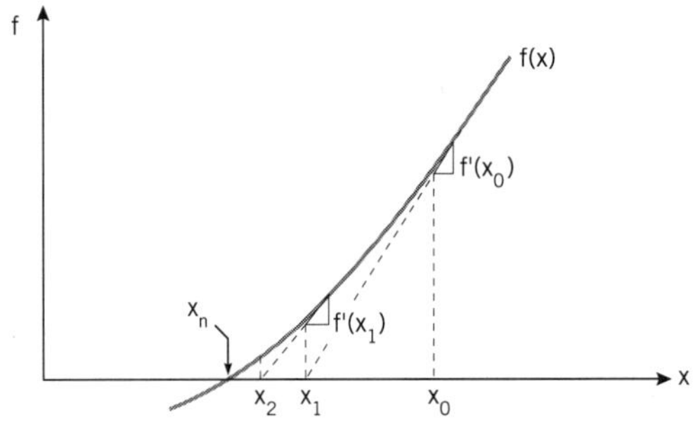

Figure 1.11 Graphical interpretation of Newton method.

The solution method is shown graphically in Figure 1.11, or it may be understood readily in terms of a **Taylor expansion** of f(x) about some trial solution x_o:

$$f(x_1) = f(x_0) + (x_1 - x_0)\, f'(x_0) \ + \ \text{higher order terms} = 0 \tag{1.40}$$

$$f(x_0) + (x_1 - x_0)\, f'(x_0) \approx 0 \tag{1.41}$$

Equation 1.41 can be used to provide a **recursive formula** for finding a new (and hopefully better) trial solution given a starting point x_0:

$$x_1 = x_0 - \frac{f(x_0)}{f'(x_0)}, \qquad x_2 = x_1 - \frac{f(x_1)}{f'(x_1)} \tag{1.42}$$

[5] The Newton method is strictly applicable only to one-variable equations, such as Equation 1.38, while the Newton-Raphson method is a straightforward generalization to functions of many variables.

Equation 1.42 and so on is the basis of the Newton method, which is used repeatedly until the solution is known to a given **tolerance**. The tolerance is usually expressed in one of two ways:

$$|f(x_n)| \leq \text{tolerance} \tag{1.43}$$

$$|x_n - x_{n-1}| \leq \text{tolerance} \tag{1.44}$$

The first of these states a tolerance for the equation, while the second is an approximate measure of the uncertainty of the value of the variable that has been found.

Exercise 1.2 **Find the solution of the following equation using Newton's method within a tolerance of 10^{-3} of f or 10^{-3} of x, whichever occurs first. Start with a trial solution of $x_o = 1$.**

$$0 = f(x) = e^x \ln x - x$$

We must first compute the derivative of f(*x*) and then proceed to apply Equation 1.42 sequentially until Equation 1.43 or 1.44 is satisfied:

$$f'(x) = e^x \ \ln \ x + \frac{e^x}{x} - 1 \qquad \text{1.2-.1}$$

Iteration number	x_i	*($f(x_i)$ (tol. = 10^{-3})*	$f'(x_i)$	x_{i+1}	$\lvert x_{i+1} - x_i \rvert$ *(tol. = 10^{-3})*
0	1	–1	1.718	1.582	0.582
1	1.582	0.649	9.771	1.516	0.066
2	1.516	0.379	3.899	1.419	0.098
3	1.419	0.027	3.359	1.411	**0.008**
4	1.411	**0.0006**	3.317	1.41082	2×10^{-4}
5	1.41082	4×10^{-5}	3.3165	1.41081	1×10^{-5}

The results for the Newton iteration shown in the table are typical. The desired tolerances (in bold) were reached at the fourth iteration, for $x_4 = 1.411$. It is interesting to note a couple of other things from this example:

- Once the procedure begins to converge, it converges very rapidly. However, there may be oscillations if the trial solution is too far away from the actual one.
- The exact tolerance is not critical because convergence proceeds very rapidly for x values close to the solution.

Let's now use Newton's method to solve the first step or two of our problem, as presented in Equations 1.38 with the boundary conditions in Equations 1.36 and 1.37. At each time step we must solve

$$0 = F\left(\dot{\varepsilon}_i^{(j)}\right) = \left\{\left[\frac{A_c}{A_i}\exp\left[\varepsilon_i^{(j)} - \varepsilon_c^{(j)}\right]\left[\frac{\varepsilon_c^{(j)}}{\varepsilon_i^{(j)}}\right]^n\right]\right\}^{1/m} \dot{\varepsilon}_c^{(j)} - \dot{\varepsilon}_i^{(j)} \tag{1.45}$$

$$\text{where} \quad \varepsilon_i^{(j)} = g\left[\dot{\varepsilon}_i^{(j)}\right] = \varepsilon_i^{(j-1)} + \dot{\varepsilon}_i^{(j)}\,\Delta t^{(j)} \tag{1.45a}$$

$$\varepsilon_c^{(j)} = g\left[\dot{\varepsilon}_c^{(j)}\right] = \varepsilon_c^{(j-1)} + k\,\Delta t^{(j)}. \tag{1.45b}$$

In order to apply Newton's method, we must find $F'\left[\dot{\varepsilon}_i^{(j)}\right]$, which involves applying the chain rule:

$$F'\left[\dot{\varepsilon}_i^{(j)}\right] = \frac{\partial f}{\partial \varepsilon_i^{(j)}}\frac{d\varepsilon_i^{(j)}}{d\dot{\varepsilon}_i^{(j)}} + \frac{\partial f}{\partial \dot{\varepsilon}_i^{(j)}} \tag{1.46}$$

where we equate the terms related to $\varepsilon_i^{(j)}$ and $\dot{\varepsilon}_i^{(j)}$ because each is known explicitly from our boundary conditions. Using either the chain rule or direct substitution obtains

$$F'\left(\dot{\varepsilon}_i^{(j)}\right) = \frac{1}{m}\left[1 - \frac{n}{\varepsilon_i^{(j)}}\right]\dot{\varepsilon}_c^{(j)}\left\{\frac{A_c}{A_i}\exp\left[\varepsilon_i^{(j)} - \varepsilon_c^{(j)}\right]\left[\frac{\varepsilon_c^{(j)}}{\varepsilon_i^{(j)}}\right]^n\right\}^{1/m} \Delta t - 1$$

$$\text{where} \quad \varepsilon_i^{(j)} = \varepsilon_i^{(j-1)} + \dot{\varepsilon}_i^{(j)}\,\Delta t \tag{1.47}$$

The reader should verify that Equation 1.47 does represent the correct derivative for the Newton procedure.

Using Equation 1.47, we could construct a table for the implicit procedure like the one presented in Table 1.1 for the explicit procedure. At each time step, we would need a table like the one presented in Exercise 1.1 in order to solve the implicit equation at that instant. As can be easily imagined, such computations are much more effectively carried out by digital computer; however, it may be instructive to carry out the first step or two by hand in order to verify and understand the procedure better. See Problem 19.

Once a numerical solution has been obtained, two tasks remain:

1. To verify that the solution is not very sensitive to the arbitrary choice of numerical parameters. For our sample calculation, the purely numerical parameters consist of

   - number of spatial elements or, equivalently, the element size
   - number of time steps or, equivalently, the time-step size
   - the tolerance allowed in solving the equilibrium equation at each time step (implicit solution only)

While there are various procedures available to test these numerical sensitivities, the best and simplest lies in choosing combinations of smaller Δl and Δt $(1/2\,x)$ (often reducing each by a factor of 2), and the convergence tolerance, then verifying that the new solution is within the desired accuracy.

2. To relate the numerical solution variables to external ones important for engineering insight into the problem. For the case we have chosen, it may be desirable to calculate and present the following kinds of results:

   - engineering strain $e = \Delta l / l_o$ for a given gage length of the original specimen)
   - engineering stress $\sigma_e = F / A_o$
   - uniform elongation (e_u = e at the point of maximum force or engineering stress)
   - total elongation (e_f = e when failure occurs) (This will require the use of a simple failure criterion because the numerical procedure does not contain information about failure.)

In order not to lose sight of the overall procedure, we close this chapter with a summary of the numerical procedure presented for analyzing a tensile test.

Summary: Numerical Tensile Test Analysis

1. Derive a differential equation satisfying equilibrium everywhere, at any given time.

$$\text{Result:} \quad \frac{dF}{dx} = \frac{d}{dx}\left\{a(x)\, K\, \varepsilon(x)^n\, \dot{\varepsilon}(x)^m \exp[-\varepsilon(x)]\right\} = 0 \tag{1.48}$$

Construct finite elements and convert Equation 1.48 to a series of regular equations at any given time (assume that stress, strain, and area are uniform in each element at a given time).

$$A_i \exp(-\varepsilon_i) K \varepsilon_i^n \dot{\varepsilon}^m = A_c \exp(-\varepsilon_c) K \varepsilon_c^n \dot{\varepsilon}_c^m \tag{1.49}$$

2. Construct a time-discretization scheme to relate the time derivatives to integral quantities.

$$\varepsilon^{(t_2)} = \varepsilon^{(t_1)} + (1-\beta)\, \dot{\varepsilon}^{(t_1)} \Delta t + \beta \dot{\varepsilon}^{(t_2)} \Delta t$$

where the choice of β determines whether the time-integration scheme is explicit, implicit, or semi-implicit.

3. Solve the regular equations at time t_i given the solution t_{i-1} and subject to certain boundary condition and initial conditions.

   a. Start from boundary conditions for a true constant strain-rate test.

$$\text{at } t = 0; \quad \varepsilon_i = 0,$$

$$\text{at } t = t_i; \quad \dot{\varepsilon}_c = k$$

   b. Use any solving method—algebraic (explicit only) trial and error, direct, Newton-Raphson— to solve for $\dot{\varepsilon}_i$ at each time.

   c. Update the integral variables (ε_i) according to the time-discretization scheme and move forward.

4. a. Calculate external quantities (e.g., engineering stress-strain curve, uniform elongation, total elongation).
   b. Check the sensitivity of your solution to mesh size (number of elements) and time-step size (Δt) to insure that they are sufficiently small to obtain the accuracy desired.

CHAPTER 1 PROBLEMS

A. Proficiency Problems

1. The following plot of load versus extension was obtained using a specimen (shown in the figure) of an alloy remarkably similar to the aluminum-killed steel found in automotive fenders, hoods, and so forth. The cross-head speed, v, was 3.3×10^{-4} inch/second. The extension was measured using an extensometer with a gage length of two inches, as shown (G). Eight points on the plastic part of the curve have been digitized for you. Use these points to help answer the following questions.

a. Determine the following quantities. Do not neglect to include proper units in your answer.

yield stress	Young's modulus
ultimate tensile strength	total elongation
uniform elongation	post-uniform elongation
engineering strain rate	

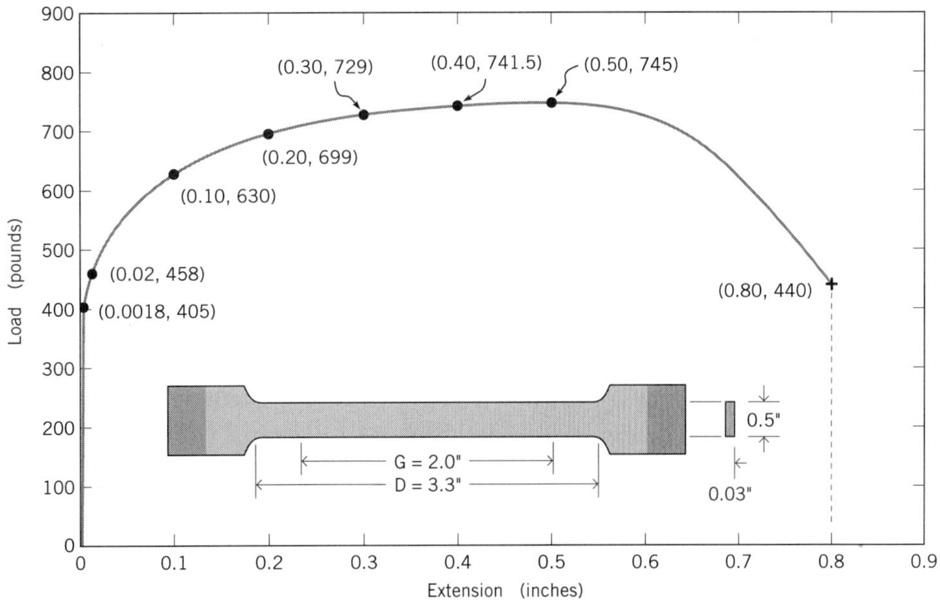

b. Construct a table with the following headings, left-to-right: extension, load, engineering strain, engineering stress, true strain, true stress. Fill in for the eight points on graph. What is the percentage difference between true and engineering strains for the first point? (i.e., % = ______ x 100) What is the percentage difference between true and engineering strains for the last point?

c. Plot the engineering and true stress-strain curves on a single graph using the same units.

d. Calculate the work-hardening rate graphically and provide the ln-ln plot along with the value of n. How does n compare with the uniform elongation in part a? Why?

e. A second tensile test was carried out on an identical specimen of this material, this time using a crosshead speed of 3.3 x 10^{-2} inch/second. The load at an extension of 0.30 inch was 763.4 lb. What is the strain-rate-sensitivity index, m, for this material?

2. Starting from the basic idea that tensile necking begins at the maximum load point, find the true strain and engineering strain where necking begins for the following material laws. Derive a general expression for the form and find the actual strains.

a.	$\sigma = k(\varepsilon + \varepsilon_o)^n$	$\sigma = 500(\varepsilon + 0.05)^{0.25}$	Swift[6]
b.	$\sigma = \sigma_o + k(\varepsilon + \varepsilon_o)^n$	$\sigma = 100 + 500(\varepsilon + 0.05)^{0.25}$	Ludwik
c.	$\sigma = \sigma_o(1 - A_e{}^{-B\varepsilon})$	$\sigma = 500[1 - 0.6\exp(-3\varepsilon)]$	Voce[7]
d.	$\sigma = \sigma_o$	$\sigma = 500$	ideal
e.	$\sigma = \sigma_o + k\varepsilon$	$\sigma = 250 + 350\varepsilon$	linear
f.	$\sigma = k\ \sin(B\varepsilon)$	$\sigma = 500\ \sin(2\pi\varepsilon)$	trig

3. What effect does a multiplicative strength coefficient (for example, k in the Hollomon law, k in Problem 2a, or σ_o in Problem 2c) have on the uniform elongation?

4. For each of the explicit hardening laws presented in Problem 2, calculate the true stress at $\varepsilon = 0.05$, 0.10, 0.15, 0.20, 0.25 and plot the results on a

---

[6] H. W. Swift, *J. Mech. Phys. Solids* 1 (1922):1.

[7] E. Voce, *J. Inst. Met* 74 (1948): 537, 562, 760.
E. Voce, *The Engineer* 195(1953): 23.
E. Voce, *Metallurgic:* 51 (1953): 219.

$\ln\sigma - \ln\varepsilon$ figure. Use the figure to calculate a best-fit n value for each material and compare this with the uniform strain calculated in Problem 2. Why are they different, in view of Equation 1.16?

5. For each of the explicit hardening laws presented in Problem 2, plot the engineering stress-strain curves and determine the maximum load point graphically. How do the results from this procedure compare with those obtained in Problems 2 and 4?

6. Tensile tests at two crosshead speeds (1 mm/sec and 10 mm/sec) can be fit to the following hardening laws:

$$\text{at } V_1 = 1\,\text{mm/sec}, \qquad \sigma = 500(\varepsilon + 0.05)^{0.25}$$

$$\text{at } V_2 = 10\,\text{mm/sec}, \qquad \sigma = 520(\varepsilon + 0.05)^{0.25}$$

What is the strain-rate-sensitivity index for these two materials? Does it vary with strain? What is the uniform strain of each, according to the Considere criterion?

7. Repeat Problem 6 with two other stress-strain curves:

$$\text{at } V_1 = 1\,\text{mm/sec}, \qquad \sigma = 550\varepsilon^{0.25}$$

$$\text{at } V_2 = 10\,\text{mm/sec}, \qquad \sigma = 500\varepsilon^{0.20}$$

Plot the stress-strain curves and find the strain-rate-sensitivity index at strains of 0.05, 0.15, and 0.25. In view of these results, does Equation 1.17 apply to this material?

8. Consider the engineering stress-strain curves for three materials labeled A, B, and C in the following illustration. Qualitatively, put the materials in order in terms of largest-to-smallest strain hardening (n-value), strain-rate sensitivity (m-value), and total ductility (formability).

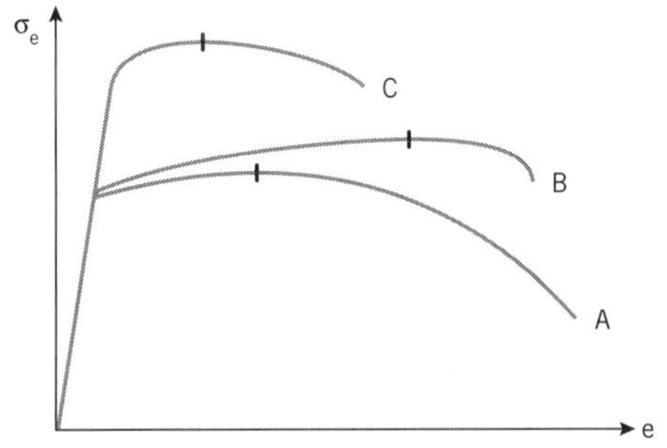

9. It is very difficult to match tensile specimens precisely. For sheet materials, the thickness, width, and strength may vary to cause a combined uncertainty

of about ±1% in stress. Considering this uncertainty of K's in Problem 6, calculate the range of n values that one might obtain if the tests were conducted at both rates several times.

10. Considering the specimen-to-specimen variation mentioned in Problem 9, it would be very desirable to test strain-rate sensitivity using a single specimen. Typically, "jump-rate tests" are conducted by abruptly changing the crosshead velocity during the test. Find the strain-rate sensitivity for the idealized result shown.

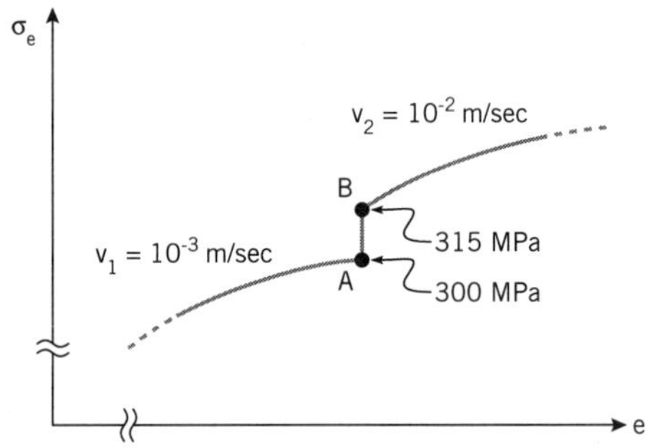

11. The procedure outlined in Problem 10, while being convenient and attractive, has its own difficulties. In order to obtain sufficient resolution of stress, it is necessary to expand the range and move the zero point. Some equipment does not have this capability. More importantly, the response shown in Problem 10 is not usual. For the two more realistic jump-rate tests reproduced in the following illustration, find m values using the various points marked.

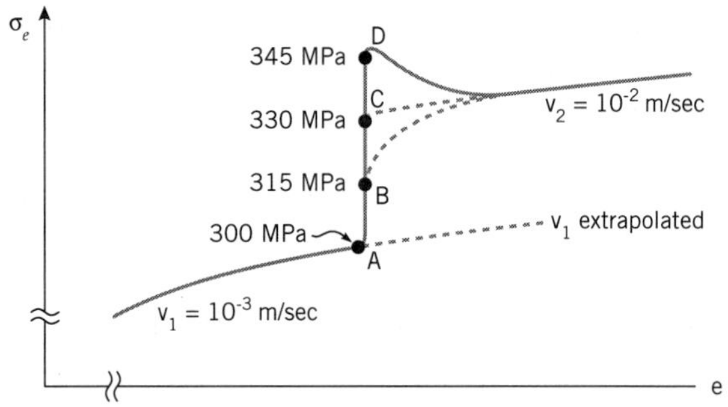

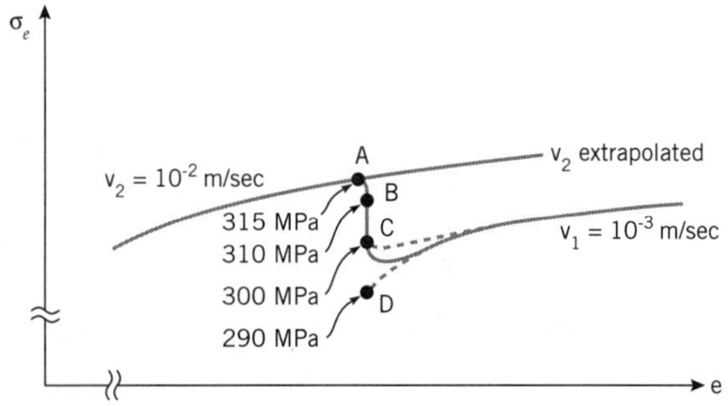

B. Depth Problems

12. If a tensile-test specimen was not exactly uniform in cross section—for example, if there were initial tapers as shown in the following illustration—how would you expect the measured true stress-strain curves to appear relative to one generated from a uniform specimen? Sketch the stress-strain curves you would expect.

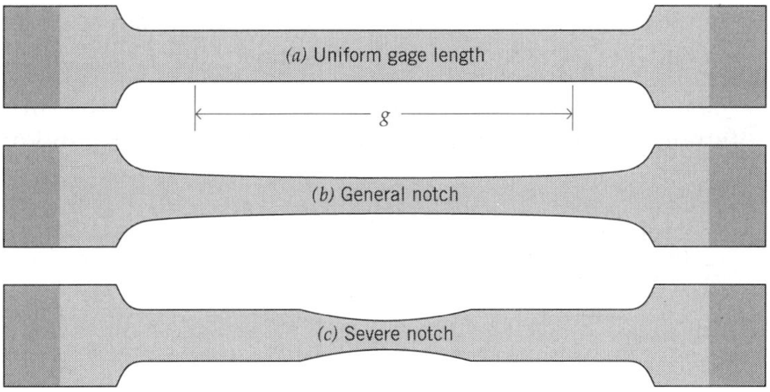

13. What is the relevance of the 0.2% offset in determining the engineering yield stress?

14. Some low-cost steels exhibit tensile stress-strain curves as shown here. What would you expect to happen with regard to necking?

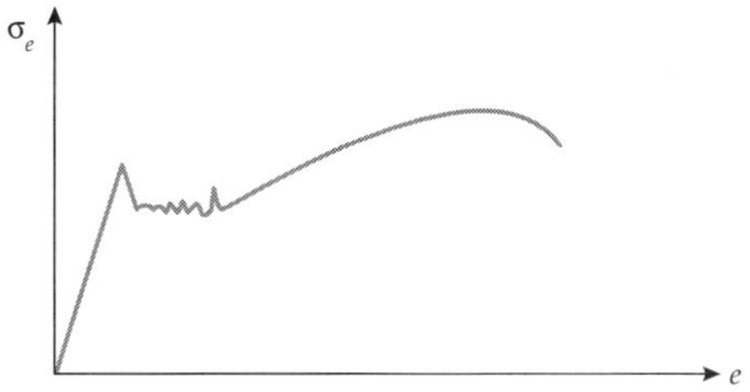

15. It has been proposed that some materials follow a tensile constitutive equation that has additive effects of strain hardening and strain-rate hardening rather than multiplicative ones:

$$\text{multiplicative:} \quad \sigma = f(\varepsilon)\, g(\dot{\varepsilon})$$

$$\text{additive:} \quad \sigma = f(\varepsilon) + g(\dot{\varepsilon})$$

In the first case, one investigates $G(\dot{\varepsilon})$ at constant ε by examining $\sigma(V_2)/\sigma(V_1)$, as we have done so far. In the second case, one would watch

$\sigma(V_2)-\sigma(V_1)$. Assume that an additive law of the following type was followed by a material

$$\sigma = 500\ \varepsilon^{0.25} + 25\left(\frac{\dot{\varepsilon}}{\dot{\varepsilon}_o}\right)^{0.03}$$

where $\dot{\varepsilon}_o$ is the base strain-rate where the strain-hardening law is determined (i.e., a tensile test conducted at a strain rate of $\dot{\varepsilon}_o$ exhibits $\sigma = 500\,\varepsilon^{0.25}$).

Given this law, determine the usual multiplicative m value at various strains from two tensile tests, one conducted at $\dot{\varepsilon}_o$ and one at $10\dot{\varepsilon}_o$.

16. Use Eq. 1.1-19 (or, equivalently, Eqs. 1.1-20 and 1.1-22) to find the plastic instability for the strain hardening [f(ε)] and strain-rate hardening [g(ε)] forms specified. In each case m = 0.02 and $\dot{\varepsilon}_o = 1/\text{sec}$.

a. $\sigma = f(\varepsilon)\, g(\dot{\varepsilon})$, f(ε) from Problem 2a, $g(\dot{\varepsilon}) = \left(\frac{\dot{\varepsilon}}{\dot{\varepsilon}_o}\right)^m$

b. $\sigma = f(\varepsilon)\, g(\dot{\varepsilon})$, f(ε) from Problem 2c, $g(\dot{\varepsilon}) = \left(\frac{\dot{\varepsilon}}{\dot{\varepsilon}_o}\right)^m$

c. $\sigma = f(\varepsilon)\, g(\dot{\varepsilon})$, f(ε) from Problem 2d, $g(\dot{\varepsilon}) = \left(\frac{\dot{\varepsilon}}{\dot{\varepsilon}_o}\right)^m$

d. $\sigma = f(\varepsilon)\, g(\dot{\varepsilon})$, f(ε) from Problem 2e, $g(\dot{\varepsilon}) = \left(\frac{\dot{\varepsilon}}{\dot{\varepsilon}_o}\right)^m$

e. $\sigma = f(\varepsilon)\, g(\dot{\varepsilon})$, f(ε) from Problem 2f, $g(\dot{\varepsilon}) = \left(\frac{\dot{\varepsilon}}{\dot{\varepsilon}_o}\right)^m$

f. $\sigma = f(\varepsilon) + g(\dot{\varepsilon})$, f(ε) and $g(\dot{\varepsilon})$ from Problem 15. (Leave Part f in equation form.)

C. Numerical Problems

17. Repeat Exercise 1.2 using a trial-and-error method. Start with $x_0 = 1$ and $x_1 = 2$ and proceed by intuition in choosing $x_2, x_3, \ldots x_n$ until the specified

tolerance is attained. How many steps were required relative to Exercise 1.2? How would you automate such a procedure for digital computation?

18. Repeat Exercise 1.1 using a direct-solving method, as follows:

Procedure: Choose trial solution, χ_o.

Calculate: $f(x_o) = x_1$ (if $x_o = x_1$, then this is the solution).

Next try: $f(x_1) = (x_2)$ and

$f(x_2) = x_3$

$f(x_3) = x_4$, etc., until $f(x_i) = x_i + 1$ and $x_i = x_i + 1$.

To check: $f(x_i) - x_{i+1} =$ error, and can solve to given tolerance; e.g., $x_{real} = x_i \pm 10^{-6}$.

19. Perform a manual implicit analysis of a tensile test with only two elements and two time increments to get a "feel" for how the computation proceeds. Set up tables like Table 1.1 and the one in Exercise 1.2 (at each time step) to carry out the calculation.

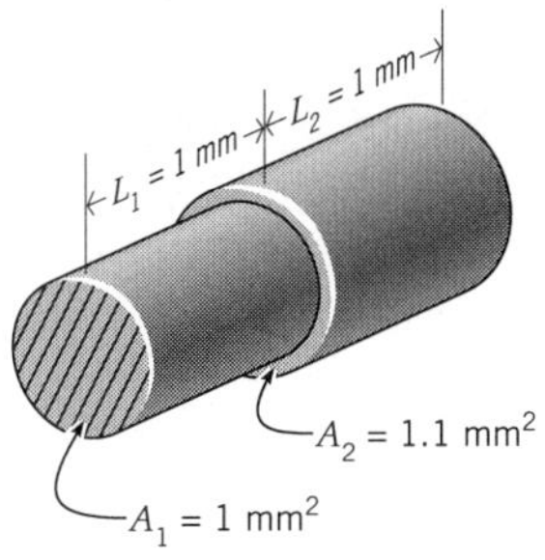

$$\sigma = 500\varepsilon^{0.25}\left(\frac{\dot{\varepsilon}}{0.1/\text{sec}}\right)^{0.05} \text{MPa}$$

$$\Delta t = 2 \text{ sec}, \qquad \dot{\varepsilon}_c = 0.1/\text{sec}$$

Find $\dot{\varepsilon}_2$ and e_2 at 2 sec, 4 sec, 6 sec.

Note: *The remaining numerical problems involve writing, modifying, and executing a short computer program in order to accomplish several objectives:*

- *understand how material behavior affects tensile-test results*
- *understand the role of specimen geometry in tensile testing*
- *understand and compare the advantages of different numerical approaches*
- *understand measures of error*

Problems 20 to 23 focus on the first two objectives, which are related to understanding physical effects, while the remaining problems focus on numerical aspects. The reader may choose to write a general program capable of handling all the variations we propose, or may wish to write shorter, more specialized programs for each problem. In either case, the procedures should be as consistent as possible so that meaningful comparisons among the numerical procedures can be made.

20. Write a computer program to carry out the *explicit* analysis of a tensile test, as outlined in this chapter. Test your programming by simulating the 2-element, 3-time test presented in Problem 19 and comparing your results to the manually calculated ones.

21. Use your program (Problem 20) to investigate the rule of material strength, strain hardening, strain-rate sensitivity, and the test shape and testing rate on the shape of the engineering stress-strain curve for a slightly tapered tension test conducted at constant true- strain rate at the center. Consider the following geometry and perform the following calculations:

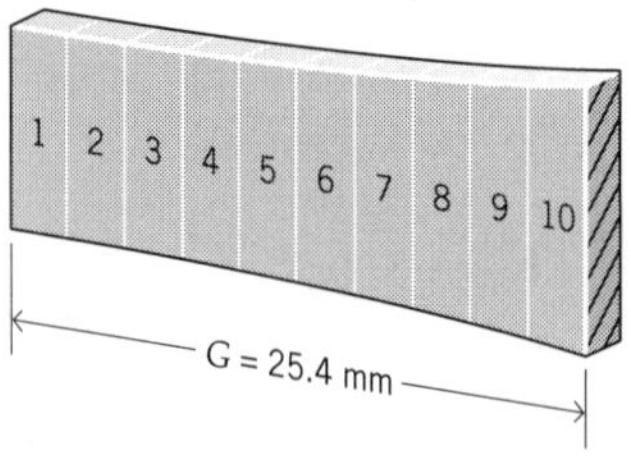

$$\Delta l_i = \frac{G}{10} = 2.54 \text{ mm} \qquad A_i = 0.990 + (0.001)i \ (\text{mm}^2)$$

Stop your calculations when $e_1 = 2.0$, at which point we will assume that fracture takes place and the total elongation has been reached. Choose a reasonable time-step size of 0.2 sec, so that there will be 100 steps at $\dot{\varepsilon}_c = 0.1/\text{sec}$ to the final center strain of 2.

a. Compare stress-strain curves (plot), uniform elongations (e_u), and total elongations (e_f) for materials with different work-hardening rates at a constant true-strain rate of 0.1/sec, as follows[8]:

$$\sigma = k\varepsilon^n \left(\frac{\dot{\varepsilon}}{0.1/\text{sec}} \right)^{0.05}, \qquad n = 0.05,\ 0.1,\ 0.2,\ 0.4,\ 0.6$$

$$k = \frac{100}{(0.01)^n}$$

Does n have a larger effect on the uniform elongation (e_u)or on the post-uniform elongation ($e_{pu} = e_f - e_u$)? (Plot e_u, e_f, and e_{pu} as functions of n to demonstrate.)

b. Compare stress-strain curves (plot), uniform elongations (e_u), total elongations (e_f), and post-uniform elongations ($e_{pu} = e_f - e_u$) at a constant true strain rate of 0.1/sec, as follows:

$$\sigma = k\varepsilon^{0.25} \left(\frac{\dot{\varepsilon}}{0.1/\text{s}} \right)^{m}, \qquad m = 0.02,\ 0.05,\ 0.1,\ 0.3,\ 0.5$$

$$k = \frac{100\ \text{MPa}}{(0.01)^{0.25}} = 316.2\ \text{MPa}$$

Does m have a larger effect on e_u, e_f, or e_{pu}? [Plot these quantities as functions of m.]

c. Compare stress-strain curves (plot), uniform elongations (e_u), total elongations (e_f), and post-uniform elongations ($e_{pu} = e_f - e_u$) at several constant true strain rates:

$$\sigma = k\varepsilon^{0.25} \left(\frac{\dot{\varepsilon}}{0.1/\text{sec}} \right)^{0.05} \qquad k = \frac{100}{(0.01)^{0.25}} 316.2\ \text{MPa}$$

[8] k is chosen such that the stress will be the same at e = 0.01 in each case.

$$\dot{\varepsilon}_c = 10^{-5}/\text{sec},\ 10^{-3}/\text{sec},\ 10^{-1}/\text{sec},\ 10^{1}/\text{sec},\ 10^{3}/\text{sec}$$

Use a time step such that each calculation will consist of 100 steps (e.g., at $\dot{\varepsilon}_c = 10^{-5}/\text{sec}$, $\Delta t = \varepsilon_f / N\dot{\varepsilon}_c = 2000\ \text{sec}$. What effect does the strain rate have on e_u, e_f, and e_{pu}?

d. Repeat part c, except scale the material strength coefficients so that the flow stress is the same at $\varepsilon = 0.01$ for each of the strain rates. How do the engineering stress-strain curves compare now?

$$k = (0.01)^{0.25} \left(\frac{\dot{\varepsilon}_c}{0.1/\text{sec}} \right)^{0.05}$$

e. Investigate the role of the taper in determining the engineering stress-strain. Consider one material law and one test rate ($\dot{\varepsilon}_c = 0.1/\text{sec}$) and several linear tapers:

$$\sigma = k\,\varepsilon^{0.25} \left(\frac{\dot{\varepsilon}_c}{0.1/\text{sec}} \right)^{0.05} \qquad k = \frac{100}{(0.01)^{0.25}}\ 316.2\ \text{MPa}$$

$$A_i = \left(\frac{1 - f_o}{N - 1} \right)(i - 1) + f_o$$

$$f_o = 0.999,\ 0.99,\ 0.9,\ 0.7,\ 0.5$$

The quantity f_o is called the **flow strength** or **weakness factor,** equal to A_c/A_N.

22. Compare the engineering stress-strain curves to failure ($\varepsilon_c = 2$) for two materials that have strain-hardening laws almost indistinguishable up to the uniform limit. Use ($\dot{\varepsilon}_c = 0.1/\text{sec}$) and the following laws:

$$\sigma = 600\,\varepsilon^{0.224} \left(\frac{\dot{\varepsilon}}{0.1/\text{sec}} \right)^{m} \ (\text{MPa})$$

$$\sigma = 430 \left(1 - 0.51\, e^{-11.3\,\varepsilon}\right) \left(\frac{\dot{\varepsilon}}{0.1/\text{sec}} \right) (\text{MPa})$$

with m = 0.01, 0.1, 0.5

23. Modify your program (Problem 20) to handle general kinds of strain-hardening laws and strain-rate-sensitivity laws via subroutines that return the required values and derivatives. Test your program using one of the laws presented in Problem 2 in conjunction with multiplicative or additive strain-rate sensitivity (See Problem 15).

24. Write a computer program to carry out the *implicit* analysis (using general formulation with P) of a tensile test, as outlined in this chapter. As was done in Problem 20, test your program against the 2-element, 3-time problem posed in Problem 19. You may wish to write a new program starting from your explicit one, or you may wish to introduce the implicit option into your existing program.

25. In order to compare the advantages of various choices among numerical procedures and parameters, it is necessary to establish a measure of numerical error, and to compare the various choices at approximately the same level of uncertainty. For our purposes, consider the following measure of error:

$$\text{error} = \sqrt{\frac{\sum_{i=2}^{N}\left(\varepsilon_i - \varepsilon_i^*\right)^2}{N-1}}, \qquad \%\ \text{error} = \frac{\text{error}}{\sqrt{\sum_{i=2}^{N}\frac{\varepsilon_i^{*2}}{N-1}}} \times 100$$

where ε_i^* is the most accurate estimate available for the strain in the ith element, and N is the number of elements. Since in general there will be no analytical result giving e_i^{exact}, we will use the most precise numerical result we have available and will compare other results with this.

In performing the following comparisons, consider the following geometric and material descriptions:

a. Find the ε_i^* at $\varepsilon_c = 2$ for a 10-element model by simulating the tensile test using a series of Δt's that get smaller until no appreciable difference in the solution is found, and record the ε_i^* in that range of Δt. Note: you must always compare solutions under exactly the same boundary conditions, so choose your Δt's by letting

$$\Delta t = \frac{\varepsilon_c}{\dot{\varepsilon}_c N_t} = \frac{20\ \text{sec}}{N_t},$$

where N_t is the integral number of time steps to load precisely at $\varepsilon_c = 2$. Compare solutions for $N_t = 50, 100, 200, 300$, and so on, until little additional accuracy is achieved.

where $f_o = 0.90$, $\dot{\varepsilon}_c = 0.1/\text{sec}$

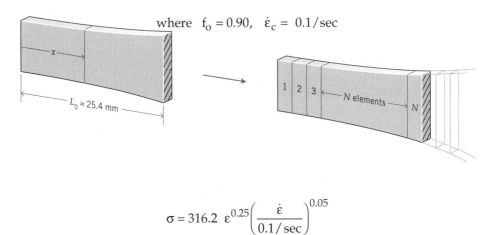

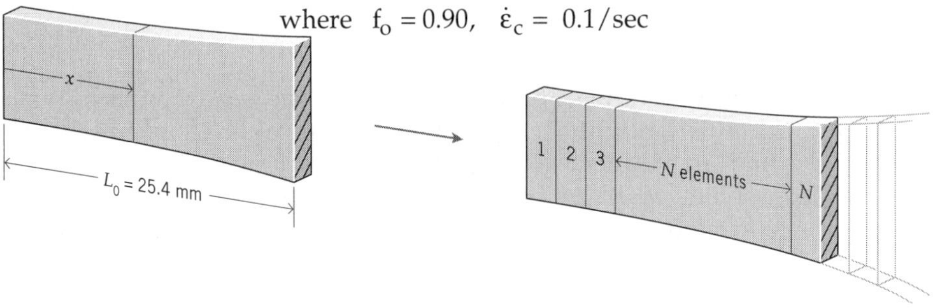

$$\sigma = 316.2\ \varepsilon^{0.25}\left(\frac{\dot{\varepsilon}}{0.1/\text{sec}}\right)^{0.05}$$

b. Carry out part a with explicit and implicit programs to verify that they give the same result for N_t large enough. Use a tolerance of 10^{-6} on the functional iteration.

c. Plot the error as a function of Δt or N for various values of β corresponding to explicit, implicit, and partially implicit numerical schemes: $\beta = 0, \frac{1}{4}, \frac{1}{2}, \frac{3}{4}, 1$. Which value of β minimizes the number of time steps needed to achieve an error of 1%? Compare computation times to achieve 1% error for the various values of β. Which value of β minimizes computer time?

26. Carry out Problem 22 with your implicit program.

27. Modify your program to handle the realistic boundary conditions for a standard tensile test; that is, that the right boundary moves at a constant speed, V. Carry out the simulations requested in Problems 21 and 22 using crosshead velocities that correspond to $\dot{\varepsilon}_c$ at the beginning of each test.

CHAPTER 2

Tensors, Matrices, Notation

This chapter has three goals: (1) to generalize the physical and mathematical concept of a vector to higher order **tensors**, (2) to review the formal mathematical manipulations for matrices and tensors, and (3) to introduce the various kinds of notation to be used throughout this book. For ease of understanding the physical nature of all such quantities, we shall avoid extensive use of curvilinear coordinate systems (but will introduce elementary concepts), instead relying on other texts to provide the detailed procedures for transforming physical laws to dynamic coordinate systems.

In fact, we view this chapter as a necessary evil. Perhaps the material could have been relegated to the appendices, with items consulted as necessary. On the other hand, it is often convenient to have all such mathematical preliminaries in one place, for easier reference. In any case, the reader should try to avoid becoming entangled in the formal manipulations—it is much more fun and rewarding, in an engineering sense, to become entangled in the physical aspects of the problems represented by the mathematics!

Although the chapter has been arranged in a text format, with problems and exercises, some readers (especially those with strong backgrounds in tensor algebra) may wish to skip this chapter entirely, referring to it only as necessary. For completeness then, we introduce the mathematical notation and manipulations in one place, while continuing to believe that this chapter can be treated as a reference, as necessary, when studying the other chapters.

2.1 VECTORS: PHYSICAL PICTURE AND NOTATION

We assume that the reader is familiar with vectors and vector notation. Therefore, we start with this familiar application to introduce the basic notions and notation from a physical perspective. In subsequent sections, we concentrate on mathematical manipulations and alternate notations.

Physically, a vector is best conceptualized as an arrow with a certain direction ("tail-to-head") and a length, or **magnitude**.[1] Because this is a completely physical picture, we are not concerned with how we describe the arrow (vector) on paper. We can easily see that the arrow (vector) exists independently of our means to observe it, describe it, rotate it, operate on it, or categorize it. This physical reality is crucial to understanding physical vectors and **vector notation**, which emphasizes that coordinate systems, observers, and so on are unimportant. Furthermore, a vector may be translated (but not rotated or stretched!) freely, without changing its meaning. Thus only the relative positions of the arrow's head and tail are important, not the position of either one independently. The vector notation (usually a bold letter or an underlined letter if bold is not available) specifies the physical nature of a vector, without regard to the observer:

$$\mathbf{a} = |\mathbf{a}| \cdot \frac{\mathbf{a}}{|\mathbf{a}|} = |\mathbf{a}|\, \hat{\mathbf{a}}, \tag{2.1}$$

where $|\mathbf{a}|$ means the length of the arrow, and $\hat{\mathbf{a}}$ is an arrow of length 1 in the direction of **a** (i.e., $\hat{\mathbf{a}}$ is a **unit vector**). While this notation carries the clearest physical picture of a vector, it is rather inconvenient for carrying out mathematical manipulations.

Addition

We start from physical vectors and later deduce the formal mathematical rules. For example, consider the vectors representing the **displacements** of an insect walking on this paper. If it walks one unit length in the **a** direction and one unit in the **b** direction, its total displacement from the beginning of its travel to the end, is shown in Figure 2.1

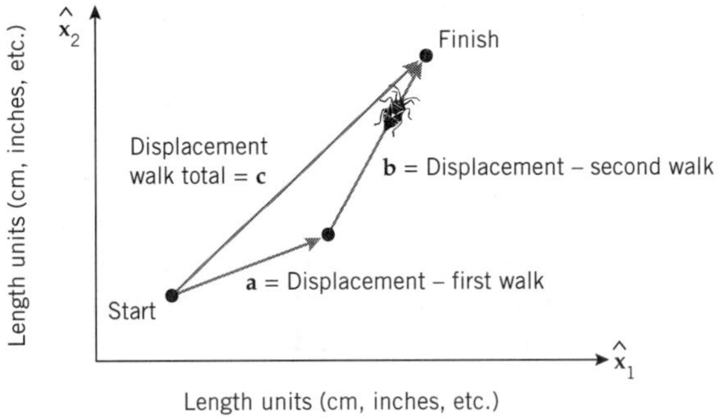

Figure 2.1 Adding two displacement vectors to obtain total displacement.

Note that the **resultant displacement vector** is *not* the total distance traveled but corresponds to the movement between the start and finish of the travel.

---

[1] The arrow itself can represent many physical quantities. As a common example, a velocity vector has its direction aimed in the direction of the motion, and its length equal to the speed of the motion.

Therefore, we can say that displacement is a two-point vector, which does not depend on the path or the period of time between the beginning and end of the observation. The vector addition of the two vector legs of the travel, **a** and **b**, is physically defined by the parallelogram rule, as shown in Figure 2.1, by placing the tail of one arrow on the tip of the other one. It can be easily seen, by a simple drawing, that the operation is **commutative,**

$$\mathbf{a}+\mathbf{b}=\mathbf{b}+\mathbf{a}=\mathbf{c}, \tag{2.2}$$

and **associative**,

$$\mathbf{a}+(\mathbf{b}+\mathbf{c})=(\mathbf{a}+\mathbf{b})+\mathbf{c}=\mathbf{a}+\mathbf{b}+\mathbf{c}. \tag{2.3}$$

Multiplication of a vector by a positive scalar changes only its magnitude, not direction. That is, in Equation 2.1, the multiplication of the vector **a** by λ becomes $\lambda\,|\mathbf{a}|\,\hat{\mathbf{a}}$, with unchanged direction of $\hat{\mathbf{a}}$ and magnitude of $\lambda\ |\mathbf{a}|$. Of course, the operation is distributive: $\lambda(\mathbf{a}+\mathbf{b})=\lambda\mathbf{a}+\lambda\mathbf{b}$. Multiplication by a negative scalar reverses the sense of the arrow $(\lambda\mathbf{a}=-\lambda\,|\mathbf{a}|\,\hat{\mathbf{a}}=-\lambda\,|\mathbf{a}|\,(-\hat{\mathbf{a}}))$ and multiplication by zero results in a **zero vector**, with no magnitude and undefined direction. The use of these definitions leads to definition of vector subtraction:

$$\mathbf{c}-\mathbf{a}=\mathbf{c}+(-1)\mathbf{a}=\mathbf{c}+(-\mathbf{a})=\mathbf{b} \tag{2.4}$$

Thus the negative of a vector is defined as one with the same magnitude but opposite sense, or direction.

There are three other ways to combine two vectors using vector operations, similar to the scalar operation of multiplication; hence the use of the word "product." Two of these operations will be defined from the physical picture in this section, and the remaining operation will be presented in the section on tensors.

Dot Product

There are two principal kinds of information that are of interest in vector operations. First, it is often necessary to know the length of a vector, as expressed along another line; that is, the orthogonal "projection" of the vector.

For example, to return to the idea of an insect traveling on this page (imagine this page to be a plane), one might ask how far did the insect *appear to travel*[2] along an arbitrary direction. Figure 2.2 illustrates the question.

This operation may be accomplished with the use of the vector operation

[2] The *appear to travel* phrase is quite revealing of the purpose of the dot product. It is typically employed to relate a vector action to a given observer's reference frame; that is, to relate a vector action to the way it appears to a given observer. In Figure 2.2, the observer sees only the motion marked by length d along the shown arbitrary direction, $\hat{\mathbf{d}}$.

called the **dot product,** the **inner product,** or the **scalar product.** Where $\hat{\mathbf{d}}$ is a unit vector along the projection line and **c** is an arbitrary vector (perhaps the resulting vector in Figure 2.1), the projection of **c** onto the line of $\hat{\mathbf{d}}$ is denoted

$$\hat{\mathbf{d}} \cdot \mathbf{c} = |\mathbf{c}| \cos\theta \tag{2.5}$$

where θ is the angle between $\hat{\mathbf{d}}$ and $\hat{\mathbf{c}}$, and $|\mathbf{c}|$ means the length of vector **c**. The more general definition of the dot product (which is not used as often) may be obtained by multiplying the unit vector $\hat{\mathbf{d}}$ (e.g., to obtain **a**) by an arbitrary length, $|\mathbf{a}|$, thereby generalizing both vectors completely:

$$\mathbf{a} \cdot \mathbf{b} = |\mathbf{a}|\,|\mathbf{b}| \cos\theta \tag{2.6}$$

where we have used **a** and **b** to represent any two arbitrary vectors.

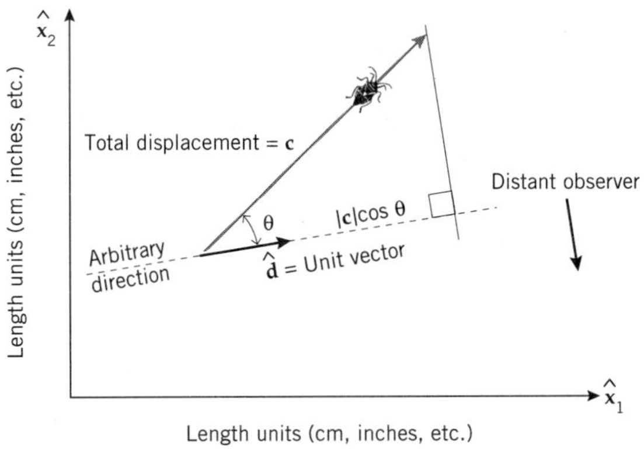

Figure 2.2 An insect's total displacement appears different to an observer at an angle to the motion. This problem defines the dot product.

Either Equation 2.5 or Equation 2.6 may be considered the vector definition of the dot product. Equation 2.6 shows that the length of a vector can be found with the aid of the dot product, since $\theta = 0$ and $\cos\theta = 1$; thus $\mathbf{a} \cdot \mathbf{a} = |\mathbf{a}|^2$ or $\hat{\mathbf{a}} \cdot \mathbf{a} = |\mathbf{a}|$. Equation 2.6 also shows most simply that the scalar product operation is commutative:

$$\mathbf{a} \cdot \mathbf{b} = \mathbf{b} \cdot \mathbf{a} = |\mathbf{a}|\,|\mathbf{b}| \cos\theta = |\mathbf{b}|\,|\mathbf{a}| \cos\theta, \tag{2.7}$$

It may also be shown that vector addition is distributive over addition:

$$\mathbf{d} \cdot (\mathbf{a} + \mathbf{b}) = (\mathbf{d} \cdot \mathbf{a}) + (\mathbf{d} \cdot \mathbf{b}), \tag{2.8}$$

while multiplication by a scalar may be deconvoluted as shown:

$$\lambda(\mathbf{a} \cdot \mathbf{b}) = (\lambda\mathbf{a}) \cdot \mathbf{b} = \mathbf{a} \cdot (\lambda\mathbf{b}). \tag{2.9}$$

Equation 2.9 may be deduced physically by looking at a figure combined of Figures 2.1 and 2.2. Clearly, no matter how the insect traveled, the combination of displacement vectors, when added, must appear the same to the far-off observer. Figure 2.3 shows the physical reality of the distributive principle, Equation 2.8.

It is convenient to treat one of the vectors as a unit vector, without altering the meaning of the equations. The equivalent to Equation 2.8 is obtained formally by writing $\mathbf{d}$ as $|\mathbf{d}|\hat{\mathbf{d}}$ and dividing both sides of the equation by the scalar length $|\mathbf{d}|$:

$$\hat{\mathbf{d}}\cdot(\mathbf{a}+\mathbf{b})=\hat{\mathbf{d}}\cdot\mathbf{c}=(\hat{\mathbf{d}}\cdot\mathbf{a})+(\hat{\mathbf{d}}\cdot\mathbf{b}) \tag{2.8a}$$

Then, Figure 2.3 illustrates the geometric equality.

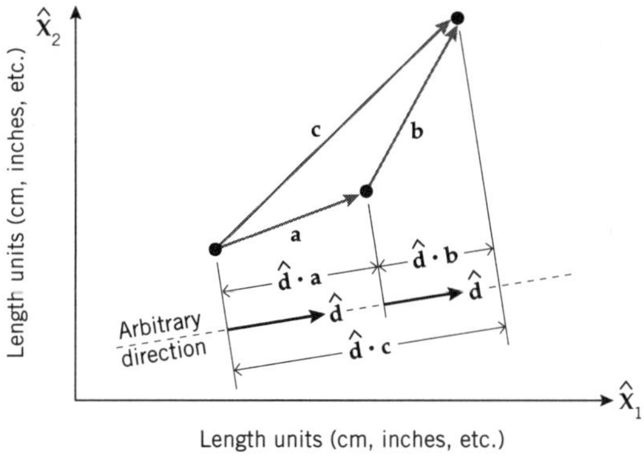

Figure 2.3 Geometric demonstration that the dot product is distributive over vector addition.

Cross Product

The other information of great interest in combining two vectors is the definition of an area in space: that is, the size and orientation of a planar region. Returning again to our friend the insect, his travels along two vectors define an area in space on a certain plane; specifically, the plane of the page. We would therefore like to define a vector operator that results in a vector representing the orientation and size of the planar region. This result will be another vector because it contains size (area) and direction information.

Therefore, we define the vector product, or cross product, to produce a third vector normal to the first two (and thus normal to the plane defined by the two) with a length equal to the area of the enclosed parallelogram, Figure 2.4

$$\mathrm{c}=\mathbf{a}\times\mathbf{b}=|\mathbf{a}|\,|\mathbf{b}|\sin\theta\,\hat{\mathrm{c}},\ \text{where } (\hat{\mathbf{c}}\perp\mathbf{a},\ \hat{\mathbf{c}}\perp\mathbf{b}), \tag{2.10}$$

i.e., the direction of $\mathbf{c}$ is normal to both $\mathbf{a}$ and $\mathbf{b}$. By convention, we choose the sense of $\mathbf{c}$ by the right-hand rule: when the fingers of the right hand lie parallel to **a** and curl up (make a fist) toward **b**, the thumb lies along the positive sense of **c**. Because of the sign convention, the cross product is not commutative; rather, it is anti-commutative:

$$\mathbf{a} \times \mathbf{b} = -(\mathbf{b} \times \mathbf{a}). \tag{2.11}$$

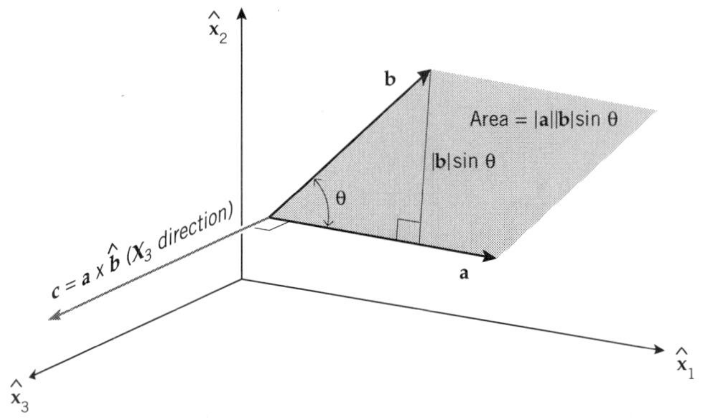

Figure 2.4 Geometric definition of the cross product.

However, it can be shown that the cross product is distributive over vector addition:

$$\mathbf{a} \times (\mathbf{b} + \mathbf{c}) = (\mathbf{a} \times \mathbf{b}) + (\mathbf{a} \times \mathbf{c}), \tag{2.12}$$

and, similar to the dot product, it may be simply deconvoluted for multiplication by a scalar:

$$\lambda(\mathbf{a} \times \mathbf{b}) = (\lambda\mathbf{a}) \times \mathbf{b} = \mathbf{a} \times (\lambda\mathbf{b}) \tag{2.13}$$

Equation 2.13 is very useful when evaluating the cross product in terms of component vectors.

2.2 BASE VECTORS

It is often convenient to decompose vectors into a linear combination of other vectors oriented along prescribed directions. This procedure makes clear the physical connection between vectors (physical "arrows") and their **components** (numbers describing the arrow to a given observer). While the physical vector in no way depends on any coordinate system, the components that describe it to a given observer very clearly depend on the observer's reference.

Any set of three non-coplanar (i.e., 3-D space filling) **base vectors** can be used to express all other vectors. Notice that although a special term is given to these constituent vectors, it applies only to their intended use and does not denote any difference in their physical reality. That is, base vectors are simply vectors, with the added intention that we make use of them for expressing other vectors in the system. (The set of three perpendicular unit vectors used with a **Cartesian coordinate system** is called an **orthonormal basis**.)

> *Note: A* ***Cartesian coordinate system*** *is one based on three mutually perpendicular axes with the same unit of measure along each axis. The corresponding* ***orthonormal basis set*** *is the three unit vectors directed along these three axes.*

Let us decompose two arbitrary vectors (for example, our insect displacements, as in Figure 2.1) into components (a_i, b_i) and base vectors ($\mathbf{e}_i$):

$$\begin{aligned} \mathbf{a} &= a_1\hat{\mathbf{e}}_1 + a_2\hat{\mathbf{e}}_2 + a_3\hat{\mathbf{e}}_3 \\ \mathbf{b} &= b_1\hat{\mathbf{e}}_1 + b_2\hat{\mathbf{e}}_2 + b_3\hat{\mathbf{e}}_3. \end{aligned} \tag{2.14}$$

Clearly, the components representing the fixed vectors **a** and **b** (a_i and b_i) depend on the choice of base vectors (or observers, as we have said), and, therefore, the generality of vector rules is obscured when written in component form. Although we have shown the base vectors as having unit length (and this is a convenient way to define them), there is no necessity to do so. Furthermore, it is often convenient to take the base vectors as an orthogonal set, thus forming a Cartesian coordinate system and orthonormal basis set. When we wish to *emphasize* that the base vectors represent an orthonormal set, we will use the labels $\hat{\mathbf{x}}_1, \hat{\mathbf{x}}_2, \hat{\mathbf{x}}_3$ in place of $\hat{\mathbf{e}}_1, \hat{\mathbf{e}}_2, \hat{\mathbf{e}}_3$ for the base vectors.

The formal connection between a set of base vectors and the corresponding components facilitates deriving and remembering of the component mathematical operations from the vector definitions. (In the next section, we will show the vector operations in component form for Cartesian coordinate systems.)

Vector operations can often be carried out most clearly by resorting to the base vector/component form. We illustrate this usage below without regard to the specific basis, which we denote by the base vectors $\hat{\mathbf{e}}_1, \hat{\mathbf{e}}_2, \hat{\mathbf{e}}_3$. For example, vector addition becomes a simple matter of multiple scalar addition when carried out in component form:

$$\begin{array}{l} \quad a = a_1\hat{e}_1 + a_2\hat{\mathbf{e}}_2 + a_3\hat{e}_3 \leftrightarrow a_i \\ +b = b_1\hat{e}_1 + b_2\hat{\mathbf{e}}_2 + b_3\hat{e}_3 \leftrightarrow b_i \\ \hline \quad c = (a_1+b_1)\hat{e}_1 + (a_2+b_2)\hat{e}_2 + (a_3+b_3)\hat{e}_3 \leftrightarrow a_i + b_i = c_i \end{array} \tag{2.15}$$

(any coordinate system)

The dot and cross products can be similarly constructed in terms of components and base vectors, by making use of their respective distributive properties over addition, Equations 2.8 and 2.12.

$$\begin{aligned}\mathbf{a}\cdot\mathbf{b} &= (a_1\hat{\mathbf{e}}_1 + a_2\hat{\mathbf{e}}_2 + a_3\hat{\mathbf{e}}_3)\cdot(b_1\hat{\mathbf{e}}_1 + b_2\hat{\mathbf{e}}_2 + b_3\hat{\mathbf{e}}_3)\\ &= a_1b_1\hat{\mathbf{e}}_1\cdot\hat{\mathbf{e}}_1 + a_1b_2\hat{\mathbf{e}}_1\cdot\hat{\mathbf{e}}_2 + a_1b_3\hat{\mathbf{e}}_1\cdot\hat{\mathbf{e}}_3\\ &+ a_2b_1\hat{\mathbf{e}}_2\cdot\hat{\mathbf{e}}_1 + a_2b_2\hat{\mathbf{e}}_2\cdot\hat{\mathbf{e}}_2 + a_2b_3\hat{\mathbf{e}}_2\cdot\hat{\mathbf{e}}_3\\ &+ a_3b_1\hat{\mathbf{e}}_3\cdot\hat{\mathbf{e}}_1 + a_3b_2\hat{\mathbf{e}}_3\cdot\hat{\mathbf{e}}_2 + a_3b_3\hat{\mathbf{e}}_3\cdot\hat{\mathbf{e}}_3\end{aligned} \tag{2.6a}$$

(any coordinate system)

The cross product may be similarly expanded, because it is also distributive over vector addition, and the resulting indicial form illustrates the definition and use of the permutation operator:

$$\begin{aligned}\mathbf{a}\times\mathbf{b} &= (a_1\hat{\mathbf{e}}_1 + a_2\hat{\mathbf{e}}_2 + a_3\hat{\mathbf{e}}_3)\times(b_1\hat{\mathbf{e}}_1 + b_2\hat{\mathbf{e}}_2 + b_3\hat{\mathbf{e}}_3)\\ &= a_1b_1\hat{\mathbf{e}}_1\times\hat{\mathbf{e}}_1 + a_1b_2\hat{\mathbf{e}}_1\times\hat{\mathbf{e}}_2 + a_1b_3\hat{\mathbf{e}}_1\mathrm{x}\hat{\mathbf{e}}_3\\ &+ a_2b_1\hat{\mathbf{e}}_2\times\hat{\mathbf{e}}_1 + a_2b_2\hat{\mathbf{e}}_2\times\hat{\mathbf{e}}_2 + a_2b_3\hat{\mathbf{e}}_2\times\hat{\mathbf{e}}_3\\ &+ a_3b_1\hat{\mathbf{e}}_3\times\hat{\mathbf{e}}_1 + a_3b_2\hat{\mathbf{e}}_3\times\hat{\mathbf{e}}_2 + a_3b_3\hat{\mathbf{e}}_3\times\hat{\mathbf{e}}_3\end{aligned} \tag{2.10a}$$

(any coordinate system)

2.3 VECTOR ALGEBRA - MATHEMATICAL APPROACH

With the introduction of base vectors and corresponding components in the previous section, it is useful to define notation representing a given vector and its components (defined in a certain basis). For this purpose, we define and use two closely related notations for vectors, which we shall call the **matrix form** and the **indicial form:**

$$\mathbf{a} \leftrightarrow [a] = \begin{bmatrix} a_1 \\ a_2 \\ a_3 \end{bmatrix} \leftrightarrow a_i \cdot \tag{2.16}$$

The matrix and indicial forms make explicit use of the **components** of the vector, that is, the individual numbers that describe its direction and length. Unfortunately, the values of the individual components depend on the choice of coordinate system, or observer, so the physical, invariant nature of the vector itself is obscured. That is, [a] or a_i may or may not represent the components of a physical vector.

Equation 2.16 requires a bit more explanation for clarity. The matrix form of representation is intermediate, with the first form, [a], emphasizing the connection with the vector label, and the columnar representation emphasizing the connection with the individual components. The indicial notation assumes that the subscript, or **index** (hence "indicial"), i, can take any of the values up to the dimension of the space, usually three. Therefore, a_i signifies one of three ordered values corresponding to directions in space and suggests that the other subscript values exist and play the same role. Thus the assembly of a_i's may be visualized as a matrix. The double arrows replace equal signs in Equation 2.16 to emphasize that a vector is not simply a set of components, but rather that the components are an alternate representation of the physical entity.

Each of the forms of notation has distinct advantages. The vector form has meaning independent of coordinate system and clearly identifies the symbol as a vector. The matrix notation is most convenient for visualizing the mathematical manipulations, but there is no assurance that the components represent a vector at all, or what coordinate system is used to define the components. The indicial notation is clearly the most compact and carries the least physical information. Unless stated elsewhere, the indicial notation does not distinguish among vectors and nonphysical quantities, and does not specify to the observer, the coordinate system upon which the components depend, or the dimensionality of the space. As we shall see later in this chapter, however, the indicial form is very convenient for computer programming because of the simple and rigid rules for manipulation of the indices. Throughout this book, we will use the form that is most convenient for a given application and will often show the alternate forms for comparison.

Addition, in Indicial and Matrix Notation

To illustrate the use of the alternate notations, we recall sample vector equations from Section 2.1:

$$\mathbf{a}+\mathbf{b}=\mathbf{b}+\mathbf{a}=\mathbf{c} \qquad \text{(commutative law)} \tag{2.2}$$

$$\mathbf{a}+(\mathbf{b}+\mathbf{c})=(\mathbf{a}+\mathbf{b})+\mathbf{c}=\mathbf{a}+\mathbf{b}+\mathbf{c} \qquad \text{(associative law)} \tag{2.3}$$

$$\mathbf{c}-\mathbf{a}=\mathbf{c}+(-1)\mathbf{a}=\mathbf{c}+(-\mathbf{a})=\mathbf{b} \qquad \text{(vector subtraction)} \tag{2.4}$$

It can easily be verified that Equations 2.2, 2.3, and 2.4 can be rewritten in matrix and indicial form by defining base vectors and looking at the physical addition process embodied in the parallelogram rule:

$$\begin{bmatrix} a_1 \\ a_2 \\ a_3 \end{bmatrix} + \begin{bmatrix} b_1 \\ b_2 \\ b_3 \end{bmatrix} = \begin{bmatrix} c_1 \\ c_2 \\ c_3 \end{bmatrix} \leftrightarrow a_i + b_i = c_i \tag{2.2a-b}$$

$$\begin{bmatrix} a_1 \\ a_2 \\ a_3 \end{bmatrix} + \left(\begin{bmatrix} b_1 \\ b_2 \\ b_3 \end{bmatrix} + \begin{bmatrix} c_1 \\ c_2 \\ c_3 \end{bmatrix} \right) = \left(\begin{bmatrix} a_1 \\ a_2 \\ a_3 \end{bmatrix} + \begin{bmatrix} b_1 \\ b_2 \\ b_3 \end{bmatrix} \right) + \begin{bmatrix} c_1 \\ c_2 \\ c_3 \end{bmatrix} = \begin{bmatrix} a_1 \\ a_2 \\ a_3 \end{bmatrix} + \begin{bmatrix} b_1 \\ b_2 \\ b_3 \end{bmatrix} + \begin{bmatrix} c_1 \\ c_2 \\ c_3 \end{bmatrix} \quad \text{or} \tag{2.3a}$$

$$a_i + (b_i + c_i) = (a_i + b_i) + c_i = a_i + b_i + c_i \tag{2.3b}$$

$$\begin{bmatrix} c_1 \\ c_2 \\ c_3 \end{bmatrix} - \begin{bmatrix} a_1 \\ a_2 \\ a_3 \end{bmatrix} = \begin{bmatrix} c_1 \\ c_2 \\ c_3 \end{bmatrix} + (-1) \begin{bmatrix} a_1 \\ a_2 \\ a_3 \end{bmatrix} = \begin{bmatrix} b_1 \\ b_2 \\ b_3 \end{bmatrix} \quad \text{or} \tag{2.4a}$$

$$c_i - a_i = c_i + (-1)\, a_i = c_i - a_i = b_i \tag{2.4b}$$

Comparison of the indicial forms with matrix forms illustrates the relative compactness and obscurity of the former.

Exercise 2.1 **Show that the numerical definition of vector addition in a Cartesian coordinate system, Equation 2.2a, agrees with the physical definition of vector addition shown in Figure 2.1.**

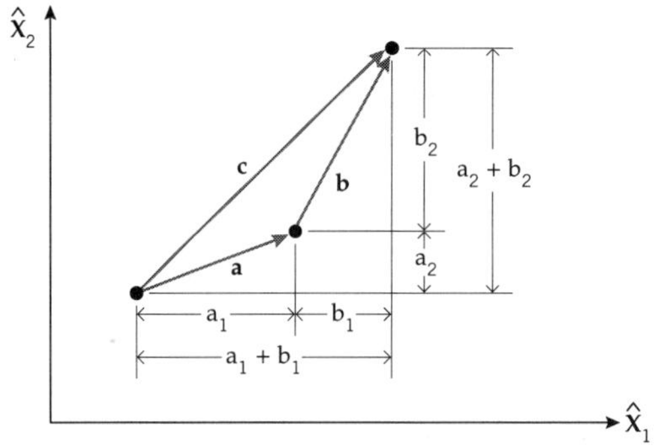

By decomposing each of the vectors to be added into components and arranging the vectors tip-to-tail, we can easily illustrate the process of vector addition:

$$\mathbf{c} = \mathbf{a} + \mathbf{b} \leftrightarrow [c] = [a] + [b] = \begin{bmatrix} a_1 + b_1 \\ a_2 + b_2 \end{bmatrix} = \begin{bmatrix} c_1 \\ c_2 \end{bmatrix} \tag{2.1-1}$$

This operation also demonstrates that vector definitions are independent of translation.

***Note**: Equations written in matrix and indicial form (e.g., Equations 2.2 through 2.4b) are valid only when the components of all vectors are defined in the same coordinate system. Again, Equations 2.2, 2.3, and 2.4 are always true because they are independent of the observer, while the forms involving components have the stated restriction.*

Manipulations to illustrate the commutative and distributive properties of the dot product and vector addition could have been performed much more succinctly using indicial or matrix notation. However, without checking, one cannot be certain that the indicial result, for example, will be valid for every coordinate system, mixed coordinate systems, and so on. Therefore, it is best to rely on **frame-invariant** results and to generalize subsequently.

Products: Summation Convention, Kronecker Delta, Permutation Operator

Introduction of a Cartesian coordinate system provides many simplifications, especially for multiplicative operations. Some new symbols and conventions make this representation even more compact. With the help of the vector definition of the dot product, the base vector dot products can easily be evaluated for a Cartesian coordinate system, because the angle between any nonidentical base vectors is $\theta = 90°$, and between identical base vectors $\theta = 0°$. Further, since unit base vectors are chosen, the results are particularly simple:

$$\hat{\mathbf{x}}_i \cdot \hat{\mathbf{x}}_j = \begin{Bmatrix} 1 \text{ if } i = j \\ 0 \text{ if } i \neq j \end{Bmatrix} \equiv \delta_{ij} \tag{2.17}$$

(Cartesian coordinate system)

and with this additional notation, Equation 2.17 may be evaluated for a Cartesian system:

$$\mathbf{a} \cdot \mathbf{b} = a_1 b_1 + a_2 b_2 + a_3 b_3 = \begin{bmatrix} a_1 a_2 a_3 \end{bmatrix} \begin{bmatrix} b_1 \\ b_2 \\ b_3 \end{bmatrix} = a_i b_i = a_i b_j \delta_{ij} \tag{2.6b}$$

The new symbol, δ_{ij}, is called the **Kronecker delta** and has the property and definition shown in Equation 2.17. Its use is shown in Equation 2.6b, where only the nonzero terms of Equation 2.6a, which have indices i = j, are retained.

Equation 2.6b embodies several conventions that must be understood in order to carry out the necessary arithmetic. The matrix notation shown implies that each element in the row of [a] is multiplied by the corresponding element in the column of [b] and the resulting terms are summed to obtain a single quantity. This is implied by the writing of the **a**-matrix and **b**-matrix adjacent to one another, without an intervening symbol. The same operation is implied in the indicial form by the $a_i b_i$. Because the index i is repeated in a single term, a summation over the index is implied. This rule is often called the **Einstein summation convention,** and it allows a wide range of systems of equations to be written clearly and compactly. An index such as i, which is repeated in a term, is called a **dummy index** because the result does not contain it. The other kind of index is called a **free index** because it may be chosen arbitrarily and appears in the result of the equation. Usually, both kinds of indices have ranges of 1 to 3 for 3-D space, but the notation is general and may be used for other kinds of sums.

Exercise 2.2 Show that Equation 2.6b (Cartesian coordinates) follows directly from the physical definition of the dot product, Equations 2.5 or 2.6.

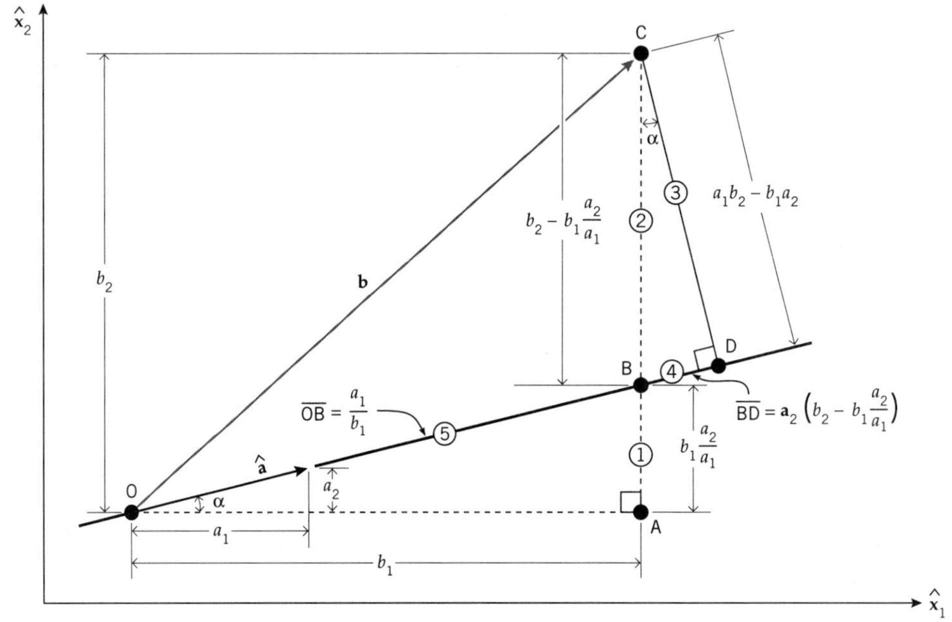

To assist in the geometry, we arrange the two 2-D vectors with the tails in common and assume that one is a unit vector (times its magnitude):

$$\mathbf{a} \cdot \mathbf{b} = |\mathbf{a}|\ \hat{\mathbf{a}} \cdot \mathbf{b} \tag{2.2-1}$$

Then we find the projection of **b** onto the line containing $\hat{\mathbf{a}}$ by applying trigonometric rules (see figure).

The five line lengths (numbered 1, 2, 3, 4, 5) are found by noting that triangles BOA, and BDC are similar to the unit vector triangle with sides shown as 1, $\hat{a}_1$, and $\hat{a}_2$, where $\hat{a}_1$ and $\hat{a}_2$ are the Cartesian components of the unit vector $\hat{\mathbf{a}}$. The result of $\hat{\mathbf{a}} \cdot \mathbf{b}$ is, physically, the projection of **b** onto the line containing **a**; that is, line segment OB. The length is simply the sum of the two line lengths 4 (segment BD) and 5 (segment OB):

$$\hat{\mathbf{a}} \cdot \mathbf{b} = \frac{b_1}{\hat{a}_1} + \hat{a}_2\left(b_2 - b_1 \frac{\hat{a}_2}{\hat{a}_1}\right) \tag{2.2-2}$$

$$= \frac{b_1}{\hat{a}_1} + \hat{a}_2 b_2 - b_1 \frac{\hat{a}_2{}^2}{\hat{a}_1} \qquad \textbf{(2.2-3)}$$

$$= \frac{b_1}{\hat{a}_1} + \hat{a}_2 b_2 - \frac{b_1}{a_1} + b_1 - a_1 \qquad \textbf{(2.2-4)}$$

$$\hat{\mathbf{a}} \cdot \mathbf{b} = \hat{a}_1 b_1 + \hat{a}_2 b_2 \qquad \textbf{(2.2-5)}$$

To find the general result, when **a** is not a unit vector, we need only to scale up the components of $\hat{\mathbf{a}}$ to **a** (i.e., to multiply each component by the magnitude of **a**). Thus $\hat{\mathbf{a}}_1$ becomes $\mathbf{a}_1$, $\hat{\mathbf{a}}_2$, becomes $\mathbf{a}_2$, and the general result is obtained:

$$\mathbf{a} \cdot \mathbf{b} = a_1 b_1 + a_2 b_2. \qquad (2.2\text{-}6)$$

The cross-product expression, Equation 2.10, can be similarly simplified by adopting a Cartesian coordinate system and noting the following:

$$\hat{\mathbf{x}}_i \times \hat{\mathbf{x}}_j = \pm \hat{\mathbf{x}}_k \qquad (2.18)$$

where $i \neq j \neq k$ and the plus sign is used if (i,j,k) takes an even permutation—(1,2,3), (2,3,1), or (3,1,2)—and the minus sign is used if (i,j,k) is an odd permutation—(1,3,2), (2,1,3), or (3,2,1). For $i = j$, $i = k$, or $j = k$, the result is a null vector.

With this simplification (arising from the orthonormal basis), we can write

$$\mathbf{c} = \mathbf{a} \times \mathbf{b} = \begin{vmatrix} \hat{\mathbf{x}}_1 & \hat{\mathbf{x}}_2 & \hat{\mathbf{x}}_3 \\ a_1 & a_2 & a_3 \\ b_1 & b_2 & b_3 \end{vmatrix} = (a_2 b_3 - a_3 b_2)\hat{\mathbf{x}}_1 - (a_1 b_3 - a_3 b_1)\hat{\mathbf{x}}_2 + (a_1 b_2 - a_2 b_1)\hat{\mathbf{x}}_3, \qquad (2.10b)$$

where | | indicates the standard **determinant,** which is used here in an unusual form involving vectors as a memory device for the expanded form of Equation 2.10b. The determinant is introduced more formally in Section 2.5.

The matrix form is shown as the result of a determinant operation—a very strange one, though, contrived of mixed components and vectors that serves as a simple guide to remember the procedure. The indicial notation relies on introducing the symbol ε_{ijk} , known as the **permutation operator**, specifically for this purpose (i.e., it allows writing of determinants indicially). ε_{ijk} has the following definition:

$$\varepsilon_{ijk} = \begin{Bmatrix} 0 & \text{unless} \quad i \neq j \neq k \\ 1 \text{ for } 1,2,3; \; 2,3,1; \; 3,1,2 \\ -1 \text{ for } 3,2,1; \; 1,3,2; \; 2,1,3 \end{Bmatrix} \qquad (2.19a)$$

so that the cross product of orthonormal base vectors can be expressed as

$$\hat{\mathbf{x}}_i \times \hat{\mathbf{x}}_j = \varepsilon_{ijk}\, \hat{\mathbf{x}}_k \,. \tag{2.19b}$$

(Cartesian coordinate system)

With the introduction of the permutation operator, Equation 2.19, the indicial form of the cross product may be written compactly:

$$c_i = \varepsilon_{ijk} a_j b_k \,. \tag{2.10c}$$

(Cartesian coordinate system)

Reference to Equations 2.6a and b and 2.10a, b, and c illustrates the natural definitions of the Kronecker delta, permutation operator, and summation convention when performing vector algebra in a Cartesian coordinate system. The summation convention allows one to consider all combinations of repeated indices while δ_{ij} and ε_{ijk} "weed out" (and assign the proper sign to) the terms that are zero in an orthonormal basis set. Note that when using a nonorthonormal basis set, the simplifications cannot be made indicially, and the complete form of the vector operations must be retained—for example, Equations 2.6 and 2.10.

Exercise 2.3 **Show that Equations 2.10b and c are equivalent to the physical definition of the cross product, Equation 2.10.**

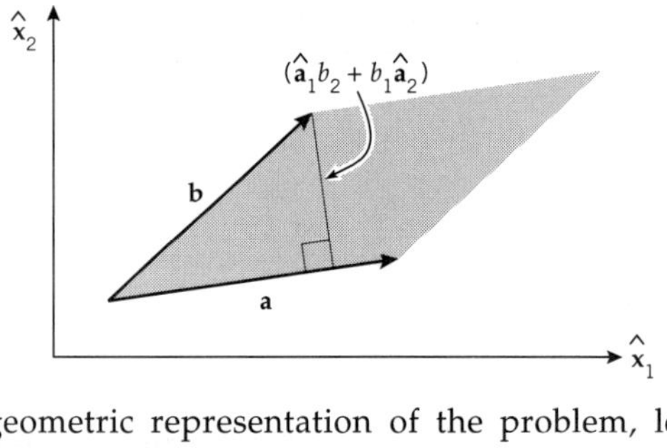

To simplify the geometric representation of the problem, let's first restrict our attention to two vectors lying in the $\hat{x}_1 - \hat{x}_2$ plane such that the resultant vector lies along the $\hat{x}_3$ axis (see figure). Using the same geometric construction as in Exercise 2.2, we find the altitude (normal to **a**), and thus the area of the parallelogram (base times altitude):

$$\text{Area} = |a|\,(\hat{a}_1 b_2 - b_1 \hat{a}_2) = a_1 b_2 - b_1 a_2 \tag{2.3-1}$$

The direction is known to be $+\hat{\mathbf{x}}_3$ by the right-hand rule and the orthogonality condition. Therefore, for this limited case, the determinant formula does indeed provide the correct result:

$$\begin{vmatrix} \hat{x}_1 & \hat{x}_2 & \hat{x}_3 \\ a_1 & a_2 & 0 \\ b_1 & b_2 & 0 \end{vmatrix} = (a_1 b_2 - b_1 a_2)\, \hat{x}_3 \tag{2.3-2}$$

To generalize this result to arbitrary vectors, it is necessary only to consider the 2-D case as a projection of 3-D vectors onto the $\hat{x}_1 - \hat{x}_2$ plane. The projected area is therefore the length along $\hat{x}_3$. Similarly, the 3-D vectors could be projected to the $\hat{x}_1 - \hat{x}_3$ and $\hat{x}_2 - \hat{x}_3$ planes to obtain the resulting vector components along the $\hat{x}_2$ and $\hat{x}_1$ axes. Thus the full result is found by summing the three components of the resultant vector:

$$\begin{vmatrix} \hat{x}_1 & \hat{x}_2 & \hat{x}_3 \\ a_1 & a_2 & a_3 \\ b_1 & b_2 & b_3 \end{vmatrix} = (a_2b_3 - a_3b_2)\hat{x}_1 + (-1)(a_1b_3 - a_3b_1)\hat{x}_2 + (a_1b_2 - a_2b_1)\hat{x}_3 \qquad (2.3\text{-}3)$$

2.4 ROTATION OF CARTESIAN AXES (CHANGE OF ORTHONORMAL BASIS)

We have emphasized the physical nature of vectors and vector operations, noting that the vectors and operations *do not* depend on the observer in any way. For mathematical manipulations, however, the *numbers* representing a particular vector depend on both the vector and the observer's coordinate system. Therefore, it is necessary to know how to relate the mathematical description of a given physical vector expressed in one coordinate system to the mathematical description of the *same* vector expressed in a second coordinate system.[3]

> ***Note***: *In the developments that follow, we consider only Cartesian coordinate systems, for simplicity. However, the relationships between vector coordinate systems are much more general and can be extended to any sort of space-filling coordinate system.*

The problem may be stated more precisely: For a given physical vector, **a**, what is the relationship between the components of **a** in two coordinate systems, labeled by their base vectors $(\hat{x}_1, \hat{x}_2, \hat{x}_3)$ and $(\hat{x}_1', \hat{x}_2', \hat{x}_3')$? We label the respective components a_1, a_2, a_3 in the original coordinate system and a_1', a_2', a_3' in the prime (second) coordinate system. In base-vector form, we can write

$$\mathbf{a} = a_1\hat{x}_1 + a_2\hat{x}_2 + a_3\hat{x}_3 = a_1'\hat{x}_1' + a_2'\hat{x}_2' + a_3'\hat{x}_3 \qquad (2.20)$$

(any coordinate systems)

where all of the unprimed quantities and the primed base vectors are known, and only the primed components a_1', a_2', a_3' are unknown. Note that Equation 2.20 is a pure expression of our physical requirement that **a** is unchanged even though our coordinate system is changed.

[3] Since a vector represents only the difference in position between its head and tail (or, alternatively, its length and direction), the translation of a coordinate system does not affect a vector's mathematical description.

There are many formal ways to approach this problem using vector forms, regular equations, or matrices. Let's approach it in a piecemeal vector fashion, which seems to make the most physical sense, and then rewrite the solution in various notations. First, let's seek a single unknown coefficient, a_1', for example, while restricting our attention to Cartesian coordinate systems.[4] We know that a_1' is the length of the orthogonal projection of **a** onto $\hat{x}_1'$. Therefore, we can make use of the dot product to find a_1':

$$a_1' = \mathbf{a}\cdot\hat{x}_1' = a_1\hat{x}_1'\cdot\hat{x}_1 + a_2\hat{x}_1'\cdot\hat{x}_2 + a_3\hat{x}_1'\cdot\hat{x}_3 \tag{2.20a}$$

(Cartesian coordinate system)

Similarly, for a_2' and a_3', following the same procedure:

$$a_2' = \mathbf{a}\cdot\hat{x}_2' = a_1\hat{x}_2'\cdot\hat{x}_1 + a_2\hat{x}_2'\cdot\hat{x}_2 + a_3\hat{x}_2'\cdot\hat{x}_3 \tag{2.20b}$$

$$a_3' = \mathbf{a}\cdot\hat{x}_3' = a_1\hat{x}_3'\cdot\hat{x}_1 + a_2\hat{x}_3'\cdot\hat{x}_2 + a_3\hat{x}_3'\cdot\hat{x}_3 \tag{2.20c}$$

Because Equation 2.20 is written in vector form, it is independent of the choice of Cartesian coordinate system. Unit vectors are indicated; thus we can easily apply the definition of the dot product to evaluate the new Cartesian components of **a**.

$$\begin{aligned} a_1' &= a_1 \cos(\hat{x}_1',\hat{x}_1) + a_2 \cos(\hat{x}_1',\hat{x}_2) + a_3 \cos(\hat{x}_1',\hat{x}_3) \\ a_2' &= a_1 \cos(\hat{x}_2',\hat{x}_1) + a_2 \cos(\hat{x}_2',\hat{x}_2) + a_3 \cos(\hat{x}_2',\hat{x}_3) \\ a_3' &= a_1 \cos(\hat{x}_3',\hat{x}_1) + a_2 \cos(\hat{x}_3',\hat{x}_2) + a_3 \cos(\hat{x}_3',\hat{x}_3) \end{aligned} \tag{2.21}$$

By defining the angles between the various pairs of base vectors in simpler notation, Equation 2.21 can be simplified in appearance:

$$\begin{aligned} a_1' &= R_{11}a_1 + R_{12}a_2 + R_{13}a_3 \\ a_2' &= R_{21}a_1 + R_{22}a_2 + R_{23}a_3 \\ a_3' &= R_{31}a_1 + R_{32}a_2 + R_{33}a_3 \end{aligned}$$

or

$$\begin{bmatrix} a_1' \\ a_2' \\ a_3' \end{bmatrix} = \begin{bmatrix} R_{11} & R_{12} & R_{13} \\ R_{21} & R_{22} & R_{23} \\ R_{31} & R_{32} & R_{33} \end{bmatrix} \begin{bmatrix} a_1 \\ a_2 \\ a_3 \end{bmatrix} \tag{2.22}$$

[4] The use of a Cartesian coordinate system allows us to make use of the dot product, which gives the **orthogonal** properties of a vector onto a given unit vector.

or

$$[a'] = [R]\,[a]$$

or

$$a_i' = R_{ij} a_j$$

where R_{ij} represents the cosine of the angle between the ith base vector in the primed system and the jth base vector in the original (unprimed) system. Although, as usual, some physical insight is lost in proceeding from vector form to matrix form to indicial form, a simple rule to remember is that the subscripts (or matrix position) must agree with the base vector subscript (or matrix position). The aids are shown below:

$$\text{old} \rightarrow$$

$$\begin{matrix}\text{n}\\\text{e}\\\text{w}\\\downarrow\end{matrix}\begin{bmatrix} a_1' \\ a_2' \\ a_3' \end{bmatrix} = \begin{matrix}\text{n}\\\text{e}\\\text{w}\\\downarrow\end{matrix}\begin{bmatrix} R_{11} & R_{12} & R_{13} \\ R_{21} & R_{22} & R_{23} \\ R_{31} & R_{32} & R_{33} \end{bmatrix} \begin{bmatrix} a_1 \\ a_2 \\ a_3 \end{bmatrix}\begin{matrix}\text{o}\\\text{l}\\\text{d}\\\downarrow\end{matrix}$$

$$a_i' = R_{ij}\, a_j \tag{2.23}$$

The arrows shown serve to remind the user of the connection between R_{ij} and two sets of vector components a_i and a_i'. By using the arrows, one can remember that R_{12} must be the cosine of the angle between $\hat{x}_1'$ and $\hat{x}_2$ (rather than $\hat{x}_1$ and $\hat{x}_2'$). There is nothing magic about the designations of the unprimed coordinate system as old and the primed one as new. One must only remember that the row number of [R] corresponds to the resulting vector components, while the column marker refers to the given vector components.

> ***Note***: *Equation 2.23 is the general form of transformation for the components of a vector between two coordinate systems. (Even more generally, Equations 2.22 and 2.23 are known as linear transformations.) The use of* R *(rather than* T*, for example) is meant to signify a rotation rather than a general linear transformation between non-orthogonal systems, or between systems having different length units (i.e., non-unitary base vectors).*

[R] has a special symmetry when both coordinate systems($\hat{x}_i$, and $\hat{x}_i'$) are orthogonal; namely, that [R] is said to be an **orthogonal matrix.** We will present in the next section the definition of an orthogonal matrix and show that [R] must be of this type. This property greatly simplifies calculations with orthogonal systems.

Exercise 2.4 Using the definition of an orthogonal matrix provided in the next section, along with the physical situation leading to Equation 2.23, show that [R] must be orthogonal.

We approach the problem in two ways. For 2-D, the geometry of the problem can be solved using trigonometric identities. For the 3-D generalizations, it is much easier to rely on indicial operations to demonstrate the orthogonality.

For a simple rotation of a Cartesian coordinate system by angle θ (counter-clockwise when looking along the $-\hat{x}_3$ axis) about $\hat{x}_3$, the rotation matrix can be found simply. Using the cosine definition of the rotation matrix (Equations 2.22 and 2.23) and the figure showing all of the important angles in terms of θ, one obtains

$$R \equiv \begin{bmatrix} \cos\hat{x}_1'\hat{x}_1 & \cos\hat{x}_1'\hat{x}_2 & \cos\hat{x}_1'\hat{x}_3 \\ \cos\hat{x}_2'\hat{x}_1 & \cos\hat{x}_2'\hat{x}_2 & \cos\hat{x}_2'\hat{x}_3 \\ \cos\hat{x}_3'\hat{x}_1 & \cos\hat{x}_3'\hat{x}_2 & \cos\hat{x}_3'\hat{x}_3 \end{bmatrix} = \begin{bmatrix} \cos\theta & \cos(90-\theta) & 0 \\ \cos(90+\theta) & \cos\theta & 0 \\ 0 & 0 & 1 \end{bmatrix}$$

$$= \begin{bmatrix} \cos\theta & \sin\theta & 0 \\ -\sin\theta & \cos\theta & 0 \\ 0 & 0 & 1 \end{bmatrix} \tag{2.4-1}$$

The last representation of the rotation matrix should be remembered — it is useful for many changes of coordinate systems (as long as the $\hat{x}_3$ axis is chosen as invariant and

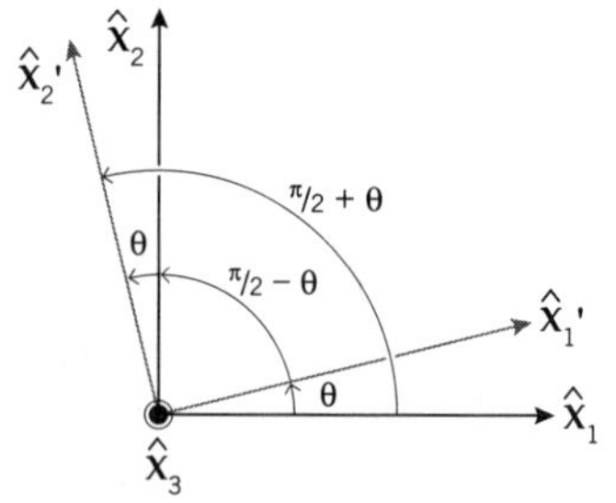

the axis rotation angle is counter-clockwise when viewed along the $-\hat{x}_3$ axis). The orthogonality of this form is easily demonstrated:

$$[R][R]^T = [I] \tag{2.4-2}$$

$$\begin{bmatrix} \cos\theta & \sin\theta & 0 \\ -\sin\theta & \cos\theta & 0 \\ 0 & 0 & 1 \end{bmatrix} \begin{bmatrix} \cos\theta & -\sin\theta & 0 \\ \sin\theta & \cos\theta & 0 \\ 0 & 0 & 1 \end{bmatrix} = \begin{bmatrix} 1 & 0 & 0 \\ 0 & 1 & 0 \\ 0 & 0 & 1 \end{bmatrix} \tag{2.4-3}$$

For the general rotation about two axes (two rotations are sufficient to define a general Cartesian coordinate system), we follow a more mathematical approach. The physical situation is that [R] operates as follows:

$$x_i' = R_{ij}\, x_j \tag{2.4-4}$$

such that the vector $\hat{\mathbf{x}}$ is unchanged. (Its components, x_i, and x_i', depend, of course, on the chosen coordinate system.) Since [R] is a pure rotation, the new basis vectors $(\hat{\mathbf{x}}_i')$ have the same unitary length as the original ones $(\hat{\mathbf{x}}_i)$ and the length of $\mathbf{x}$ is unchanged. Furthermore, since both $\hat{\mathbf{x}}_i$ and $\hat{\mathbf{x}}_i'$ are orthogonal basis sets, the squared length of $\hat{\mathbf{x}}$ is given in both coordinate systems by

$$\mathbf{x}\cdot\mathbf{x} = |\mathbf{x}|^2 = \hat{x}_i\, \hat{x}_i = \hat{x}_j'\, \hat{x}_j', \text{ or } |\mathbf{x}|^2 = \hat{x}_i\, \hat{x}_j\, \delta_{ij} = \hat{x}_i'\, \hat{x}_j'\, \delta_{ij}. \tag{2.4-5}$$

Substituting Equation 2.4-3 into 2.4-4, we obtain a necessary condition on [R] to ensure that the vector length of $\mathbf{x}$ doesn't change:

$$\delta_{ij}\, x_i\, x_j = (R_{ij}\hat{x}_j)\,(R_{ik}x_k)$$

$$\delta_{ij}\, x_i\, x_j = R_{ij}\, R_{ik}\, \hat{x}_j\, x_k$$

$$\delta_{jk} = R_{ij}\, R_{ik} \tag{2.4-6}$$

Equation 2.4-5 may be written in matrix or component form to further illustrate the connection with the definition of an orthogonal matrix:

$$\delta_{jk} = R_{ij}\, R_{ik} \leftrightarrow [R]\,[R]^T = [I], \tag{2.4-7}$$

or

$$\begin{array}{ll} R_{11}^2 + R_{12}^2 + R_{13}^2 = 1 & R_{11}R_{21} + R_{12}R_{22} + R_{13}R_{23} = 0 \\ R_{11}R_{31} + R_{12}R_{32} + R_{13}R_{33} = 0 & R_{21}^2 + R_{22}^2 + R_{23}^2 = 0 \\ R_{21}R_{31} + R_{22}R_{32} + R_{23}R_{33} = 0 & R_{31}^2 + R_{32}^2 + R_{33}^2 = 0 \end{array}$$

Thus [R] must be orthogonal by the usual definition.

2.5 MATRIX OPERATIONS

We have "secretly" introduced matrices into the notation of this chapter in a natural manner, to emphasize the use of the particular notation. For example, Equation 2.22 used the correspondence among systems of equations, matrices, and indices to present a change of orthonormal basis.

Instead of trying to develop all matrix rules by such correspondences, we review below the structure and uses of the basic notation. We have found linear algebra to be one of the most easily learned subjects in mathematics, and therefore present the principal results briefly. For those readers who have not been exposed to linear algebra, reference to any of the standard texts on the subject is recommended.[5]

A matrix is an array of numbers, often two-dimensional ($n \times m$, e.g.), but having any number of dimensions greater than zero.[6] Two-dimensional matrices are the most popular because they are the largest ones simply written on paper and because one-dimensional matrices can usually be replaced by a more convenient notation, such as vectors, to advantage. By convention (amazingly, this is one convention that is often obeyed), the elements of a 2-D matrix are labeled with the first subscript corresponding to the row number and the second corresponding to the column number:

$$[A] = \begin{bmatrix} A_{11} & A_{12} & \cdots & A_{1m} \\ A_{21} & A_{22} & \cdots & A_{2m} \\ \vdots & \vdots & & \vdots \\ A_{n1} & A_{n2} & \cdots & A_{nm} \end{bmatrix} \leftrightarrow A_{ij} \tag{2.24}$$

One-dimensional matrices are called **column matrices** or **row matrices,** depending on their orientation.

Thus Equation 2.24 illustrates various ways to write an array of $n \times m$ numbers in 2-D. (Note: We use the brackets shown to emphasize the matrix form of the array. A single letter may also be occasionally used for compactness).

---

[5] For example, we find the book by I.S. Sokolnikoff and R.M. Redheffer, *Mathematics of Physics and Modern Engineering* (New York, McGraw-Hill, 1958, 1966), presents the principles of a wide range of engineering mathematics in compact form.

[6] The dimensions of a matrix should not be confused with spatial dimensions. Some authors prefer the word "rank" to dimension, but in any case we refer to the number of subscripts needed to represent a general element. Matrices of dimension larger than two are usually modified for representation on paper (two mathematical dimensions), but the n-dimensional representation is often more convenient for writing computer programs.

Matrices may be added or subtracted if they are of the same size:

$$[A] \pm [B] = \begin{bmatrix} A_{11} \pm B_{11} \ A_{12} \pm B_{12} \dots A_{1m} \pm B_{1m} \\ \cdot \\ \cdot \\ \cdot \\ A_{n1} \pm B_{n1} \ A_{n2} \pm B_{n2} \dots A_{nm} \pm B_{nm} \end{bmatrix} \leftrightarrow A_{ij} \pm B_{ij} \qquad (2.25)$$

or, in a more general procedure, by filling out the smaller matrix with zeroes to create the same size artificially. Clearly, matrix addition is **commutative,** because each element is generated by real addition.

Matrix multiplication by a scalar is performed element by element ($\alpha[A] = [\alpha \mathbf{A}] \leftrightarrow \alpha A_{ij}$), while multiplication of two matrices can be performed only when the matrices are said to **conform**, or to be **conformable**. That is, the number of rows in the pre-multiplication matrix must equal the number of columns in the post-multiplication matrix. The multiplication convention is:

$$[A][B] = \begin{bmatrix} A_{11} & A_{12} & A_{13} & \dots \\ A_{21} & A_{22} & A_{23} & \dots \\ A_{31} & A_{32} & A_{33} & \dots \\ \cdot & \cdot & \cdot & \\ \cdot & \cdot & \cdot & \\ \cdot & \cdot & \cdot & A_{nm} \end{bmatrix} \begin{bmatrix} B_{11} & B_{12} & B_{13} & \dots \\ B_{21} & B_{22} & B_{23} & \dots \\ B_{31} & B_{32} & B_{33} & \dots \\ \cdot & \cdot & \cdot & \\ \cdot & \cdot & \cdot & \\ \cdot & \cdot & \cdot & B_{mp} \end{bmatrix} = \begin{bmatrix} C_{11} & C_{12} & C_{13} & \dots \\ C_{21} & C_{22} & C_{23} & \dots \\ C_{31} & C_{32} & C_{33} & \dots \\ \cdot & \cdot & \cdot & \\ \cdot & \cdot & \cdot & \\ \cdot & \cdot & \cdot & C_{np} \end{bmatrix} = [C]$$

$$\text{where: } C_{ij} = A_{ik} B_{kj} = \sum_{k=1}^{m} A_{ik} B_{kj} \qquad (2.26)$$

$$\text{e.g., } \quad C_{23} = A_{21} B_{13} + A_{22} B_{23} + A_{23} B_{33} + \dots + A_{2m} B_{m3}$$

As introduced earlier, it is unnecessary to show the summation explicitly because of the Einstein summation convention. It can easily be seen that matrix multiplication is *not commutative*. For example, unless the matrices are *square* ($n \times n$), multiplication is possible only in one order (and even then is not commutative).

> ***Note***: *Although not formally correct, it is possible to enlarge nonconforming matrices (by introducing additional elements equal to zero) to allow them to conform. This procedure, while appearing strange, and disallowed by mathematical purists, is sometimes convenient in representing an operation. For example, two 1-D matrices could be "multiplied" in two ways, obtaining quite different results (and further demonstrating the lack of a commutation property):*

$$\begin{bmatrix} a_1 \; a_2 \; a_3 \end{bmatrix} \begin{bmatrix} b_1 \\ b_2 \\ b_3 \end{bmatrix} = c \text{ (a scalar)} = a_1 b_1 + a_2 b_2 + a_3 b_3 \tag{2.27}$$

or using generalized multiplication:

$$\begin{bmatrix} a_1 \\ a_2 \\ a_3 \end{bmatrix} \begin{bmatrix} b_1 b_2 b_3 \end{bmatrix} \Rightarrow \begin{bmatrix} a_1 & 0 & 0 \\ a_2 & 0 & 0 \\ a_3 & 0 & 0 \end{bmatrix} \begin{bmatrix} b_1 & b_2 & b_3 \\ 0 & 0 & 0 \\ 0 & 0 & 0 \end{bmatrix} = \begin{bmatrix} a_1 b_1 & a_1 b_2 & a_1 b_3 \\ a_2 b_1 & a_2 b_2 & a_2 b_3 \\ a_3 b_1 & a_3 b_2 & a_3 b_3 \end{bmatrix} \tag{2.28}$$

(a 3 × 3 matrix)

(Equation 2.28 is the component form of the generalized tensor product of two vectors, $a \otimes b$.

Exercise 2.5 **Three mutually orthogonal unit vectors are expressed below in a standard Cartesian coordinate system $\hat{x}_i$.**

$$\begin{aligned} \hat{x}_1{}' &= 0.707,\ 0.707,\ 0.0 \\ \hat{x}_2{}' &= -0.707,\ 0.707,\ 0.0 \\ \hat{x}_3{}' &= 0.000,\ 0.000,\ 1.0 \end{aligned} \tag{2.5-1}$$

Using simple matrix manipulations, find the rotation matrix, [R], which transforms components from the original coordinate system to the primed coordinate system based on the three $\hat{x}_i{}'$ Also, find the transformation of components from the $\hat{x}_i{}'$ system to the $\hat{x}_i$ system.

Without thinking too deeply about the problem, let's just remember the purpose of [R] and what it accomplishes. For example, if we would like the vector $\hat{x}_i{}'$ to become the new $\hat{x}_i$ coordinate axis (there is no necessity to do this; any of the $\hat{x}_i{}'$ could become the first coordinate axis), then our [R] must transform components as follows:

$$\begin{bmatrix} 1.0 \\ 0.0 \\ 0.0 \end{bmatrix} = \begin{bmatrix} R_{11} & R_{12} & R_{13} \\ R_{21} & R_{22} & R_{23} \\ R_{31} & R_{32} & R_{33} \end{bmatrix} \begin{bmatrix} 0.707 \\ 0.707 \\ 0.000 \end{bmatrix} \tag{2.5-2}$$

because we want the components of $\hat{x}_i{}'$ expressed in the original coordinate system (0.707, 0.707, 0.0) to become the $\hat{x}_1$ axis (i.e., 1, 0, 0) in the new coordinate system we are constructing. Similarly, $\hat{x}_2{}'$ and $\hat{x}_3{}'$ will become the $\hat{x}_2$ and $\hat{x}_3$ axes in the new coordinate system. (Again, the choice is arbitrary, with the only restriction that the new order of coordinate axes represent a right-handed system.)

$$\begin{bmatrix} 0.0 \\ 1.0 \\ 0.0 \end{bmatrix} = \begin{bmatrix} R_{11} & R_{12} & R_{13} \\ R_{21} & R_{22} & R_{23} \\ R_{31} & R_{32} & R_{33} \end{bmatrix} \begin{bmatrix} -0.707 \\ 0.707 \\ 0.000 \end{bmatrix} \tag{2.5-3}$$

$$\begin{bmatrix} 0.0 \\ 0.0 \\ 1.0 \end{bmatrix} = \begin{bmatrix} R_{11} & R_{12} & R_{13} \\ R_{21} & R_{22} & R_{23} \\ R_{31} & R_{32} & R_{33} \end{bmatrix} \begin{bmatrix} 0.0 \\ 0.0 \\ 1.0 \end{bmatrix} \tag{2.5-4}$$

Equations 2.5-2 through 2.5-4 can be combined into a single system of equations as follows:

$$\begin{bmatrix} 1.0 & 0.0 & 0.0 \\ 0.0 & 1.0 & 0.0 \\ 0.0 & 0.0 & 1.0 \end{bmatrix} = \begin{bmatrix} R_{11} & R_{12} & R_{13} \\ R_{21} & R_{22} & R_{23} \\ R_{31} & R_{32} & R_{33} \end{bmatrix} \begin{bmatrix} 0.707 & -0.707 & 0.0 \\ 0.707 & 0.707 & 0.0 \\ 0.000 & 0.000 & 1.0 \end{bmatrix} \tag{2.5-5}$$

$$\text{or} \quad [I] = [R]\,[A]$$

where we have simply noted that [R] is the same in each equation, and each column vector is operated on independently, with an independent result. Therefore, combining Equations 2.5-2 to 2.5-4 lost no information and did not specialize the representation in any way.

> ***Note**: We could have selected the three equations corresponding to the three column vectors in any order, but the way we've done it is particularly convenient because the left-hand side of Equation 2.5-5 becomes the identity matrix.*

To find [R], we simply note that the last matrix in Equation 2.5-5 ([A]) is orthogonal, such that the inverse is equal to the transpose. We then perform simple matrix manipulations and use the identity property of [I]:

$$\text{or} \quad [I] = [R]\,[A]$$

$$[I]\,[A]^T = [R]\,[A]\,[A]^T = [R]\,[I] \tag{2.5-6}$$

$$[R] = [A]^T = \begin{bmatrix} 0.707 & 0.707 & 0.0 \\ -0.707 & 0.707 & 0.0 \\ 0.000 & 0.000 & 1.0 \end{bmatrix} \tag{2.5-7}$$

and thus the original problem is solved. In order to find the inverse transformation (from components expressed in the primed system to ones in the unprimed system), we simply invert [R]; that is, take its transpose, since it is orthogonal:

$$[R]^T = \begin{bmatrix} 0.707 & -0.707 & 0.0 \\ 0.707 & 0.707 & 0.0 \\ 0.000 & 0.000 & 1.0 \end{bmatrix} \tag{2.5-8}$$

Matrix multiplication introduces the possibility of one of the most important uses for matrices: representing sets of linear equations. For example, if x_1, x_2, ...x_n are *n* unknowns, and there are *n* equations of the form,

$$A_1x_1 + A_2x_2 + A_3x_3 + \ldots A_nx_n + = r_1, \quad \cdot \\ \begin{matrix} \cdot & \cdot & \cdot & \cdot & \cdot \\ \cdot & \cdot & \cdot & \cdot & \cdot \\ \cdot & \cdot & \cdot & \cdot & \cdot \end{matrix} \tag{2.29}$$

the set of equations can be written more compactly using matrix notation (and *much* more compactly using indicial notation):

$$\begin{bmatrix} A_{11} & A_{12} & A_{13} & \cdots & A_{1n} \\ A_{21} & A_{22} & A_{23} & \cdots & A_{2n} \\ \cdot & \cdot & \cdot & & \cdot \\ \cdot & \cdot & \cdot & & \cdot \\ \cdot & \cdot & \cdot & & \cdot \\ A_{n1} & A_{n2} & A_{n3} & \cdots & A_{nn} \end{bmatrix} \begin{bmatrix} x_1 \\ x_2 \\ \cdot \\ \cdot \\ \cdot \\ x_n \end{bmatrix} = \begin{bmatrix} r_1 \\ r_2 \\ \cdot \\ \cdot \\ \cdot \\ r_n \end{bmatrix} \tag{2.29a}$$

$$[A][x] = [r]$$

$$A_{ij}\, x_j = r_i$$

It can be shown that such a set of equations has a solution only if each one is independent of all the others. This condition may be checked by evaluating the **determinant**[7] of the matrix, written $|\mathbf{A}|$. If the determinant is zero, the equations are not linearly independent, the matrix [A] is said to be singular, and a unique solution cannot be found. The determinant is evaluated by summing a series of cofactors (the rule for a 3×3 matrix is shown below), but the procedure is complex and cumbersome for large matrices:

$$\begin{vmatrix} A_{11} & A_{12} & A_{13} \\ A_{21} & A_{22} & A_{23} \\ A_{31} & A_{32} & A_{33} \end{vmatrix} =$$

$$A_{11}(A_{22}A_{33} - A_{32}A_{23}) - A_{12}(A_{21}A_{33} - A_{23}A_{31}) + A_{13}(A_{21}A_{32} - A_{31}A_{22}) \tag{2.30}$$

---

[7] Our discussion of determinants is by necessity brief. For further information on the properties and evaluation of determinants see, for example, I. S. Sokolnikoff and R. M. Redheffer, *Mathematics of Physics and Modern Engineering*, 7th ed. (New York: McGraw-Hill), 702-707 (Appendix A).

or,
$$|A| = \frac{1}{6} A_{ij} A_{k1} A_{mn} \varepsilon_{ikm} \varepsilon_{j1n}$$

where, as before, ε_{ijk} is the permutation operator, as defined in connection with the determinant form of the vector cross product, Equation 2.10b. The quantities inside the parentheses in Equation 2.30 are called the cofactors of the elements multiplying them. There is no escaping the "ugly" appearance of equations involving determinants or permutation operators — they work and are conceptually the simplest route to accomplishing the goal, but they do not have the elegant appearance of much of linear algebra.

> ***Note:*** *Equation 2.30 clearly shows that only the determinant of square matrices can be nonzero, and therefore only equations with a square coefficient matrix (e.g., Equation 2.29a) can be solved for a unique set of* x_i, *the unknowns. Therefore, even by using our proposed method for "filling out" matrices, the determinant will be zero (because at least one row or column in* [A] *will be zero) and the set of equations will be insoluble.*

Exercise 2.6 **Show that the determinant of a 3 × 3 matrix must be non-zero if the three equations represented by the rows of the matrix are linearly independent.**

Assume that the three equations represented by the rows of [A] are not independent; that is:

$$\begin{aligned} A_{11} &= \alpha A_{21} + \beta A_{31} \\ A_{12} &= \alpha A_{22} + \beta A_{32} \\ A_{13} &= \alpha A_{23} + \beta A_{33} \end{aligned} \qquad (2.6\text{-}1)$$

Simply insert this assumption into the definition of the determinant [Equation 2.33] and simplify:

$$\begin{aligned} |A| = &(\alpha A_{21} + \beta A_{31})(A_{22}A_{33} - A_{32}A_{23}) - (\alpha A_{22} + \beta A_{32})(A_{21}A_{33} - A_{23}A_{31}) + (\alpha A_{23} + \beta A_{33})(A_{21}A_{32} - A_{31}A_{22}) = \\ &+\alpha A_{21}A_{22}A_{33} - \alpha A_{21}A_{32}A_{23} + \beta A_{31}A_{22}A_{33} - \beta A_{31}A_{32}A_{23} \\ &-\alpha A_{22}A_{21}A_{33} + \alpha A_{22}A_{31}A_{23} - \beta A_{32}A_{21}A_{33} + \beta A_{32}A_{31}A_{23} \\ &+\alpha A_{23}A_{21}A_{32} + \alpha A_{23}A_{31}A_{22} - \beta A_{33}A_{21}A_{32} + \beta A_{33}A_{31}A_{22} = 0 \end{aligned} \qquad (2.6\text{-}2)$$

Therefore, since Equation 2.6-1 represents the most general linear dependence, the determinant of [A] is shown to be zero for all singular sets of equations. Similarly, it is easy to prove that the value of the determinant of the [A] matrix is not changed when a row is transformed by adding a linear combination of the other rows, and only the sign could be changed when the order of rows or columns is changed.

Now if the determinant of [A] is different from zero we have to prove that necessarily the rows are linearly independent. To obtain this result, the [A] matrix is transformed into matrice with the same determinant, according to our previous results. The algorithm is the Gaussian elimination which can be used for solving linear systems.

If $A_{11} \neq 0$, then replace rows 2 and 3 by

$$A_{2k} \rightarrow A_{2k} - A_{1k}A_{21} / A_{11}$$

and

$$A_{3k} \rightarrow A_{3k} - A_{1k}A_{31} / A_{11}$$

If $A_{11} = 0$ and $A_{12} \neq 0$, then permute columns 1 and 2 and perform the same transformation as previously on the new matrix.

If $A_{11} = 0 = A_{12}$, then permute columns 1 and 3 and perform the same transformation as previously on the new matrix.

$A_{11} = A_{12} = A_{13} = 0$ is not possible, as the determinant would be equal to zero.

Call [A′] the new matrix, the determinant of which is the same (except a possible change of sign due to column permutation). It is easy to verify that the structure of [A′] is:

$$[A'] = \begin{bmatrix} A'_{11} & A'_{12} & A'_{13} \\ 0 & A'_{22} & A'_{23} \\ 0 & A'_{32} & A'_{33} \end{bmatrix}$$

If $A'_{22} \neq 0$, then replace row 3 by

$$A'_{3k} \rightarrow A'_{3k} - A'_{2k}A'_{32} / A'_{22}$$

If $A'_{22} = 0$, then permute colums 2 and 3 and perform the same transformation as previously on the new matrix.

$A'_{22} = A'_{23} = 0$ is not possible as the determinant would be equal to zero.

Call [A″] the new matrix. We observe that:

$$\det[A''] = \det\begin{bmatrix} A''_{11} & A''_{12} & A''_{13} \\ 0 & A''_{22} & A''_{23} \\ 0 & 0 & A''_{33} \end{bmatrix} = A''_{11}A''_{22}A''_{33} = \pm\det[A]$$

We must have $A''_{11} \neq 0$, $A''_{22} \neq 0$ and $A''_{33} \neq 0$ so that we see that the rows of [A″] are linearly independent. But it is obvious that the rows of [A″] are linear combinations of rows of [A] so that we can conclude that the rows of [A] are linearly independent. The same method is easily generalized to $n \times n$ matrice.

With reference to Equation 2.29 (a general linear system of equations), we have now *demonstrated* the condition for the system to be soluble. The method of solution of the system is another matter. The formal method depends on finding the **inverse** of [A], which we define as

$$[A]\,[A]^{-1} = [A]^{-1}\,[A] = [I]$$

$$A_{ij}A_{jk}' = A_{ij}'A_{jk} = \delta_{ik} \tag{2.31}$$

where$[A]^{-1}$ (with components A_{ij}') is the **inverse matrix** (or the indexed components), and [I] (with δ_{ij}) is the **identity matrix,** which has all diagonal elements equal to 1 and all nondiagonal elements equal to zero. [I] is called the identity matrix because it exhibits the identity property for matrix multiplication (just as 1 does for real arithmetic):

$$[A][I]=[I][A]=[A] \tag{2.32}$$

Equation 2.31 may be solved to find the inverse of [A], with the result as ugly as the determinant. The result for the inverse matrix is

$$\begin{bmatrix} A_{11} & A_{12} & A_{13} \\ A_{21} & A_{22} & A_{23} \\ A_{31} & A_{32} & A_{33} \end{bmatrix}^{-1} = \frac{\begin{bmatrix} +(A_{22}A_{33}-A_{32}A_{23}) & -(A_{12}A_{33}-A_{32}A_{13}) & +(A_{12}A_{33}-A_{32}A_{13}) \\ -(A_{21}A_{33}-A_{31}A_{23}) & +(A_{11}A_{33}-A_{31}A_{13}) & -(A_{11}A_{23}-A_{21}A_{12}) \\ +(A_{21}A_{32}-A_{31}A_{22}) & -(A_{11}A_{32}-A_{31}A_{12}) & +(A_{11}A_{22}-A_{21}A_{12}) \end{bmatrix}}{|A|}$$

$$[A]^{-1} = \frac{\begin{bmatrix} \text{signed} \\ \text{cofactor} \\ \text{matrix} \end{bmatrix}^T}{|A|} \tag{2.33}$$

$$A'_{ij} = \frac{A_{kl}\,A_{mn}\,\varepsilon_{jkm}\,\varepsilon_{jln}}{2\,|A|}$$

The superscript T on the matrix indicates the transpose, where rows and columns have been exchanged, equivalent to reflecting the matrix elements across the diagonal. Each element of the inverse matrix is thus the signed transpose cofactor of the original element, divided by the original determinant. As with the determinant, the inverse for large matrices becomes complex and inefficient for computation. Numerical methods more suited to digital computation will be presented in later sections on numerical methods.

Now that we have defined the inverse (and determinant) of a matrix, we are in a position to accomplish our original goal of solving Equation 2.29, shown below in compact matrix form. Remember that the matrices [A] and [r] are known constants, and the column matrix [x] represents the unknowns. We use the properties of the inverse matrix of [A] (known, in principle, via Equation 2.32) to solve this problem:

$$[A]\,[x]=r \tag{2.34}$$

(Original Equation)

$$[A]^{-1}[A][x]=[A]^{-1}[r] \tag{2.34a}$$

(Pre-multiplication by the inverse of [A])

$$[I][x]=[A]^{-1}[r] \tag{2.34b}$$

(Replacement of $[A]^{-1}[A]$ by [I])

$$[x]=[A]^{-1}[r] \tag{2.34c}$$

Thus the elements of the unknown column matrix [x] are found by a purely formal process, the only restriction being the equivalent conditions that the inverse of [A] exist, the determinant of [A] be nonzero (i.e., that [A] be nonsingular), or that the system of equations be linearly independent. In such cases (i.e., when these conditions are violated), there is no unique solution for the x values, or no solution at all.

Exercise 2.7 **Solve Equation 2.29 in 3×3 form by obtaining [A] in *upper triangular* form using *Gaussian reduction.***

> ***Note:*** *Gaussian reduction involves transforming a set of equations by sequential matrix operations to a form where all elements below the diagonal are zero. Then, starting from the last equation, the set of equations is solved by substituting the known values of the variables into higher equations. The method is illustrated as follows:*

$$\begin{bmatrix} A_{11} & A_{12} & A_{13} \\ A_{21} & A_{22} & A_{23} \\ A_{31} & A_{32} & A_{33} \end{bmatrix}\begin{bmatrix} x_1 \\ x_2 \\ x_3 \end{bmatrix}=\begin{bmatrix} r_1 \\ r_2 \\ r_3 \end{bmatrix} \tag{2.7-1}$$

The first row stands as shown, and we use it to reduce the first column (beneath the diagonal) to zero. To accomplish this, we multiply the first row by a coefficient and subtract from all lines below. For example, to reduce the second row, we multiply the first row by A_{21}/A_{11} (if $A_{11}\neq 0$) and subtract. For the third line, we use a factor of A_{31}/A_{11}. The system row appears as follows:

$$\begin{bmatrix} A_{11} & A_{12} & A_{13} \\ 0 & A_{22}-\dfrac{A_{21}}{A_{11}}A_{12} & A_{23}-\dfrac{A_{21}}{A_{11}}A_{13} \\ 0 & A_{32}-\dfrac{A_{31}}{A_{11}}A_{12} & A_{33}-\dfrac{A_{31}}{A_{11}}A_{13} \end{bmatrix}\begin{bmatrix} x_1 \\ \underline{x_2} \\ \underline{x_3} \end{bmatrix}=\begin{bmatrix} r_1 \\ r_2-\dfrac{A_{21}}{A_{11}}r_1 \\ r_3-\dfrac{A_{31}}{A_{11}}r_1 \end{bmatrix} \tag{2.7-2}$$

or

$$\begin{bmatrix} A_{11} & A_{12} & A_{13} \\ 0 & A_{22}' & A_{23}' \\ 0 & A_{32}' & A_{33}' \end{bmatrix} \begin{bmatrix} x_1 \\ x_2 \\ x_3 \end{bmatrix} = \begin{bmatrix} r_1 \\ r_2' \\ r_3' \end{bmatrix}. \tag{2.7-3}$$

We continue the same procedure, now operating on the second row, to eliminate the second column below the diagonal:

$$\begin{bmatrix} A_{11} & A_{12} & A_{13} \\ 0 & A_{22}' & A_{23}' \\ 0 & 0 & A_{33}' - \frac{A_{32}'}{A_{22}'} A_{23} \end{bmatrix} \begin{bmatrix} x_1 \\ x_2 \\ x_3 \end{bmatrix} = \begin{bmatrix} r_1 \\ r_2' \\ r_3' - \frac{A_{32}'}{A_{22}'} r_2' \end{bmatrix}, \qquad 2.74)$$

or

$$\begin{bmatrix} A_{11} & A_{12} & A_{13} \\ 0 & A_{22}' & A_{23}' \\ 0 & 0 & A_{33}'' \end{bmatrix} \begin{bmatrix} x_1 \\ x_2 \\ x_3 \end{bmatrix} = \begin{bmatrix} r_1 \\ r_2' \\ r_3'' \end{bmatrix}. \tag{2.7-5}$$

To solve for the values of x, one simply begins with the last row and back substitutes to sequentially solve for x_3, x_2, x_1

$$x_3 = \frac{r_3''}{A_{33}''} = \frac{r_3' - \frac{A_{32}'}{A_{22}'} r_2'}{A_{33}' - \frac{A_{32}'}{A_{22}'} A_{23}'} = \tag{2.7-6}$$

$$\frac{r_3 - \frac{A_{31}}{A_{11}} r_1 - \left(r_2 - \frac{A_{21}}{A_{11}} r_1\right) \frac{\left(A_{23} - \frac{A_{21}}{A_{11}} A_{12}\right)}{\left(A_{22} - \frac{A_{21}}{A_{11}} A_{12}\right)}}{A_{33} - \frac{A_{31}}{A_{11}} A_{13} - \frac{\left(A_{23} - \frac{A_{21}}{A_{11}} A_{13}\right)}{\left(A_{22} - \frac{A_{21}}{A_{11}} A_{12}\right)}} \tag{2.7-7}$$

$$x_2 = \frac{r_2' - A_{23}'' X_3}{A_{22}'} \tag{2.7-8}$$

$$x_1 = \frac{r_1 - r_2' X_2 - r_3' X_3}{A_{11}} \tag{2.7-9}$$

We have shown only x_3 in terms of the original coefficients, A_{ij}, to illustrate the rapid growth in the number of terms in a closed-form, explicit solution. Fortunately, such a procedure is not necessary because only the final coefficients are needed to obtain the solution for x_j, and they have already been obtained as part of the reduction process.

The Gaussian reduction method is immediately useful for larger systems of equations by following the procedure given. Compare this with the determination of an inverse for a matrix larger than 3×3 (which follows the spirit of Equation 2.7-7 rather than Equation 2.7-6) and one can readily see why reduction methods are used almost exclusively for solving large systems of equations with digital computers.

Another class of important problems involves finding column vectors that are changed only in magnitude, but not in direction[8], when multiplied by a given matrix. In fact, it can be easily seen that this property is a function only of the matrix itself. The basic equation for this problem is as follows:

$$[A]\,[x] = \lambda[x] \tag{2.35}$$

where [x] is the unknown column vector and λ is an unknown scalar ratio of vector lengths between the **x** vector on the left-hand side and the result of the transformation (right-hand side). Since λ is an unknown, arbitrary constant, the condition that [x] be parallel to λ[x] is applied without concern about the lengths.

Equation 2.35 is called the **characteristic equation,** and [x] and λ are called the **eigenvector** and **eigenvalue**, respectively.

The solution of Equation 2.35 usually proceeds in two steps. Before beginning the solutions, let's write it in simpler form:

$$[A]\,[x] - \lambda[x] = 0 \tag{2.35a}$$

$$[A]\,[x] - [\lambda I]\,[x] = 0 \tag{2.35b}$$

$$[A - \lambda I]\,[x] = 0 \tag{2.35c}$$

[8] Formally, the column matrix has the same ratios among all elements, but the magnitude of any element is unknown. This procedure is almost always performed for vectors, lines, and similar spatial quantities, hence our use of the term "vector." Formally, there is no need for [x] to be a physical vector.

We introduced the identity matrix [I] in Equation 2.35b to bring the second term in the equation up to the same measure as the first term, so that the two expressions could be subtracted. The only interesting solution to Equation 2.35c requires that $|A-\lambda I| = 0$. Otherwise, [x] = [0] (where [0] = the **null matrix**), and only the null vector satisfies the equation. Therefore, Equation 2.35c can first be approached by using that condition:

$$|A-\lambda I| = 0$$

or, for a 3×3 matrix,

$$\begin{vmatrix} A_{11}-\lambda & A_{12} & A_{13} \\ A_{21} & A_{22}-\lambda & A_{23} \\ A_{31} & A_{32} & A_{33}-\lambda \end{vmatrix} = 0. \tag{2.36}$$

By using the expressions for the determinant of a 3×3 symmetric matrix[9], Equation 2.30, we find the equation for λ:

$$-\lambda^3 + \lambda^2 \left[A_{11} + A_{22} + A_{33}\right] + \lambda\left[-A_{11}A_{22} - A_{22}A_{33} - A_{11}A_{33} + A_{23}^2 + A_{13}^2 + A_{12}^2\right] +$$

$$\left[A_{11}(A_{22}A_{33} - A_{32}A_{23}) - A_{12}(A_{21}A_{33} - A_{31}A_{23}) + A_{13}(A_{21}A_{32} - A_{31}A_{22})\right] \text{ or,} \tag{2.37}$$

$$\lambda^3 - J_1\lambda^2 - J_2\lambda - J_3 = 0,$$

where J_1, J_2, J_3 are conventionally called the **invariants** of matrix [A]. In particular, the invariants are

$$J_1 = A_{ij}\,\delta_{ij} = \text{ trace}\left[A\right] = A_{11} + A_{22} + A_{33} \tag{2.38}$$

$$J_2 = \frac{1}{2}\left[A_{ij}A_{ij} - (A_{ij}\,\delta_{ij})^2\right] = \text{second invariant of } \left[A\right] =$$

$$-A_{11}A_{22} - A_{22}A_{33} - A_{11}A_{33} + A_{23}^2 + A_{13}^2 + A_{12}^2$$

$$J_3 = |A| = \text{determinant}\left[A\right] =$$

$$A_{11}(A_{22}A_{33} - A_{32}A_{23}) - A_{12}(A_{21}A_{33} - A_{31}A_{23}) + A_{13}(A_{21}A_{32} - A_{31}A_{22})$$

---

[9] The expressions given in Eqs. 2.37–2.38 assume a symmetric matrix, but the corresponding non-symmetric case may easily be written following Eq. 2.36.

For a diagonal matrix (defined later in this section), the invariants take a simpler form:

$$J_1 = A_{11} + A_{22} + A_{33} \text{ (same)}$$
$$J_2 = -(A_{11}A_{22} + A_{22}A_{33} + A_{11}A_{33})$$
$$J_3 = A_{11}A_{22}A_{33}$$

Equation 2.37 is a normal cubic equation in λ, and therefore must have three roots that satisfy it; say, λ_1, λ_2, and λ_3 (in any order). It may easily be shown[10] that these roots will be real if [A] is real and symmetric, the case of most interest in mechanics. Once λ_1, λ_2, and λ_3 are known, they may be substituted into the original equation (Equation 2.35c) individually to obtain the conditions leading to the three corresponding vectors **x**, **x′**, **x″**:

$$\begin{bmatrix} A_{11}-\lambda_1 & A_{12} & A_{13} \\ A_{21} & A_{22}-\lambda_1 & A_{23} \\ A_{31} & A_{32} & A_{33}-\lambda_1 \end{bmatrix}\begin{bmatrix} x_1 \\ x_2 \\ x_3 \end{bmatrix} = \begin{bmatrix} A_{11}-\lambda_2 & A_{12} & A_{13} \\ A_{21} & A_{22}-\lambda_2 & A_{23} \\ A_{31} & A_{32} & A_{33}-\lambda_2 \end{bmatrix}\begin{bmatrix} x_1' \\ x_2' \\ x_3' \end{bmatrix} \tag{2.39}$$

$$\begin{bmatrix} A_{11}-\lambda_3 & A_{12} & A_{13} \\ A_{21} & A_{22}-\lambda_3 & A_{23} \\ A_{31} & A_{32} & A_{33}-\lambda_3 \end{bmatrix}\begin{bmatrix} x_1'' \\ x_2'' \\ x_3'' \end{bmatrix} = \begin{bmatrix} 0 \\ 0 \\ 0 \end{bmatrix}$$

Because each set of equations is not sufficient to find a unique [x] (because the equations are equally valid for any length of **x**), we can complete the set by specifying that $x_1^2 + x_2^2 + x_3^2 = 1$. That is, that [x] is a unit vector. There are then sufficient equations to find the three sets of eigenvalue-eigenvector pairs:

$$\lambda_1 \leftrightarrow x_1, x_2, x_3, \leftrightarrow x,$$

$$\lambda_2 \leftrightarrow x_1', x_2', x_3', \leftrightarrow x', \tag{2.40}$$

$$\lambda_3 \leftrightarrow x_1'', x_2'', x_3'', \leftrightarrow x'',$$

and the original problem is solved. λ is the ratio of length between the vectors on the left and right sides of Equation 2.35, while the three sets of (x_1, x_2, x_3) are the particular directions of the three lines that do not change direction when multiplied by [A]. The physical consequences and use of eigenvector-eigenvalue equations will be made clear in the next chapter, when we approach the physical problem of principal stresses.

[10] See, for example, L. E. Malvern: *Introduction to the Mechanics of a Continuous Medium,* (Englewood Cliffs, N.J., 1969),Prentice-Hall, Inc., p. 45.

Note, however, that **symmetric matrices** can be shown always to possess three **real eigenvalues** and three **mutually perpendicular eigenvectors**.

Exercise 2.8 **Find the eigenvalues and eigenvectors of the matrix shown below.**

$$[A] = \begin{bmatrix} 1.312 & 0.541 & 0.217 \\ 0.541 & 1.940 & 0.375 \\ 0.217 & 0.375 & 2.748 \end{bmatrix}$$

Note: [A] *has been chosen symmetric so that the eigenvectors will be real and orthogonal.*

$$\text{Physical Requirement} \quad \begin{bmatrix} A_{11} & A_{12} & A_{13} \\ A_{21} & A_{22} & A_{23} \\ A_{31} & A_{32} & A_{33} \end{bmatrix} \begin{bmatrix} x_1 \\ x_2 \\ x_3 \end{bmatrix} = \begin{bmatrix} \lambda x_1 \\ \lambda x_2 \\ \lambda x_3 \end{bmatrix} \tag{2.8-1}$$

$$\mathbf{A} \cdot \mathbf{x} = \lambda \mathbf{x} \ \text{ or } \ [A - \lambda I]x = 0$$

$$\text{Characteristic Equation} \quad |A_{ij} - \lambda\delta_{ij}| = \begin{bmatrix} 1.312 - \lambda & 0.541 & 0.217 \\ 0.541 & 1.940 - \lambda & 0.375 \\ 0.217 & 0.375 & 2.748 - \lambda \end{bmatrix} = 0 \tag{2.8-2}$$

$$\begin{aligned} \text{Cubic Polynomial Form} \quad & (1.312 - \lambda)\left[(1.940 - \lambda)(2.748 - \lambda) - (0.375)^2\right] \\ & -0.541[(0.541)(2.748 - \lambda) - (0.217)(0.375)] \\ & +0.217[(0.541)(0.375) - (0.217)(1.940 - \lambda)] \\ & = 0 \end{aligned} \tag{2.8-3}$$

In terms of invariants:

$$\lambda^3 - 6\lambda^2 + 11\lambda - 6 = 0 \quad \text{where} \quad \begin{aligned} J_1 &= 6 \\ J_2 &= -11 \\ J_3 &= 6 \end{aligned} \tag{2.8-4}$$

To avoid introducing a numerical technique yet, we try the root $\lambda = 1$ and discover that Equation 2.8-3 can be factored:

$$(\lambda - 1)(\lambda^2 - 5\lambda + 6) = (\lambda - 1)(\lambda - 2)(\lambda - 3) = 0 \tag{2.8-5}$$

So the three eigenvalues have been *formed* equal to $\lambda_1 = 1$, $\lambda_2 = 2$, and $\lambda_3 = 3$. In order to find the *eigenvectors*, we must return to the physical requirement of [x] (Equation

2.8-1). Remember, the characteristic equation, Equation 2.2-2, only assures that the problems have a nontrivial solution. Now, substitute one of the required λ values into Equation 2.8-1 and solve for the corresponding eigenvector:

$$\text{for } \lambda_1 = 1: \begin{bmatrix} 0.312 & 0.541 & 0.217 \\ 0.541 & 0.940 & 0.375 \\ 0.217 & 0.375 & 1.748 \end{bmatrix} \begin{bmatrix} x_1 \\ x_2 \\ x_3 \end{bmatrix} = \begin{bmatrix} 0 \\ 0 \\ 0 \end{bmatrix} \qquad (2.8\text{-}6)$$

We require that **x** be a unit vector such that

$$x_1{}^2 + x_2{}^2 + x_3{}^2 = 1 \text{ or, } \mathbf{x} \cdot \mathbf{x} = 1. \qquad (2.8\text{-}7)$$

(This condition, or a similar one, is necessary because Equation 2.8-6 is a requirement only on the direction of **x**, not its length. Mathematically, this condition is obviously necessary because the determinant of the matrix on the left side of Equation 2.8-6 is singular. In fact, this is the condition we used to find this matrix!)

Equations 2.8-6 and 2.8-7 can be solved in a variety of ways. Let's eliminate the third equation (since it depends on the first two anyway) and assume[11] for the moment that $x_1 = 1$ After we have found x_2 and x_3, we will impose the unit vector condition, Equation 2.8-7, by scaling the three components. Equation 2.8-6 then becomes

$$\begin{bmatrix} 0.940 & 0.375 \\ 0.375 & 1.748 \end{bmatrix} \begin{bmatrix} x_2 \\ x_3 \end{bmatrix} = \begin{bmatrix} -0.541 \\ -0.217 \end{bmatrix} \qquad (2.8\text{-}8)$$

where we have assumed that $x_1 = 1$ and moved the A_{21} and A_{31} coefficients to the right side of the equation. (If this is difficult to visualize, write out Equation 2.8-6 in equation form, assume $x_1 = 1$, and subtract the constants from both sides of each equation.)

The solution for Equation 2.8-8 is found using simple algebra:

$$x_1 = 1.00 \text{ (given)} \quad x_2 = -0.57 \quad x_3 = 0.0$$

or, after scaling to form a unit vector,

$$\hat{x}\,(\lambda_1 = 1) = (0.87,\ -0.50,\ 0.0).$$

Following the same procedure for the other two eigenvalues $(\lambda_2 = 2,\ \lambda_3 = 3)$, we find that the corresponding results are

$$\hat{x}'(\lambda_2 = 2) = (-\,0.43,\ -\,0.75,\ 0.50)$$
$$\hat{x}''(\lambda_3 = 3) = (0.25,\ 0.43,\ 0.87)$$

[11] Equation 2.8-2 is homogeneous and therefore unchanged by multiplication by an arbitrary scalar. The only exception is that multiplication by zero or infinity is not allowed, so one must always check the validity of the results to determine that the component set equal to one should not be zero. If it is, another component of **x** should be chosen for elimination.

The signs of the components of the eigenvectors is arbitrary because the line direction can be represented equally well by $+\hat{\mathbf{x}}$ or $-\hat{\mathbf{x}}$. However, the $\hat{\mathbf{x}}, \hat{\mathbf{x}}'$ and $\hat{\mathbf{x}}''$, should form a right-handed set, so we verify that $\hat{\mathbf{x}} = \hat{\mathbf{x}}' \times \hat{\mathbf{x}}''$:

$$\begin{vmatrix} \hat{x} & \hat{x}' & \hat{x}'' \\ -0.43 & -0.75 & 0.50 \\ 0.25 & 0.43 & 0.87 \end{vmatrix} = \begin{matrix} \hat{x}\left[(-.75)(.87)-(.43)(.50)\right] \\ -\hat{x}'\left[(-.43)(.87)-(.25)(.50)\right] \\ +\hat{x}''\left[(-.43)(.43)-(.25)(-.75)\right] \end{matrix} = \begin{bmatrix} -0.87 \\ 0.50 \\ 0.00 \end{bmatrix} \tag{2.8-9}$$

So, in order to form a right-handed system, the signs for $\hat{\mathbf{x}}'$ should be reversed, with the final result that

$$\begin{aligned} \lambda_1 &= 1: \ \hat{x} = (-0.87, \ +0.50, \ 0.00) \\ \lambda_2 &= 2: \ \hat{x}' = (-0.43, \ -0.75, \ +0.50) \\ \lambda_3 &= 3: \ \hat{x}'' = (+0.25, \ +0.43, \ +0.87) \end{aligned} \tag{2.8-10}$$

The most important operations and nomenclature for matrices have now been presented. There are, however, many additional descriptive words involving matrices and matrix operations, a few of which are as follows:

A **symmetric matrix** is one for which $A_{ij} = A_{ji}$, that is, that a line drawn through the diagonal elements (A_{ii}) is a mirror plane of symmetry.

The **transpose** $[A]^T$ of a matrix [A] is obtained by exchanging the rows and columns of the matrix (i.e., $A_{ij}^T = A_{ji}$).

A **diagonal matrix** is one for which all elements not on the diagonal (i.e., not A_{ii}) are zero. The identity matrix, [I], is one example.

A **triangular matrix** is one that has nonzero elements appearing only in the diagonal positions, and either above or below (but not both) the diagonal positions. The following matrix is **upper triangular,** for example:

$$\begin{bmatrix} A_{11} & A_{12} & A_{13} \\ 0 & A_{22} & A_{23} \\ 0 & 0 & A_{33} \end{bmatrix}$$

Two matrices [A] and [B] are said to be **similar** if $[B] = [T]^{-1}[A][T]$, where [T] is any nonsingular matrix. The transformation shown is called a **similarity transformation**.

An **orthogonal matrix** is one for which its inverse and transpose are identical. The condition may be expressed as follows:

$$[A] \text{ orthogonal if } [A][A]^T = [A]^T[A] = [I], \tag{2.41a}$$

such that

$$A_{ik}\,A_{jk} = A_{ki}\,A_{kj} = \delta_{ij}. \tag{2.41b}$$

2.6 TENSORS

In order to introduce the idea of tensors, we will once again resort to physical reasoning and logical extension of the concepts introduced above. The reader is referred to the excellent book by Nye[12], especially the first two chapters, in which the physical basis of tensors is discussed in detail. Much of our presentation follows these ideas, with updated notation to follow current usage.

Tensor is a general word, referring to any physical quantity. Familiar examples of tensors include **scalars** and **vectors**, which we have already discussed in some detail. **Scalars** are represented by a single number that is independent of the coordinate system chosen by an observer. For example, the temperature at a point in space (in particular, the number assigned to the measurement) does not depend on the direction from which one observes it. *Scalars are tensors of* ***rank zero***.

Vectors, as introduced above, consist of a set of three quantities describing a length and orientation which exist independently of the observer, but which have a numerical representation that depends on the observer and coordinate system. Thus the usual physical concept of an arrow with length and direction will *look* different to each observer (in a completely determined way) although it retains the same orientation and length in space. Familiar vector quantities include displacement, velocity, force, and gradients of scalar fields (for example, the thermal gradient down which heat flows). *Vectors are tensors of* ***rank one***.

Tensors of order higher than one do not have special names, but are referred to as **tensors of the second rank, tensors of the third rank,** etc. For simplicity, we will concentrate on tensors of the second rank in this introduction, but the principles are general and apply to all orders. We shall provide more discussion of the various tensors useful in continuum mechanics as we use them later in the book, but will concentrate on the principles for now. Second-rank tensors are quantities that connect two vectors, such as stress (force and area), strain (displacement and length), electrical conductivity (voltage and current), and so on. Tensors of the second rank contain nine components in 3-D space, to account for all of the cross terms between the components of the two constituent vectors. For example, one can see that the general relationship between two vectors (voltage and current density, for example) must contain nine components:

$$\begin{bmatrix} j_1 \\ j_2 \\ j_3 \end{bmatrix} = \begin{bmatrix} S_{11} & S_{12} & S_{13} \\ S_{21} & S_{22} & S_{23} \\ S_{31} & S_{32} & S_{33} \end{bmatrix} \begin{bmatrix} v_1 \\ v_2 \\ v_3 \end{bmatrix} \tag{2.42}$$

[12] J. F. Nye, *Physical Properties of Crystals* (Oxford: Clarendon Press, 1957).

$$\mathbf{j} = \mathbf{S}\ \mathbf{v}$$

where [S] is the conductivity, **v** the applied voltage (the electric field), and **j** the current density.

where **S** is the conductivity, **v** the applied voltage (the electric field), and **j** the current density.

Higher-order tensors follow the same pattern, with the connection among other vectors and tensors apparent. Table 2.1 lists examples of tensors of various ranks, their related tensor quantities, and the rules for transforming their components.

Equation 2.42 illustrates the connection between a second-rank tensor and a **linear vector operator**. To test for tensor character of an arbitrary 3×3 matrix, one must check only whether the argument and resultant are truly vectors (i.e., that their meanings are independent of the observer). This is done by testing that the orientation and length of the vectors are independent of the choice of coordinate system. Using Equation 2.22—the vector component transformation for a rotation of orthonormal basis—the vector components for an unchanged vector may be found in a new (primed) coordinate system $(\hat{x}_1', \hat{x}_2', \hat{x}_3')$. Then, in the new coordinate system, we can rewrite Equation 2.42:

$$\begin{bmatrix} j_1' \\ j_2' \\ j_3' \end{bmatrix} = \begin{bmatrix} S_{11}' & S_{12}' & S_{13}' \\ S_{21}' & S_{22}' & S_{23}' \\ S_{31}' & S_{32}' & S_{33}' \end{bmatrix} \begin{bmatrix} v_1' \\ v_2' \\ v_3' \end{bmatrix} \leftrightarrow [j]' = [S]'[v]'$$

$$\text{where:} \begin{bmatrix} j_1' \\ j_2' \\ j_3' \end{bmatrix} = \begin{bmatrix} R_{11} & R_{12} & R_{13} \\ R_{21} & R_{22} & R_{23} \\ R_{31} & R_{32} & R_{33} \end{bmatrix} \begin{bmatrix} j_1 \\ j_2 \\ j_3 \end{bmatrix} \leftrightarrow [j]' = [R]\,[j] \tag{2.43}$$

$$\begin{bmatrix} v_1' \\ v_2' \\ v_3' \end{bmatrix} = \begin{bmatrix} R_{11} & R_{12} & R_{13} \\ R_{21} & R_{22} & R_{23} \\ R_{31} & R_{32} & R_{33} \end{bmatrix} \begin{bmatrix} v_1 \\ v_2 \\ v_3 \end{bmatrix} \leftrightarrow v_i' = R_{ij} v_j \leftrightarrow [v]' = [R]\,[v]$$

With the use of Equation 2.43, the direct transformation of tensor components from S_{ij} to S_{ij}' (expressed in the new coordinate system) can be found:

$$[j]' = [S]'[v]' \tag{2.43a}$$

$$[R]\,[j] = [S']\,[R]\,[v] \tag{2.43b}$$

$$[R]^T [R]\,[j] = [R]^T [S]'\,[R]\,[v] \tag{2.43c}$$

TABLE 2.1 Typical Tensor Notation and Component Transformations*

Rank	*Name*	*Example*	*Component Transformations*	*Use*	*Examples*
0	Scalar	t	$t' = t$	$\frac{3}{M}$	temperature (T), time (t), distance (x), speed (v), mass (m), pressure (p), density (r), area (a)
1	Vector	**x**, x_i	$x_i = R_{ij}x_j$	$\mathbf{x} = x_1\hat{\mathbf{x}}_1 + x_2\hat{\mathbf{x}}_2 + x_3\hat{\mathbf{x}}_3$	position (**x**), displacement (**u**), velocity (**v**), force (**f**), area (**a**), temperature gradient (—t), electric field(**v**), electric current density (**j**)
2	Tensor	**S**, S_{ij}	$S'_{ij} = R_{ik}R_{jl}S_{kl}$	$\mathbf{j}=\mathbf{S}\cdot\mathbf{v}$, $j_i=S_{ij}v_j$	electrical conductivity (**S**), stress (σ), strain (**E,** ε), deformation gradient (**F**), thermal conductivity (**K**)
3	Tensor	**D**, D_{ijk}	$D'_{ijk}=R_{il}R_{jm}R_{kn}D_{lmn}$	$\mathbf{P}=\mathbf{D}\cdot\sigma$, $P_i=D_{ijk}\sigma_{jk}$	piezoelectric moduli (**D**)
4	Tensor	**C**, C_{ijkl}	$C'_{ijkl}=R_{im}R_{jn}R_{ko}R_{lp}C_{mnop}$	$\sigma=\mathbf{C}\cdot\varepsilon$, $\sigma_{ij}=C_{ijkl}\varepsilon_{kl}$	elastic constants (**C**), elastic compliances (**S**)

*Cartesian coordinate systems are assumed throughout.

$$\text{so: } [j] = [R]^T [S]' [R] [v] \tag{2.43d}$$

(because $[R]^T [R] = [I]$ for orthogonal [R])

$$\text{so: } [S] = [R]^T [S]' [R] \text{, or} \tag{2.43e}$$

$$\text{so: } [S]' = [R] [S] [R]^T \tag{2.44}$$

Therefore, Equation 2.44 shows how the components of a tensor are related in the two coordinate systems. Note that we made use of the fact that [R] is orthogonal in Equation 2.43c, such that $[R]^T [R] = [I]$. Equations 2.44 and 2.43e represent the transformation of tensor components back and forth between the original and primed coordinate system in terms of the rotation matrix, [R], defined as in Equation 2.43. That is, [R] is the cosine matrix that transforms vector components in the original coordinate system to those in the primed coordinate system. These component transformations are most easily remembered in indicial form:

$$S_{ij}' = R_{ik} R_{jl} S_{kl} \tag{2.45a}$$

$$S_{ij} = R_{ik}' R_{jl}' S_{kl} \tag{2.45b}$$

where [R]′ transforms components from the primed coordinate system to the original one, and is therefore the inverse (and transpose) of [R]. Here, [R]′ is the transpose of [R], so, for example, $R_{12}' = R_{21}$.)

> ***Note:*** *Equations 2.43e through 2.45b represent a specific example of* ***similarity transformations*** *(this name is used mainly in matrix algebra, where matrices [S] and [S]′ are known as* ***similar matrices****). In general, a similarity transformation has the form $[S]'=[T]^{-1}[S][T]$. In our example, T is orthogonal, so $[T]^{-1}=[T]^T$. The usefulness of similarity transformations is shown precisely by the tensor transformation problem.*

Equation 2.45 represents one formal mathematical definition of a tensor. That is, any 3×3 matrix of numbers that obeys these transformation rules is a second-rank tensor. The physical perspective on this definition states that a tensor is a physical relationship between two vectors and therefore exists independent of the observer. Thus a tensor's components must follow Equation 2.43 in order for the tensor itself to be independent of the coordinate system. The physical picture insists on the basic, physical nature of the tensor independent of observer, exactly following the physical description of a vector (a first-rank tensor).

An alternative and equivalent definition of a second-rank tensor can be based on its use as a **linear vector operator**—an operator that operates on a vector and has a vector result [$\mathbf{y} = L(\mathbf{x})$], and that satisfies the following two properties:

1. $L(\lambda \mathbf{a}) = \lambda L(\mathbf{a})$
2. $L(\mathbf{a}+\mathbf{b}) = L(\mathbf{a}) + L(\mathbf{b})$

for any real vectors **a**, **b**; and real number λ. The linearity ensures that L is, in fact, a tensor whose components transform according to Equations 2.42–2.43.

2.7 TENSOR OPERATIONS AND TERMINOLOGY

The language of tensors follows precisely the matrix notation we introduced in Section 2.5. This makes sense, because a second-rank tensor is simply a 3×3 matrix that obeys the component transformations shown in Equation 2.42. Therefore, second-rank tensors are a subset of 3×3 matrices, **symmetric tensors** are a subset of symmetric matrices, and so on.

Exercise 2.9 **Demonstrate that [R], the rotation matrix, is not a tensor.**

This is a difficult question, which may nonetheless illuminate the meaning of tensors. The standard textbook answer is, for example, that [R] cannot be a tensor because its role is in relating two sets of components (expressed in two Cartesian coordinate systems) for the *same* vector. For example, Nye says (using $[T_{ij}]$ to denote a tensor)

> ... $[R_{ij}]$ *is an array of coefficients relating two sets of axes;* $[T_{ij}]$ *is a physical quantity that, for one given set of axes, is represented by nine numbers.* $[R_{ij}]$ *straddles, as it were, two sets of axes;* $[T_{ij}]$ *is tied to one set at a time. One cannot speak, for instance, of transforming* $[R_{ij}]$ *to another set of axes, for it would not mean anything.*[13]

This reply relies on the concept of a tensor as a physical reality, probably the clearest way to approach the question. (Mathematical tests of "tensorness" perform the same function but more obscurely, still relying on the user's knowledge of whether the tensor transformation rules can be utilized in a meaningful way.) If we can assign an appropriate meaning to $[R_{ij}]$ that can be transformed from one coordinate system to another, then in fact $[R_{ij}]$ could be considered a tensor.

It is easily shown that there are four closely related matrices:

[R], which transforms *components* of a given vector or tensor from one coordinate system to another (let's say original to primed system).

[13] J. F. Nye, *Physical Properties of Crystals* (Oxford: Clarendon Press, 1957),p. 15.

$[R]^T$, the transpose and inverse of [R], which therefore transforms *components* of a vector expressed in the primed coordinate system to those exposed in the unprimed one.

$[r] = [R]^{T,}$ which rotates a *vector* or *tensor* to a new position in space. The vector rotation is carried out such that it is by the two angles specified for the coordinate system transformation [R].

$[r]^T = [R]$, which rotates a *vector* to a new position in space by angles exactly equivalent to the coordinate system transformation $[R]^T$.

The meaning of these quantities is as follows: The linear vector operator [r] is defined such that its resultant vector components, expressed in the primed coordinate system, are the same as those for the original vector expressed in the original coordinate system. That is,

$$\hat{\mathbf{y}} = \mathbf{r} \cdot \hat{\mathbf{x}} \tag{2.9-1}$$

$$x_i = R_{ij}\, y_j \tag{2.9-2}$$

in matrix form:

$$[x] = [r]\,[R]\,[x] = [R]\,[r]\,[x] \tag{2.9-3}$$

Although the foregoing may seem obvious to anyone who has used matrix algebra to rotate vectors, the conclusion is that [R] could easily be defined by $[r]^T$. Furthermore, since $[r]^T$ is clearly a linear vector operator, relating two physically meaningful vectors, $[r]^T$ must be a tensor. Thus [R] can be considered a tensor without trying to ascribe a special meaning to it in terms of the component transformation definition.

Similarly, tensor operations follow matrix algebra precisely. **Tensor multiplication** follows matrix multiplication:

$$\mathbf{U} = \mathbf{ST} = \mathbf{S} \cdot \mathbf{T} \leftrightarrow U_{ik} = S_{ij}\, T_{jk}, \tag{2.46}$$

(Cartesian coordinate systems)

whereas the **inner product** of two tensors follows the definition of the dot product for vectors:

$$U \text{ (a scalar)} = \mathbf{S} \cdot \mathbf{T} = S_{ij}\, T_{ij}. \tag{2.47}$$

(Cartesian coordinate systems)

Alternatively, we can represent other possible forms of tensor multiplication using the basic ones defined in Equations 2.47 and 2.48 and the notation for transposes:

$$\mathbf{U}' = \mathbf{S}\mathbf{T}^{\mathrm{T}} = \mathbf{S}\mathbf{T}^{\mathrm{T}}\ \mathrm{U}_{ik}' = \mathrm{S}_{ij}\ \mathrm{T}_{kj} \tag{2.48}$$

Alternative tensor multiplications

$$\mathbf{U}'' = \mathbf{S}^{\mathrm{T}}\mathbf{T} \leftrightarrow \mathrm{U}_{jk}'' = \mathrm{S}_{ij}\ \mathrm{T}_{ik} \tag{2.49}$$

$$\mathbf{U}''' = \mathbf{S}^{\mathrm{T}}\mathbf{T}^{\mathrm{T}} \leftrightarrow \mathrm{U}_{jk}''' = \mathrm{S}_{ij}\ \mathrm{T}_{ki} \tag{2.50}$$

$$\mathrm{U}' = \mathbf{S}\cdot\mathbf{T}^{\mathrm{T}} \leftrightarrow \mathrm{S}_{ij}\ \mathrm{T}_{ji} \tag{2.51}$$

As with first-rank tensors (vectors), the physical operations of tensor product and inner product are defined without regard to coordinate systems, but the mathematical manipulations take on simple forms, Equation 2.46 and 2.47, when carried out for Cartesian coordinate systems. However, unlike the situation with vectors, it is difficult to draw the tensors and visualize the geometric operation.

The dot product is also defined between vectors and tensors, with a simpler physical significance. Equation 2.42, which we used to define a tensor, is an example of this kind of dot product. As with the vector dot product, it is simpler to visualize the operation for a unit vector:

$$\mathbf{t} = \mathbf{S}\,\hat{\mathbf{n}} \tag{2.52}$$

In this case, **t** represents the projection of **S** along a line whose direction is represented by the unit vector $\hat{\mathbf{n}}$. This projection contains three components and is therefore a vector, contrary to the vector dot product, which yields a scalar projection. The physical significance of tensors and tensor operations is better explained by examples of specific physical tensors, introduced in the next two chapters.

Because tensor algebra follows matrix algebra so closely, we reproduce only a few common tensor properties here, without amplification:

$$\mathbf{TI} = \mathbf{IT} = \mathbf{T} \tag{2.53}$$

(identity tensor)

$$\mathbf{TS} \neq \mathbf{ST} \tag{2.54}$$

(not commutative)

$$(\mathbf{TS})\mathbf{U} = \mathbf{S}(\mathbf{TU}) \tag{2.55}$$

(associative)

$$\mathbf{T}(\mathbf{S}+\mathbf{U}) = \mathbf{TS}+\mathbf{TU} \tag{2.56}$$

(distributive over addition)

$$\mathbf{T}\cdot\mathbf{U} = \mathbf{U}\cdot\mathbf{T} \tag{2.57}$$

(commutative)

$$\mathbf{T}\cdot(\mathbf{U}+\mathbf{S})=(\mathbf{T}\cdot\mathbf{U})+(\mathbf{T}\cdot\mathbf{S}) \tag{2.58}$$
(distributive over addition)

$$\mathbf{T}+\mathbf{0}=\mathbf{T} \tag{2.59}$$
(identity element = null tensor)

Other definitions follow directly from matrix algebra. For example, the transpose of a tensor is one that has its rows and columns interchanged:

$$T^{T}{}_{ij}=T_{ji} \tag{2.60}$$

The use of tensors will be illustrated more precisely in the following three chapters.

CHAPTER 2 PROBLEMS

A. Proficiency Problems

1. Perform the indicated vector operations using the vector components provided:

$\mathbf{a}\leftrightarrow(1,\ 1,\ 1)$ $\mathbf{b}\leftrightarrow(1,\ 2,\ 3)$ $\mathbf{c}\leftrightarrow(-1,\ 1,-1)$

$\mathbf{a}\cdot\mathbf{b}$	$\mathbf{a}\times\mathbf{b}$	$\mathbf{a}\cdot(\mathbf{b}\times\mathbf{c})$
$\mathbf{a}\cdot\mathbf{c}$	$\mathbf{a}\times\mathbf{c}$	$(\mathbf{a}\times\mathbf{b})\cdot(\mathbf{a}\times\mathbf{c})$
$\mathbf{b}\cdot\mathbf{c}$	$\mathbf{b}\times\mathbf{c}$	$\mathbf{a}\cdot(\mathbf{b}+\mathbf{c})$
$\mathbf{a}+\mathbf{b}$	$\mathbf{b}\times\mathbf{a}$	$\mathbf{a}\cdot\mathbf{b}+\mathbf{a}\cdot\mathbf{c}$
$\mathbf{a}+\mathbf{c}$	$\mathbf{c}\times\mathbf{a}$	$\mathbf{a}\times(\mathbf{b}+\mathbf{c})$
$\mathbf{b}+\mathbf{c}$	$\mathbf{c}\times\mathbf{b}$	$(\mathbf{a}\times\mathbf{b})+(\mathbf{a}\times\mathbf{c})$

2. Perform the indicated vector operations.

   a. Write the given vectors in terms of the base vectors provided:

$$\hat{x}_1{}'\leftrightarrow(0.866,\ 0.500, 0.000)$$
$$\hat{x}_2{}'\leftrightarrow(-0.500,\ 0.866, 0.000),$$
$$\hat{x}_3{}'\leftrightarrow(0.000,\ 0.000, 1.000)$$

   where the components of these base vectors are expressed in the original coordinate system:

$\mathbf{a}\leftrightarrow(1,\ 1,\ 1)$ $\mathbf{b}\leftrightarrow(1,\ 2,\ 3)$ $\mathbf{c}\leftrightarrow(-1,\ 1,-1)$

as expressed in the original orthonormal bases set $\hat{\mathbf{x}}_1, \hat{\mathbf{x}}_2, \hat{\mathbf{x}}_3$

i.e., $\mathbf{a} = \hat{\mathbf{x}}_1 + \hat{\mathbf{x}}_2 + \hat{\mathbf{x}}_3 \qquad \mathbf{b} = \hat{\mathbf{x}}_1 + 2\hat{\mathbf{x}}_2 + 3\hat{\mathbf{x}}_3 \qquad \mathbf{c} = -\hat{\mathbf{x}}_1 + \hat{\mathbf{x}}_2 - \hat{\mathbf{x}}_3$

b. Perform the following operations using the components of **a, b, c** expressed in the new (primed) basis:

$$\begin{array}{lll} \mathbf{a}\cdot\mathbf{b} & \mathbf{a}\cdot(\mathbf{b}\times\mathbf{c}) & \mathbf{a}\times(\mathbf{b}+\mathbf{c}) \\ \mathbf{a}\times\mathbf{b} & (\mathbf{a}\times\mathbf{b})\cdot(\mathbf{a}\times\mathbf{c}) & (\mathbf{a}\times\mathbf{b})+(\mathbf{a}\times\mathbf{c}) \\ \mathbf{a}+\mathbf{b} & & \end{array}$$

c. Construct the rotation matrix [R] to transform components from the original coordinate system to the primed coordinate system. Is [R] orthogonal? Find the inverse of [R] in order to transform components expressed in the primed coordinate system back to the original, unprimed coordinate system.

d. Transform the components of the results found in part b to the unprimed coordinate system and compare the results with the equivalent operations carried out in part a (using components expressed in the original coordinate system).

3. Find the rotation matrices for the following operations:

a. Rotation of axes (i.e., component transformation) 45° about $\hat{\mathbf{x}}_3$ in a right-handed sense (counterclockwise when looking anti-parallel along $\hat{\mathbf{x}}_3$).

b. Rotation of a physical vector 45° about $\hat{\mathbf{x}}_3$ in a right-handed sense (i.e., the *vector* moves counterclockwise when looking antiparallel along $\hat{\mathbf{x}}_3$).

c. Rotation of axes (i.e., component transformation) 30° about $\hat{\mathbf{x}}_2$ in a right-handed sense (i.e., counterclockwise when viewed antiparallel to $\hat{\mathbf{x}}_2$).

d. Rotation of a physical vector 30° about $\hat{\mathbf{x}}_2$ in a right-handed sense (i.e., the vector moves counterclockwise when looking antiparallel along $\hat{\mathbf{x}}_2$).

4. Perform the matrix manipulations shown.

a. Find the determinants and inverses of the following matrices:

$$[A] = \begin{bmatrix} 1 & 2 & 3 \\ 4 & 5 & 6 \\ 7 & 8 & 9 \end{bmatrix} \quad [B] = \begin{bmatrix} 7 & 8 & 9 \\ 1 & 2 & 3 \\ 4 & 5 & 6 \end{bmatrix} \quad [C] = \begin{bmatrix} 1 & 1 & 1 \\ -1 & 2 & 3 \\ 3 & 1 & -1 \end{bmatrix}$$

b. Multiply $[A][A]^{-1}$, $[B][B]^{-1}$, and $[C][C]^{-1}$ to verify that the inverse has been correctly obtained.

5. The following sets of basis vectors are presented in a standard Cartesian coordinate system $(\hat{x}_1, \hat{x}_2, \hat{x}_3)$.

$$\text{Set (1)}\quad \begin{array}{lrrr} \hat{x}_1^{(1)} \leftrightarrow & 0.707, & 0.707, & 0.000 \\ \hat{x}_2^{(1)} \leftrightarrow & -0.500, & 0.500, & 0.707 \\ \hat{x}_3^{(1)} \leftrightarrow & 0.500, & -0.500, & 0.707 \end{array}$$

$$\text{Set (2)}\quad \begin{array}{lrrr} \hat{x}_1^{(2)} \leftrightarrow & 0.750, & 0.433, & 0.500 \\ \hat{x}_2^{(2)} \leftrightarrow & -0.500, & 0.866, & 0.000 \\ \hat{x}_3^{(2)} \leftrightarrow & -0.433, & -0.250, & 0.866 \end{array}$$

$$\text{Set (3)}\quad \begin{array}{lrrr} \hat{x}_1^{(3)} \leftrightarrow & 0.866, & 0.500, & 0.354 \\ \hat{x}_2^{(3)} \leftrightarrow & 0.500, & 0.866, & 0.354 \\ \hat{x}_3^{(3)} \leftrightarrow & 0.000, & 0.000, & 0.866 \end{array}$$

a. Using vector operations, determine which of the basis sets are orthogonal.

b. Determine the transformation matrices to transform components presented in the original coordinate system $(\hat{x}_1, \hat{x}_2, \hat{x}_3)$ to those in each of the other basis systems.

c. Which of the transformation matrices in part b are orthogonal? Does this agree with part a?

d. Find the transformation matrix to transform components provided in coordinate system (1) to components expressed in coordinate system (2). Is the transformation matrix orthogonal?

6. Solve the sets of equations presented below by finding the inverse of the coefficient matrix. (Note that part b will require extension of the inversion formula to matrices of size greater than 3×3.)

a.

$$\begin{aligned} X_1 + 2X_2 + 3X_3 &= 10 \\ X_1 + 5X_2 - X_3 &= 12 \\ X_1 + 3X_2 + X_3 &= 14 \end{aligned}$$

b.

$$\begin{aligned} X_1 + 2X_2 + 3X_3 + 4X_4 &= 10 \\ X_1 + 5X_2 - X_3 + 14X_4 &= 12 \\ X_1 + 3X_2 + X_3 + X_4 &= 14 \\ X_1 + 4X_2 - 2X_3 - 2X_4 &= 16 \end{aligned}$$

7. Perform the following operations related to eigenvector-eigenvalue problems.

   a. Find the eigenvalues and the associated eigenvectors for the following matrices:

$$\begin{bmatrix} 1 & 2 \\ 3 & 1 \end{bmatrix} \quad \begin{bmatrix} 1 & 2 & 3 \\ 2 & 4 & 5 \\ 3 & 5 & 6 \end{bmatrix} \quad \begin{bmatrix} 1 & -1 & 2 \\ -1 & 2 & -3 \\ 2 & -3 & 3 \end{bmatrix}$$

   b. Find the transformation matrices which change components expressed in the original coordinate system to ones expressed using the eigenvectors as base vectors. Choose the direction associated with the maximum eigenvalue to be the new $\hat{x}_1'$, the second one $\hat{x}_2'$, and the third one $\hat{\mathbf{x}}_3'$.

   c. Treating the columns of the matrices in part a as vectors, find the equivalent components expressed in the eigenvector bases from part b [i.e., use the transformation matrices found in part c to find the new components of the tensors in part a, expressed in the **principal** coordinate system.]

8. Find the new components of the tensors provided below if the coordinate system change corresponds to a rotation of 30° about $\hat{x}_3'$:

$$[R] = \begin{bmatrix} 0.866 & 0.500 & 0.000 \\ -0.500 & 0.866 & 0.000 \\ 0.000 & 0.000 & 1.000 \end{bmatrix}$$

$$\mathbf{T}_1 \leftrightarrow \begin{bmatrix} 1 & 2 & 3 \\ 2 & 4 & 5 \\ 3 & 5 & 6 \end{bmatrix} = [\mathrm{T}_1], \quad \mathbf{T}_2 \leftrightarrow \begin{bmatrix} 1 & 2 & 3 \\ 4 & 5 & 6 \\ 7 & 8 & 9 \end{bmatrix} = [\mathrm{T}_2]$$

9. In calculating contact conditions at an interface, it is often necessary to find the unit vector that represents the projection of a given vector (usually the displacement of a material point) onto a plane tangent to the interface. If the normal to the tangent plane is denoted $\hat{\mathbf{n}}$, and the arbitrary vector is **a**, find $\hat{\mathbf{t}}$, the unit tangent vector corresponding to the material displacement. (Express the result in terms of **a**, $\hat{\mathbf{n}}$, and simple vector operations.)

B. DEPTH PROBLEMS

10. Perform the following operations related to component transformations.

    a. Find the transformation for components from basis set (2) to basis set (3) in Problem 5.

    b. Find the inverse transformation; that is, one that expresses components in basis set (2) if they are given in basis set (3).

    c. Using the approach shown in Exercise 2.5, verify that transformation matrices found in parts a and b do, indeed, perform the indicated transformations.

    d. Show the matrix form of the tensor transformation for components given in basis set (2) to those in basis set (3).

11. Find the rotation matrix for the double rotation of coordinate axes: rotate 90° about $\hat{\mathbf{x}}_1$, and then 90° about $\hat{\mathbf{x}}_3$.

12. A cylindrical coordinate system is one that rotates according to the coordinates of the point in question. Typically, r, θ, z represent the coordinates of a point, with the base vectors given as $\hat{\mathbf{r}}, \hat{\boldsymbol{\theta}}, \hat{\mathbf{z}}$, for example. The diagram shows such a coordinate system and a superimposed Cartesian coordinate system that coincides when $\theta = 0$.

    a. Find the transformation matrix to change components expressed in $\hat{\mathbf{x}}_1$, $\hat{\mathbf{x}}_2$, $\hat{\mathbf{x}}_3$ to ones expressed in $\hat{\mathbf{r}}, \hat{\boldsymbol{\theta}}, \hat{\mathbf{z}}$.

    b. Find the cylindrical components of the following vectors expressed in the $\hat{\mathbf{x}}_1$, $\hat{\mathbf{x}}_2$, $\hat{\mathbf{x}}_3$ system:

a $\leftrightarrow$ (10.000, 0.000, 0.000)
b $\leftrightarrow$ (0.866, 0.500, 0.000)
c $\leftrightarrow$ (1.000, 1.000, 0.000)
d $\leftrightarrow$ (1.000, 1.000, 1.000)
e $\leftrightarrow$ (0.000, 1.000, 0.000)
f $\leftrightarrow$ (−1.000, 1.000, 1.000)

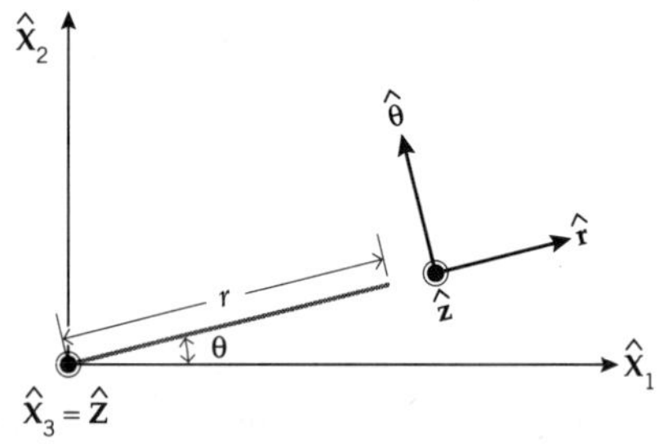

c. Find the magnitudes of the vectors given in part b, first using the Cartesian coordinate system components, then using the cylindrical coordinate system. How is the magnitude of a vector computed in cylindrical coordinates?

d. Is [R] orthogonal for this transformation? Physically, why or why not?

13. Perform the indicated operations related to equation solving.

a. Solve the equations given in Problem 6 by using Gaussian reduction instead of by finding the inverse. Which do you prefer for large matrices?

b. Given the solutions obtained in part a, find the inverse of the coefficient matrix.

c. For larger sets of equations, why is it easier to solve by a reduction method?

C. NUMERICAL AND COMPUTATIONAL PROBLEMS

14. Write a main program for handling input and output and subroutines to perform the following basic vector and matrix operations (for dimensions up to 10×10, for example). Use Problems 1, 2, and 3 to test the validity of your program.

a. Adding and subtracting vectors and matrices

b. Dot product

c. Cross product (3-D only)

d. Multiplication of two matrices

e. Multiplication of a matrix and vector

f. Test for the orthogonality of a 3×3 matrix

15. Using your subroutines from Problem 1, construct a program for rotating the components of a second-rank tensor.

16. Create and test a subroutine.

    a. Write a subroutine for solving a system of linear equations based on the Gaussian reduction method. Verify your program by comparing with the manual operation requested in Problem 12, or the inverse method requested in Problem 6.

    b. Use your subroutine from part a to find the inverse of a given matrix. Compare results with Problems 4 and 6 to verify your method.

17. Use the Newton-Raphson method (see Chapter 1) to solve a general cubic polynomial equation with real roots. Write a subroutine to solve this problem. Verify your program by comparing with Problem 7.

18. Write a program to find the eigenvalues and eigenvectors for a 3×3 symmetric matrix. Use the subroutine constructed in Problem 17 to find the eigenvalues from the cubic equation. Use the method presented in Exercise 2.8 to find the eigenvectors, but check to make certain that the component eliminated in the method is not equal to zero. Verify your program by comparison with Problem 7.

CHAPTER 3

Stress

We choose to consider the description of forces acting in and on a body before discussing deformation of the body. We adopt this approach because the concept of force is usually easier to understand than the concept of deformation. Forces are used directly in all mechanical branches of physics; that is, those utilizing Newton's laws. We begin with simple physical concepts and then go on to develop a more precise mathematical approach to the definitions. Along the way, we find the need to discuss and define the continuum (which few discussions of continuum mechanics actually do). We note in passing that the fundamental concepts in continuum mechanics often cause the most trouble and should not be avoided simply because they are not purely mathematical.

3.1 PHYSICAL MOTIVATION FOR STRESS DEFINITIONS (1-D)

Let's start the discussion by considering a long, slender body loaded axially (picture, for example, a wire or standard tensile specimen), as in Figure 3.1.

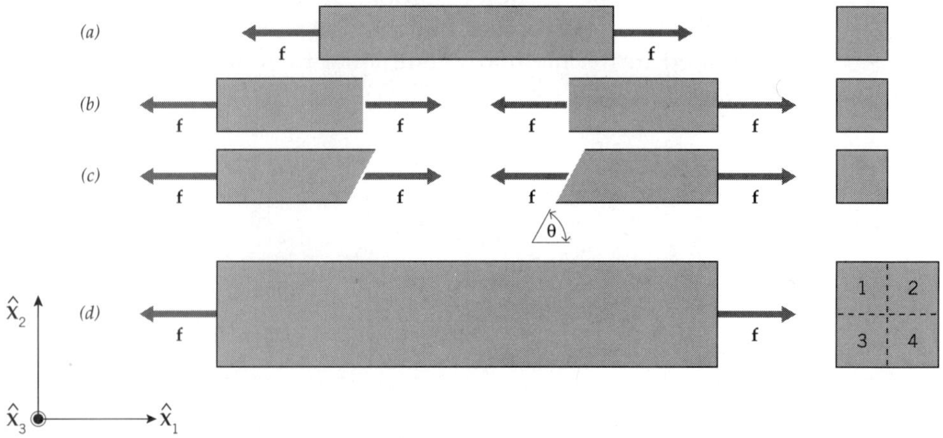

Figure 3.1 The force transmitted by a long, slender rod.

For equilibrium (i.e., so the rod doesn't accelerate), the **external forces** acting on each end of the rod must be equal. To probe the **internal forces** acting through the material rod, we conceptually "cut" the material of the rod into two pieces[1] and then apply a fictitious force on each part such that the remaining bodies do not accelerate. Of course, there is no reason that we could not do this operation in reality, but it is obvious that the result would be the same, if we are to believe Newton. By physically or conceptually cutting the specimen, we can convert the existing internal forces to external, observable forces, so we should realize that we need not be overly concerned about the distinction between the two terms.[2]

By comparing the perpendicular cut, Figure 3.1(*b*), with a cut at an arbitrary angle, Figure 3.1(*c*), one can easily see that the force passing through the cut surfaces does not depend on the cut orientation.[2]

Similarly, by comparing Figures 3.1(*c*) and (*d*), it is clear that the force passing through the cut plane does not depend on the size of the body, either. In Figure 3.1(*d*), we see that four rods like the one in Figure 3.1(*a-c*) could be put together (each loaded with $\mathbf{f}/4$) without changing the external or internal forces.

To generalize our limited observations from the slender rod, we can state the following, for cuts separating a body into two distinct pieces:

1. Internal forces can be examined by making mathematical or physical cuts *through* a body, thereby converting the internal forces to external, observable ones.
2. The internal forces are independent of the size and orientation of the cut *through* the body used to probe them as long as the reorientation of the cut does not include any additional boundary forces.

These two general observations simply mean that the internal forces transmitted across internal surfaces are as real and physical as the external ones, and that they are unrelated to the structure and properties of the body, or to the technique we use to reveal them.

Let's now compare Figures 3.1(*c*) and (*d*) in more detail. Figure 3.1(*d*) represents a bar with dimensions double that of the others shown. If we make the *trivial* cuts shown by dashed lines (trivial because no forces pass through those surfaces since the force vector lies in those planes), we can see that the larger bar is equivalent to four of the smaller ones, each loaded with a force of

[1] We restrict our discussion to cuts that separate the body into two pieces. Other cuts do not clearly make the distinction between internal and external parts of the body. In addition, we assume that a narrow cut indeed separates the two halves completely, in that there are no long-range forces from regions remote from the cut, nor are there local couples operating across surface regions.

[2] There are exceptions to this, depending on how the cut is taken. The force will depend on the cut orientation if, by rotating the cut, additional external forces are added to the body. Equivalently, any forces acting in the cut plane are not "seen" until the cut plane is rotated, because these forces are not transmitted by the cut surface. This is important because it implies that we must always examine three non-coplanar planes to reveal the full range of forces in a body. In the rod geometry, we avoid these difficulties because only a single external force acts along the axis, and we exclude cuts that contain this axis.

$\mathbf{f}/4$. Therefore, from *a material point of view*,[3] bar (*d*) is loaded only one-quarter as heavily as the bars in (*a-c*), although the same force is transmitted through it. That is, the loading felt by the material is reduced proportionally as the area of the surface (at a fixed orientation) carrying a fixed load is increased. For material purposes, then, it is more useful to think about the average intensity of the force transmitted by the area, that is, the force transmitted through a surface divided by the area of that surface. While this definition solves the problem of different-size bars, it introduces a further complication because the orientation of the cut plane, while not influencing the transmitted force, does influence the *magnitude* of computed force intensity.[4] Therefore, the vector force intensity, usually called the **stress vector** or **traction vector**, depends on the orientation of the surface across which it is transmitted. For a given surface, the transmitted force vector, **f**, is found by multiplying the stress vector, **t**, by the scalar surface area a:

$$\mathbf{f}_{\mathbf{transmitted}} = \mathbf{t}\,\mathrm{a} \tag{3.1}$$

Equation 3.1 shows that the idea of internal force per unit area must be linked to both the force and the area used to probe it. This is a simple consequence of this geometry, because the force transmitted by the plane is independent of orientation while the area of the plane varies with orientation.[5] Equation 3.1 also shows in a simple way that the maximum force per unit area will be found on planes normal to the force passing through them. These particular directions (parallel to force direction, normal to the plane) are called the **principal stress directions**, and the values of the force area intensity (a maximum) on these planes are called the **principal stresses**. (Since the force vector lies normal to the plane, only a scalar length is needed to describe a principal stress.)

The direction of the **stress vector** lies in the direction of the force passing through the plane, and its magnitude is equal to the magnitude of the force divided by the area of the cut plane used to measure the force. The magnitude is a scalar measure of the **stress** acting on the plane.

As a final note in this thought experiment, let's consider what we have neglected so far. We know, for example, that all planes containing the vector **f**—that is, all those containing the axis of the rod, like the cuts shown in Figure

---

[3] The material in each bar is presumed to have identical properties, independent of the number of bars selected or their sizes. Thus we are able to assign the force **f** equally to the four bars. A more precise statement of this appears in the next section.

[4] Again, this applies to cuts made across the whole body and whose reorientation doesn't pass through additional external forces.

[5] An equivalent, and more usual, way to look at stress is to choose a unit area (internal) of arbitrary orientation and to ask what force is transmitted by this trial area. This procedure will be followed in subsequent, more formal sections. In either case, three non-coplanar planes are required to probe all of the possible transmitted forces.

3.1(*d*)—are unaffected by **f**. Similarly, all forces lying in our chosen mathematical (or real) cut plane will be unobservable. Therefore, with a little geometric visualization, it should be clear that we must make at least three arbitrary (not coplanar) cuts in order to probe all of the possible forces passing through the body. This can be easily seen by considering two arbitrary planes. They intersect in a line that is contained in both planes. Therefore, forces transmitted in this line's direction are invisible on both of the two planes. The third plane, being noncoplanar, will have a nonzero projected area normal to the intersection line and will therefore "feel" the remaining force direction. As usual, the argument can be simplified if we think of orthogonal cuts. In Figure 3.1(*b*), for example, the first cut can be taken normal to $\hat{\mathbf{x}}_1$ and the second might be normal to $\hat{\mathbf{x}}_2$. Then, forces acting along the $\hat{\mathbf{x}}_3$ direction are untested, so we choose our third cut normal to $\hat{\mathbf{x}}_3$. (This system has a further advantage for the rod geometry—the area of the cross-sectional cut is minimum, and the stress vector magnitude is maximum. The two normal cuts experience no transmitted forces. That is, the principal stresses are obtained on the three surfaces chosen, two of which are zero.)

The remaining limitation to our thought experiment lies in our assumption that the material of the bar(s) is uniform. We used this assumption when we assigned a force of $\mathbf{f}/4$ to each quarter of the bar in Figure 3.1(d). For example, when we cut the bar into quarters, we could have found that one piece was made of steel, one of mercury, one of butter, and one of air! In this case, nearly all of **f** would be transmitted through the steel quarter while the other parts would be nearly unloaded. Thus, it would make no sense to assume that the *material* in Figure 3.1(*d*) was loaded one-quarter as heavily as that in Figure 3.1(*c*). Three-quarters would be nearly load-free, and one quarter would be nearly the same as the bar in Figure 3.1(*c*); thus any attempt to describe the force intensity by a single quantity would be meaningless. The transmitted force across the cut is, of course, the same because Newton can't be cheated, but the idea of a force-per-area intensity would require a more careful approach. In the next section, we discuss the basic assumptions in defining limitations on our picture of the material (or on our way of looking at it) in order to resolve this problem with our physically motivated definitions of stress.

3.2 GLIMPSE OF A CONTINUUM

The problem raised at the end of the last section is central to the field of **continuum mechanics**; namely, what assumptions do we require about our material behavior[6] such that we can define materially useful quantities such as stress? In the discussion of Figure 3.1, we avoided the problem by quietly assuming that the force on the rod was transmitted *uniformly* through the rod for purposes of defining an intensity of force, a reasonable assumption for a

[6] Material is defined generally here to include solids, liquids and gases.

long, slender geometry of homogeneous material (which we sneakily introduced without comment). We later showed that the idea of a single intensity on the cut surface was meaningless unless the bar could be subdivided and a part of the transmitted force assigned according to the cross-sectional area of the part.

In fact, the problem is much more serious for a generally loaded body because the forces are varying with position on each internal and external surface, with the stress varying from point to point and orientation to orientation. Therefore, to talk about a materially meaningful force/area in the general case, we need to restrict our attention to a surface area that is much smaller and/or much larger than the scales of variation for the stress itself. In the case of the long, slender rod with constant cross-section (Figure 3.1), we can see intuitively that the transmitted force/area will be nearly uniform for any area far removed from the ends of the rod, as long as the material itself is homogeneous.

The idea of **length scales** mentioned in the preceding paragraph pervades the fields of continuum mechanics and materials science. We use the idea without real precision, much like an *order-of-magnitude* estimate. Inhomogeneities in mechanical field variables (stress or strain, e.g.) arise not only because of external loading, but also from internal variations in material properties. It is necessary to distinguish which variations are of important length scales and which can safely be ignored. For example, if the length scale of interest for a given problem is $d \sim 10^{-3}$ m, one can often safely average over variations over a scale of $d \sim 10^{-6}$ m or 10^{-5} m and neglect far-removed sources of variation (10^{-1} m, e.g.), by replacing the actual distribution of forces with a statically equivalent set with the characteristics sought. The principle does not apply to the directional dependence induced by structures of various length scales, which persists at much greater length scales via long-range symmetry. For example, the atomic ($d \sim 10^{-10}$ m) arrangements in a crystal determine its elastic symmetry, up to any size single crystal ($d \sim 10^{-5}$ m). Conversely, a random assortment of such crystals in a polycrystal can have isotropic elastic properties because the local atomic symmetry is lost at large length scales.

In order to provide a crude but instructive example of typical length scales, we consider a large structure, a bridge, from the perspective of several mechanical scales. At the largest scale, the bridge might be considered as a unit ($d \sim 10^2$-10^3 m) to assess relative movements of the ends by seismic activity. For structural analysis, each member is often modeled by considering statically equivalent pin-on-rod assemblies such as the I-beam shown in Figure 3.2 ($d \sim 10^1$ m in length). The structural analysis for such a structure prior to construction would normally be carried out at this scale. Separately, the use of I-beams would be justified by more detailed structural analysis based on their actual shape, on a length scale corresponding to parts of the smallest dimension ($d \sim 10^{-2}$-10^{-3} m). At this length scale, the continuum approximation is quite good because the internal material inhomogeneities are usually much smaller and may be treated in an average way. At $d \sim 10^{-5}$ m, the typical **grains** of the steel represent marked inhomogeneities, and there may be second-phase particles **(precipitates)** of $d \sim 10^{-6}$ m inside the grains. Inside the grains and between the precipitates, the material

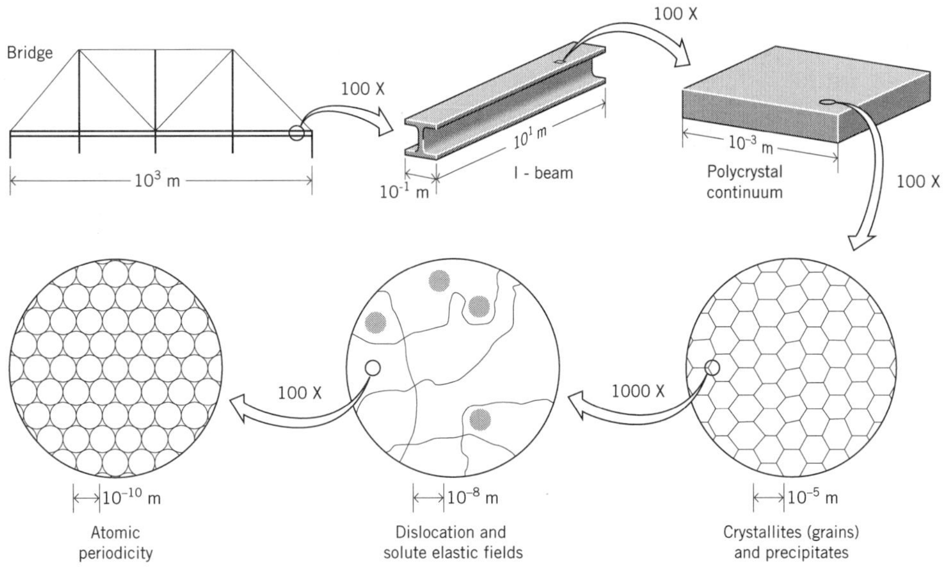

Figure 3.2 Length scales for a typical structural material spanning thirteen orders of magnitude.

can be treated homogeneously until a length scale of d ~ 10^{-8} m is approached, where the stress fields of **dislocations** are resolvable. Again, between dislocation cores (and not crossing phase or grain boundaries), the continuum model is very accurate. Finally, the continuum idea breaks down conceptually (along with the idea of a material) at the atomic level, d ~ 10^{-10} m, where many-body interactions occur via the complexities of atomic bonding.

> ***Summary:*** *A continuum is a smoothly behaving mass of material that exhibits no "lumps," "folds," or "holes," such that the limit defined in Equation 3.1 exists. The connection to real materials involves the choice of length scale (d) such that a modified limit like Equation 3.1 exists for surface areas of a scale of approximately $d \cdot d$.*

With this melange of material and loading inhomogeneities, how do we go about defining a useful force intensity? The usual procedure is to invoke, arbitrarily, the idea of a **continuum** (hence *continuum* mechanics). A continuum is an idealized sort of material where, at a sufficiently small length scale, the material behaves *smoothly*;[7] that is, all of the field variables change much more slowly (i.e., on a much larger length scale) than the material properties. For

[7] In this general discussion, we will avoid defining *smooth* in a precise mathematical way but will defer that to subsequent sections.

example, we restrict our attention to sufficiently small regions so that we can avoid boundaries between crystals and dissimilar phases. As a corollary of this intuitive idea of **smoothness**, we will not allow holes to open up or material to overlap. That is, two bits of material can never occupy the same point in space (or conversely, one bit occupy two points in space), for these actions would certainly result in some abrupt and nonsmooth changes in intensities, and so on.

With our definition of a continuum, we are free to define a meaningful force intensity (or stress) by considering an infinitesimal area of surface (internal or external), which must be at a much smaller scale than the variations in field quantity:

$$\text{stress} = \lim_{a \to 0} \frac{\text{Force transmitted through area } \mathbf{a}}{\text{Area of } \mathbf{a}} \qquad (3.2)$$

$$\left(\text{For a specified orientation of } \hat{\mathbf{a}}\right)$$

This one-dimensional view of stress corresponds to the discussion of the slender rod, Figure 3.1. Equation 3.2 must exist without regard to how the limit is taken if we want a good definition of stress at a point, hence the need to disallow **voids** or **folds**.

How useful is this idea of a continuum? It is extremely useful (although some of the most interesting research focuses on regions where the idea breaks down). Virtually all **structural analysis** and design relies completely on the continuum concept. The most serious deficiency occurs at the smallest length scales, approaching the interatomic distances (10^{-10} m). A limit like the one shown in Equation 3.2 has no meaning when the limiting procedure involves passing through periodic arrays of nuclei and electron clouds that can affect the result significantly. Therefore, we should think of the limit of Equation 3.2 as being one where "a" becomes very small relative to some variations (those from external loading, second-phase particles, etc.) while remaining very large relative to others (the interatomic distance).

3.3 DEFINITION OF STRESS: INTUITIVE APPROACH

We now proceed to generalize the idea of stress (force intensity) introduced conceptually in Section 3.1. In order to do this formally, we rely on the concept of a continuum discussed in Section 3.2.

For a general 3-D definition, let's consider cutting out a *small*[8] piece of material from a statically loaded body. (Note that we now do not require that the cuts extend across the whole body, and thus we cannot know, a priori, the force transmitted across the plane.) In order to keep the element from accelerating or deforming after the connections to the adjoining material are cut, we must impose forces equivalent to those imposed by the rest of the body on each cut

[8] The use of the word *small* refers to its dimensions relative to variations of force intensity. For the formal continuum, of course, we can ensure this condition by making the piece infinitesimal.

surface. The small piece may be cut out in any shape; however, it is convenient to use a right parallelepiped such that the edges lie along the axes of a Cartesian coordinate system. (The directions of these axes are chosen arbitrarily as long as they form an orthogonal set).

Therefore, let's consider the parallelepiped and the general set of forces that pass through its surfaces, Figure 3.3.

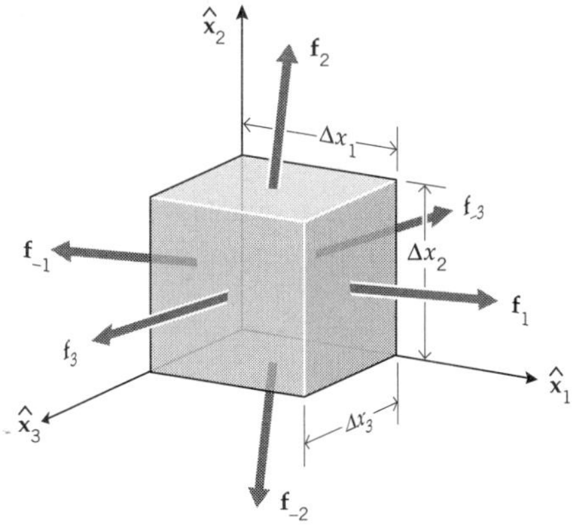

Figure 3.3 The forces acting on an element of material cut from a loaded body.

We label the forces acting on the material by the plane normal to that surface (i.e., $\mathbf{f}_{-1}$ is the force acting on the plane whose outward normal is $-\hat{\mathbf{x}}_1$). Following our 1-D example, we define the average force intensity on each face by dividing by the surface area:

$$\text{force intensities}: \quad \frac{\mathbf{f}_{\pm i}}{a_{\pm i}} = \frac{\mathbf{f}_{\pm i}}{\Delta x_j \Delta x_k} \tag{3.3}$$

where Δx_j and Δx_k are the lengths of the two sides defining a_i. In order to define stress vectors more precisely, we take the limit of Equation 3.3 as a_i approaches zero (while the orientation of the plane remains fixed), a limit that we know must exist because of our continuum assumption:

$$t_{\pm i} = \lim_{a_i \to 0} \frac{\mathbf{f}_{\pm i}}{a_i} = \lim_{\substack{\Delta x_{j \to 0} \\ \Delta x_{k \to 0}}} \frac{\mathbf{f}_i}{\Delta x_j \Delta x_k}. \tag{3.4}$$

Where $\mathbf{t}_{\pm i}$ represents the six stress vectors (i = 1, 2, 3) operating on an infinitesimal element of material via the six faces.

With our knowledge of a continuum's properties, it is easy to show that only three of the $\mathbf{t}_{\pm i}$ are independent. Since the limit in Equation 3.4 must exist independently of the path taken as $a_i \to 0$, we can consider a parallelepiped of the proportions shown in Figure 3.4 as we carry out the limit. That is, because we are dealing with a continuum, we are free to consider a_2 and a_3 second order relative to a_1. Then, by summing forces on the element, we find that $\mathbf{f}_2$ (= $\mathbf{t}_2 a_2$) and $\mathbf{f}_3$ (= $\mathbf{t}_3 a_3$) are second order and therefore do not contribute. Thus,

$$\mathbf{t}_{-1} = - \mathbf{t}_1 \text{ (for equilibrium).}$$

By taking the shape of the parallelepiped differently, such that other pairs of faces do not contribute to the equilibrium, we can obtain similar relations for the other forces:

$$\mathbf{t}_{-2} = - \mathbf{t}_2$$

$$\mathbf{t}_{-3} = - t_3,$$

such that it is shown that only three stress vectors are needed to describe the entire area-intensity of forces passing through a continuum material point.

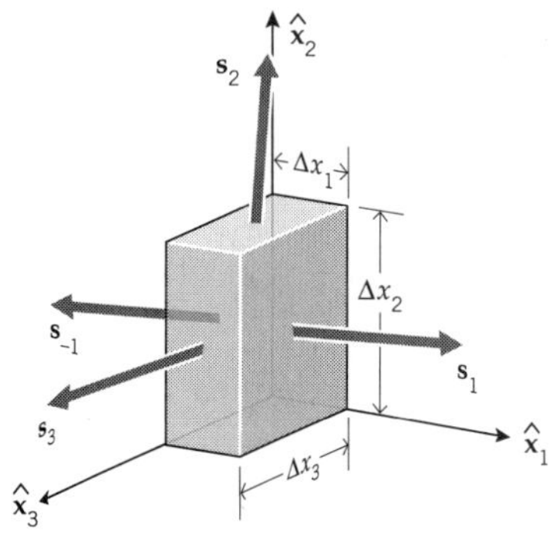

Figure 3.4 Stress vectors acting on the faces of right parallelepiped for $\Delta x_1 << \Delta x_2, \Delta x_3$.

It is convenient to arrange the components of the remaining stress vectors in a matrix for further manipulation, by making $\mathbf{t}_1$, $\mathbf{t}_2$, and $\mathbf{t}_3$ into three column vectors, part of a 3×3 matrix:

$$\begin{bmatrix} t_1(x_1) & \vdots & t_2(x_1) & \vdots & t_3(x_1) \\ t_1(x_2) & \vdots & t_2(x_2) & \vdots & t_3(x_2) \\ t_1(x_3) & \vdots & t_2(x_3) & \vdots & t_3(x_3) \end{bmatrix} = \begin{bmatrix} \sigma_{11} & \sigma_{12} & \sigma_{13} \\ \sigma_{21} & \sigma_{22} & \sigma_{23} \\ \sigma_{31} & \sigma_{32} & \sigma_{33} \end{bmatrix} = [\sigma] \tag{3.5}$$

where we have labeled the components of the stress vectors σ_{ij} as follows: i refers to the component direction $\hat{\mathbf{x}}_i$ of the force acting on the plane normal to $\hat{\mathbf{x}}_j$. Thus, the components of the stress vectors $\mathbf{t}_1$, $\mathbf{t}_2$, and $\mathbf{t}_3$ are the columns of $[\sigma]$. Figure 3.5 shows the locations and directions in which each of the components of $[\sigma]$ operates (three faces only are shown).

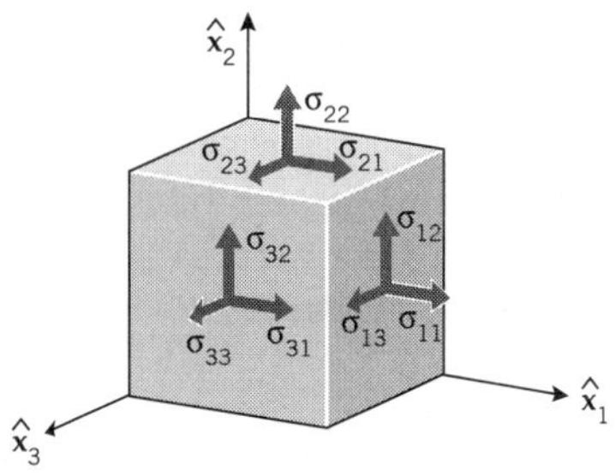

Figure 3.5 Stress vector components acting on the front of a cube (rear faces are not shown).

Although we considered the linear equilibrium of the element of material in reducing from six to three independent stress vectors, we have not yet considered the rotary equilibrium of the element. As an example, consider the view of the element along $-\hat{\mathbf{x}}_3$, Figure 3.6, and sum the moments about the $\hat{\mathbf{x}}_3$ axis in a right-handed sense:

$$\sum M_{x_3} = 2\sigma_{12}\Delta x_2 \Delta x_3 \left(\frac{\Delta x_1}{2}\right) - 2\sigma_{21}\,\Delta x_1\,\Delta x_3 \left(\frac{\Delta x_2}{2}\right) = 0 \tag{3.6}$$

Thus $\sigma_{12} = \sigma_{21}$ (by removing a common factor of $\Delta x_1 \Delta x_2 \Delta x_3$). Similarly, by considering rotary equilibrium about the $\hat{\mathbf{x}}_2$ and $\hat{\mathbf{x}}_1$ axes, one finds that $\sigma_{13} = \sigma_{31}$ and $\sigma_{23} = \sigma_{32}$. Therefore, the matrix representation introduced in Equation 3.5 is symmetric, and the distinction about whether the stress vectors are arranged as columns or rows is lost:

$$[\sigma] = \begin{bmatrix} \sigma_{11} & \sigma_{12} & \sigma_{13} \\ \sigma_{12} & \sigma_{22} & \sigma_{23} \\ \sigma_{13} & \sigma_{23} & \sigma_{33} \end{bmatrix} \tag{3.7}$$

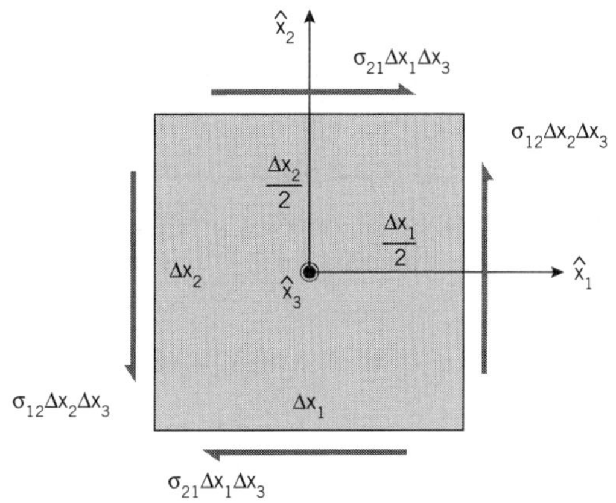

Figure 3.6 Moments acting about the $\hat{\mathbf{x}}_3$ axis through a material element (the element has thickness Δx_3 into the page).

Therefore, using Newton's laws, we have demonstrated that only six of the components of the **stress tensor** (Eq. 3.7) are independent (although we have not yet demonstrated that $[\sigma]$ represent the components of a tensor). The existence of only six independent components is obvious for the Cartesian coordinate system and the definition of stress presented here (following the original development of Cauchy and Euler), because $[\sigma]$ is symmetric. However, there are other possible definitions of stress for which the independence of only six components will not be so apparent.

Exercise 3.1 **Show that $[\sigma]$, as defined in Equation 3.7, is a tensor.**

We make a cut through the corners of the parallelepiped that we used to define σ_{ij}, thus forming a tetrahedron (this procedure follows Cauchy's original). We denote the cut surface area by a and its outward normal $\hat{\mathbf{n}}$, as shown in the figure below.

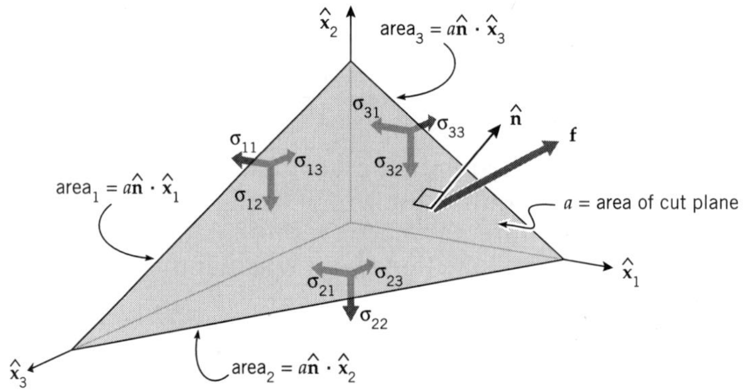

Since the new element is required to be in equilibrium, we can find the forces transmitted by the cut plane. We sum the forces on the tetrahedron in one coordinate direction at a time, noting the necessary signs to follow our outward-normal sign convention on the cut surface:

$$\sum f_{x_1} = 0 \rightarrow f_1(\text{on the cut plane}) = \sigma_{11}(\hat{\mathbf{n}}\cdot\hat{x}_1)a + \sigma_{12}(\hat{\mathbf{n}}\cdot\hat{x}_2)a + \sigma_{13}(\hat{\mathbf{n}}\cdot\hat{x}_3)a \quad (3.1\text{-}1)$$

where the area of each coordinate-plane face is simply equal to $(\hat{\mathbf{n}}\cdot\hat{x}_1)a$, obtained by projecting $\hat{\mathbf{n}}a$ along the three coordinate axes. Of course, $\hat{\mathbf{n}}\cdot\hat{x}_1$ is simply n_i, the ith component of $\hat{\mathbf{n}}$, so that Equation 3.1-1 can be rewritten as follows (dividing by a):

$$t_1 = \frac{f_1}{a} = \sigma_{11}n_1 + \sigma_{12}n_2 + \sigma_{13}n_3 \quad (3.1\text{-}1a)$$

Similarly, for equilibrium in the other two directions,

$$t_2 = \frac{f_2}{a} = \sigma_{21}n_1 + \sigma_{22}n_2 + \sigma_{23}n_3 \quad (3.1\text{-}1b)$$

$$t_3 = \frac{f_3}{a} = \sigma_{31}n_1 + \sigma_{32}n_2 + \sigma_{33}n_3 \tag{3.1-1c}$$

where t_1, t_2, and t_3 are the components of the stress vector acting on the cut plane with plane normal $\hat{\mathbf{n}}$. By combining Eqs. 3.1-1 and by noting that **t** and $\hat{\mathbf{n}}$ are physical vectors, we can write:

$$\begin{bmatrix} t_1 \\ t_2 \\ t_3 \end{bmatrix} = \begin{bmatrix} \sigma_{11} & \sigma_{12} & \sigma_{13} \\ \sigma_{21} & \sigma_{22} & \sigma_{23} \\ \sigma_{31} & \sigma_{32} & \sigma_{33} \end{bmatrix} \begin{bmatrix} n_1 \\ n_2 \\ n_3 \end{bmatrix} \leftrightarrow \mathbf{t} = \boldsymbol{\sigma} \cdot \hat{\mathbf{n}} \leftrightarrow t_i = \sigma_{ij} n_j \tag{3.1-2}$$

Thus σ is shown to be a linear vector operator (and thus a tensor) associated with the two vectors $\hat{\mathbf{n}}$ and **s** for a given plane. With a simple change to represent the cut-plane orientation and area by a single vector **a** $(\mathbf{a} = a\hat{\mathbf{n}})$, Equation 3.1-2 becomes:

$$\begin{bmatrix} f_1 \\ f_2 \\ f_3 \end{bmatrix} = \begin{bmatrix} \sigma_{11} & \sigma_{12} & \sigma_{13} \\ \sigma_{21} & \sigma_{22} & \sigma_{23} \\ \sigma_{31} & \sigma_{32} & \sigma_{33} \end{bmatrix} \begin{bmatrix} a_1 \\ a_2 \\ a_3 \end{bmatrix} \leftrightarrow \mathbf{f} = \boldsymbol{\sigma} \cdot \mathbf{a} \leftrightarrow f_i = \sigma_{ij}\, a_j \tag{3.1-3}$$

Exercise 3.2 **Show what happens to the properties of the stress tensor if body forces and body torques are allowed in the material element.**

Body forces and torques may be caused by forces that operate throughout a material because of its volume or mass. Examples include gravitational and electromagnetic forces.

Body forces can be shown not to affect the reduction of 18 possible stress components to 9 by again considering Figure 3.4. As Δx_1 approaches zero (while Δx_2 and Δx_3 remain constant), the volume of material upon which a body force can operate approaches zero. Therefore, along this limiting path (remember, all limiting paths are allowed by our definition of a continuum), the body force is second order to the surface force, and the opposite sides of an infinitesimal element must exhibit equal and opposite stress vectors. Thus the 18 possible components of $[\sigma]$ become 9. Moreover, $[\sigma]$ is still symmetric, so only 6 components are independent.

Body torques, conversely, are not second order to the torques introduced by the surface forces. Figure 3.6 shows that the volume of material in the element cannot approach zero in such a way that body torques become second order. Therefore, the existence of body torques destroys the symmetry of the stress tensor:

$$\sigma_{ij} \neq \sigma_{ji} \text{(with body torques)}$$

3.4 CONCISE DEFINITION OF THE STRESS TENSOR

We have concentrated so far on presenting a physical interpretation of stress. However, it is simpler and more concise to define the stress tensor as a vector operator, to show that the operator is linear (and thus a tensor), and to proceed from this vantage point.

To perform this definition, we consider an element of planar surface inside a loaded, *continuum* body.[9] The normal to the surface is defined by the unit vector $\hat{\mathbf{n}}$, and the average stress vector acting through this surface is **t**. (Equivalently, we could multiply **t** by the area of the surface, a, to obtain the total force acting on the surface.) Then, to obtain the value of **t** at a point, we perform a limit as the area a approaches zero, while keeping the normal ($\hat{\mathbf{n}}$) constant:

$$\mathbf{t}\,(\text{at a point}) = \lim_{a \to 0} \mathbf{t}\,(\text{average}) = \lim_{a \to 0} \frac{\mathbf{f}}{a} \tag{3.8}$$

This limit must exist because of our assumption of a continuum. The value of **t** becomes independent of the size of the test area, but does depend, in general, on the position in the body and the direction of $\hat{\mathbf{n}}$. Therefore, at a single point, we can write that **t** (or **f**) is a vector function of $\hat{\mathbf{n}}$:

$$\mathbf{t} = \mathcal{L}\left(\hat{\mathbf{n}}\right) \text{ or } d\mathbf{f} = \mathcal{L}\left(\hat{\mathbf{n}}\right) da \tag{3.9}$$

where, by the limit shown in Equation 3.8, **t** = d**f**/da (which is evaluated at constant **n**). Considering Equation 3.9, it is convenient to define an **area vector**; that is, one that has a direction normal to the surface area and a magnitude equal to the area itself:

$$\mathbf{a} = a\hat{\mathbf{n}}, \ d\mathbf{a} = \hat{\mathbf{n}}da \tag{3.10}$$

Note that we defined d**a** to conform with the limiting procedure of Equation 3.8. With this understanding of the limiting procedure (which shows that $\mathcal{L}$ is a constant for a given $\hat{\mathbf{n}}$), we can rewrite Equation 3.9 more simply:

$$\mathbf{t} = \mathcal{L}(\hat{\mathbf{n}}), \ d\mathbf{f} = \mathcal{L}(d\mathbf{a}) \tag{3.11}$$

Therefore, one of the tests that $\mathcal{L}$ be a *linear* vector operator (and thus a tensor) is satisfied:

$$\lambda d\mathbf{f} = \mathcal{L}(\hat{\mathbf{n}})\lambda da = \mathcal{L}(\lambda\hat{\mathbf{n}})da = \lambda\mathcal{L}(d\mathbf{a}) \tag{3.12}$$

[9] We restrict ourselves to an internal, planar surface to avoid ambiguities of the normal direction, and to eliminate surface-induced stresses such as occur at interfaces between dissimilar media, and which can depend on material properties and surface curvature, independent of the stresses transmitted through internal, imaginary surfaces.

The other, vector test has already been established in Exercise 3.1 using Cauchy's tetrahedron. For a more intuitive picture, we consider here a 2-D development in order to demonstrate that equilibrium requires that $\mathcal{L}(\mathbf{a}+\mathbf{b}) = \mathcal{L}(\mathbf{a}) + \mathcal{L}(\mathbf{b})$. Consider an infinitesimal element of rectangular cross-section (for simplicity!) with unit thickness along $\hat{\mathbf{x}}_3$ (Figure 3.7) with lengths of sides dx_1 and dx_2. Also consider a cut along the diagonal (which is in an arbitrary direction because dx_1 and dx_2 are arbitrary) and the equilibrium of one of the remaining triangles.

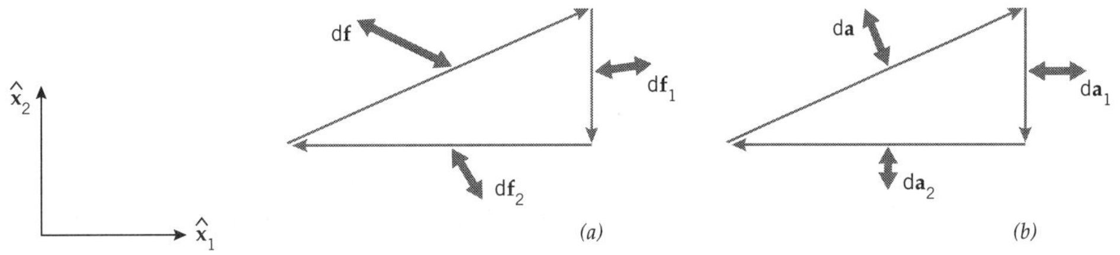

Figure 3.7 Force and area vector for a triangular element of material (unit thickness into the page).

Because the triangular element must be in equilibrium, the forces $\mathbf{df}_1$, $\mathbf{df}_2$, and $\mathbf{df}$ are related:

$$-\mathbf{df} = +(\mathbf{df}_1 + \mathbf{df}_2) \tag{3.13}$$

or, using our definition of the vector operator $\mathcal{L}$ (Equation 3.11),

$$-L(\mathbf{da}) = +[L(\mathbf{da}_1) + L(\mathbf{da}_2)] \tag{3.14}$$

But $\mathbf{da}_1$ and $\mathbf{da}_2$ are just the components (with a sign change because of the outward normal convention) of $\mathbf{da}$. That is, $\mathbf{da} = -(\mathbf{da}_1 + \mathbf{da}_2)$ and $\mathbf{da}_1 = (\mathbf{da}\cdot\hat{\mathbf{x}}_1)\hat{\mathbf{x}}_1$, $\mathbf{da}_2 = -(\mathbf{da}\cdot\hat{\mathbf{x}}_2)\hat{\mathbf{x}}_2$. Therefore, with the help of Equation 3.12, we can write

$$\mathcal{L}(\mathbf{da}_1 + \mathbf{da}_2) = \mathcal{L}(\mathbf{da}_1) + \mathcal{L}(\mathbf{da}_2) \tag{3.15}$$

and L has been shown to be a linear operator, and therefore a tensor, that we conventionally call the **Cauchy stress tensor**, σ. Following the arguments in the previous section, the Cartesian components of σ can be shown to be symmetric.

Note: *The foregoing development and the Cauchy tetrahedron of Exercise 3.1 depend on three assumptions that are usually observed to be true in common materials:*

1. The force acting through an internal surface is very short range in nature so that the entire interaction can be considered mechanically at the surface, without long-range, multi-body interaction.

2. The forces needed to maintain equilibrium on each side of the mathematical cut are equal and opposite.
3. The force acting through an internal surface depends only on the area enclosed and the surface orientation. Higher-order dependencies, curvatures, couple stresses, and chemical effects are neglected.

3.5 MANIPULATION OF THE STRESS TENSOR

Now that we have established that σ is a second-rank, symmetric tensor, we can use all the formalisms introduced in the second chapter. For example, the change of components upon a rotation of Cartesian coordinate system is

$$\sigma_{ij}' = R_{ik}R_{jl}\sigma_{kl} \leftrightarrow [\sigma]' = [R]^T[\sigma][R] \tag{3.16}$$

By the definition of σ, we understand that the force acting on any plane is simply the projection of σ onto d**a**, the vector representing the area in question:

$$\mathbf{df} = \boldsymbol{\sigma} \cdot \mathbf{da} \leftrightarrow \begin{bmatrix} df_1 \\ df_2 \\ df_3 \end{bmatrix} = \begin{bmatrix} \sigma_{11} & \sigma_{12} & \sigma_{13} \\ \sigma_{12} & \sigma_{22} & \sigma_{23} \\ \sigma_{13} & \sigma_{23} & \sigma_{33} \end{bmatrix} \begin{bmatrix} da_1 \\ da_2 \\ da_3 \end{bmatrix} \tag{3.17}$$

In a homogeneous stress field, there is no need to be concerned about the absolute size of **a**, and the differentials in Equation 3.17 can be dropped.

There is a particular orientation of coordinate axes for which the stress tensor becomes diagonal; that is, no shear stresses appear. The physical picture is that there are three orthogonal planes for which the stress vectors (or force vectors) are normal to these surfaces. In order to find these directions, we write the mathematical condition that the stress vector operating on an area d**a** (with normal $\hat{\mathbf{n}}$; i.e., $\hat{\mathbf{n}}$ is a unit vector parallel to d**a**) is parallel to $\hat{\mathbf{n}}$:

$$\mathbf{s} = \lambda\mathbf{n} \leftrightarrow \tag{3.18}$$

$$\sigma \cdot \mathbf{n} = \lambda\mathbf{n} \leftrightarrow \tag{3.19}$$

$$\begin{bmatrix} \sigma_{11} & \sigma_{12} & \sigma_{13} \\ \sigma_{12} & \sigma_{22} & \sigma_{23} \\ \sigma_{13} & \sigma_{23} & \sigma_{33} \end{bmatrix} \begin{bmatrix} n_1 \\ n_2 \\ n_3 \end{bmatrix} = \begin{bmatrix} \lambda n_1 \\ \lambda n_2 \\ \lambda n_3 \end{bmatrix} \tag{3.20}$$

$$\begin{bmatrix} \sigma_{11}-\lambda & \sigma_{12} & \sigma_{13} \\ \sigma_{12} & \sigma_{22}-\lambda & \sigma_{23} \\ \sigma_{13} & \sigma_{23} & \sigma_{33}-\lambda \end{bmatrix} \begin{bmatrix} n_1 \\ n_2 \\ n_3 \end{bmatrix} = 0 \tag{3.21}$$

As we showed in Chapter 2, Eqs. 3.20 and 3.21 are examples of standard eigenvalue and eigenvector problems. The problem has a nontrivial solution only when

$$\begin{vmatrix} \sigma_{11}-\lambda & \sigma_{12} & \sigma_{13} \\ \sigma_{12} & \sigma_{22}-\lambda & \sigma_{23} \\ \sigma_{13} & \sigma_{23} & \sigma_{33}-\lambda \end{vmatrix} = 0 \tag{3.22}$$

which yields a cubic equation in λ:

$$\lambda^3 - J_1\lambda^2 - J_2\lambda - J_3 = 0 \tag{3.23}$$

where

$$\begin{aligned} J_1 &= \text{First Stress Invariant (trace)} = \sigma_{11} + \sigma_{22} + \sigma_{33} \\ J_2 &= \text{Second Stress Invariant (quadratic invariant)} = \\ &\quad -(\sigma_{11}\sigma_{22} + \sigma_{22}\sigma_{33} + \sigma_{33}\sigma_{11}) + \sigma^2{}_{23} + \sigma^2{}_{31} + \sigma^2{}_{12} \\ J_3 &= \text{Third Stress Invariant} = \det \sigma = \\ &\quad \begin{vmatrix} \sigma_{11} & \sigma_{12} & \sigma_{13} \\ \sigma_{12} & \sigma_{22} & \sigma_{23} \\ \sigma_{13} & \sigma_{23} & \sigma_{33} \end{vmatrix} \end{aligned}$$

Equation 3.23 has real **roots** only when $[\sigma]$ is symmetric, as we have shown for the stress tensor. Therefore, Equation 3.23 will yield three values of λ, which, according to Equation 3.18, represent the magnitudes of the stress vectors operating on the three orthogonal planes where no shear components operate. That is, the three values of —say, λ_1, λ_2, λ_3—represent the three **principal stresses** (σ_1, σ_2, σ_3), without regard to the orientation of the special directions $\hat{\mathbf{n}}_1$, $\hat{\mathbf{n}}_2$, and $\hat{\mathbf{n}}_3$ along which they are oriented. Thus Equation 3.23 is independent of the orientation of the coordinate system, hence the name "invariants" for the coefficients J_1, J_2 and J_3. To put it another way, either set—(J_1, J_2, J_3) or (σ_1, σ_2, σ_3)—describes the state of stress, without reference to its orientation. These representations are convenient when dealing with isotropic materials, where the orientation does not influence the material response. The set σ_i is convenient for describing the state of stress in the principal coordinate system, while J_i ($i = 1, 2, 3$) is convenient for understanding which parts of the arbitrary stress tensor components, $[\sigma]$, don't change upon coordinate system rotation.

Once Equation 3.23 is solved and σ_1, σ_2, and σ_3 are obtained, it remains necessary only to find the **principal directions** $\hat{\mathbf{n}}_1$, $\hat{\mathbf{n}}_2$, and $\hat{\mathbf{n}}_3$ using the same methods introduced in Chapter 2 for eigenvalue/eigenvector problems. The principal directions are parallel to the three principal stress vectors, normal to the three principal planes that transmit the principal stresses.

As outlined in Chapter 2, it is frequently convenient to know the rotation matrix to change components from an arbitrary coordinate system to the **principal coordinate system**, or vice versa. If **n**, **m**, and **p** are the respective base vectors of the principal coordinate system, (i.e., $\hat{\mathbf{n}} = \hat{\mathbf{n}}_1$, $\hat{\mathbf{m}} = \hat{\mathbf{n}}_2$, $\hat{\mathbf{p}} = \hat{\mathbf{n}}_3$), then it is easily shown that the transformation must perform the following:

$$\begin{bmatrix} 1 \\ 0 \\ 0 \end{bmatrix} = \begin{bmatrix} R_{11} & R_{12} & R_{13} \\ R_{21} & R_{22} & R_{23} \\ R_{31} & R_{32} & R_{33} \end{bmatrix} \begin{bmatrix} n_1 \\ n_2 \\ n_3 \end{bmatrix} \tag{3.24a}$$

$$\begin{bmatrix} 0 \\ 1 \\ 0 \end{bmatrix} = \begin{bmatrix} R_{11} & R_{12} & R_{13} \\ R_{21} & R_{22} & R_{23} \\ R_{31} & R_{32} & R_{33} \end{bmatrix} \begin{bmatrix} m_1 \\ m_2 \\ m_3 \end{bmatrix} \tag{3.24b}$$

$$\begin{bmatrix} 0 \\ 0 \\ 1 \end{bmatrix} = \begin{bmatrix} R_{11} & R_{12} & R_{13} \\ R_{21} & R_{22} & R_{23} \\ R_{31} & R_{32} & R_{33} \end{bmatrix} \begin{bmatrix} p_1 \\ p_2 \\ p_3 \end{bmatrix} \tag{3.24c}$$

such that, by arranging the three equations as columns, we arrive at a matrix of principal unit vectors and the identity matrix:

$$\begin{bmatrix} 1 & 0 & 0 \\ 0 & 1 & 0 \\ 0 & 0 & 1 \end{bmatrix} = \begin{bmatrix} R_{11} & R_{12} & R_{13} \\ R_{21} & R_{22} & R_{23} \\ R_{31} & R_{32} & R_{33} \end{bmatrix} \begin{bmatrix} n_1 & m_1 & p_1 \\ n_2 & m_2 & p_2 \\ n_3 & m_3 & p_3 \end{bmatrix} \leftrightarrow [I]=[R][P] \tag{3.25}$$

Since we know that both coordinate systems are orthogonal (and right-handed), [R] must be an orthogonal matrix that is easily found from Equation 3.25, because $[R]^{-1} = [R]^T$:

$$[I][P]^T = [R][P][P]^T, \text{ so } [R]=[P]^T \tag{3.26}$$

$$\begin{bmatrix} R_{11} & R_{12} & R_{13} \\ R_{21} & R_{22} & R_{23} \\ R_{31} & R_{32} & R_{33} \end{bmatrix} = \begin{bmatrix} n_1 & n_2 & n_3 \\ m_1 & m_2 & m_3 \\ p_1 & p_2 & p_3 \end{bmatrix} \tag{3.27}$$

where n_i, m_i, and p_i are the components of the principal directions $\hat{\mathbf{n}}$, $\hat{\mathbf{m}}$, $\hat{\mathbf{p}}$ (or $\hat{\mathbf{n}}_1$, $\hat{\mathbf{n}}_2$, $\hat{\mathbf{n}}_3$) expressed in the same coordinate system as the stress tensor σ.

Exercise 3.3 **Find the stress invariants, the principal stresses, the principal directions, and the component rotation matrix to the principal coordinate system for the following stress tensor:**

$$\sigma = \begin{bmatrix} 1 & 2 & 3 \\ 2 & 2 & 4 \\ 3 & 4 & 3 \end{bmatrix}$$

In finding the component rotation matrix, choose $\hat{x}_1', \hat{x}_2', \hat{x}_3'$ along with principal axes having the maximum, medium, and minimum principal stresses, respectively.

With reference to Chapter 2 (eigenvalue and eigenvector problems), we write the characteristic equation $\left(|\sigma - I\lambda| = 0\right)$ in terms of the stress invariants:

$$\lambda^3 - J_1\lambda^2 - J_2\lambda - J_3 = 0 \tag{3.3-1}$$

where

$$J_1 = 6$$
$$J_2 = 18$$
$$J_3 = 8$$

To solve for the roots of Eq. 3.3-1, we resort to a Newton-Raphson procedure for the first root and then use the quadratic formula to find the other two. Following Appendix 4, we seek a numerical solution within an approximate accuracy of ±0.05 for the root, and we choose a starting, trial value for the root equal to $J_1 = (\sigma_{11} + \sigma_{22} + \sigma_{33})$:

Iteration number	*Trial* λ	$\phi(\lambda)$	$\phi'(\lambda)$	$\Delta\lambda$
0	6.0000	-116.0000	18.0000	
1	12.4400	765.0000	297.0000	-2.5800
2	9.8600	190.0000	155.0000	-1.2200
3	8.6400	34.0000	102.0000	-0.3300
4	8.3100	1.9000	89.4000	-0.0200
5	8.2880	-0.0200	88.60000	0.0002

After five iterations, we find that one root is approximately $\lambda = 8.288$, with an estimated uncertainty of 0.0002 ($\Delta\lambda$), and an estimated error of the characteristic equation $[\phi(\lambda)]$ of approximately 0.02.

We now need to find the other two roots of the characteristic equation. It would be possible to continue with the Newton-Raphson procedure, starting from different trial roots, until all three are found. This procedure is cumbersome (because the first root may be obtained repeatedly) and unnecessary (because the quadratic formula is available and convenient for quadratic equations). Therefore, we perform **synthetic long division** of the original equation by the known root to obtain the quadratic equation (and verify our Newton-Raphson solution).[10]

$$
\begin{array}{rrrrr}
 & & \lambda^2 & +2.288\lambda & +0.963 \\
\hline
(\lambda - 8.288)\,\big) & \lambda^3 & -6\lambda^2 & -18\lambda & -8.000 \\
 & \lambda^3 & -8.288\lambda^2 & & \\
 & & 2.288\lambda^2 & -18\lambda & \\
 & & 2.288\lambda^2 & -18.963\lambda & \\
 & & & 0.963\lambda & -8.00 \\
 & & & 0.963\lambda & -7.98 \\
 & & & & -0.02 \text{ (remainder / error)}
\end{array}
$$

[10] For programming purposes, the long division may be carried out symbolically, with the result that the residual quadratic equation $(a\lambda^2 + b\lambda + c = 0)$ has coefficients a, b, c as follows: $a = 1$, $b = \lambda_1 - J_1$, $c = b\lambda_1 - J_2 = \lambda_1^2 - J_1\lambda_1 - J_2$

Thus we have verified that l = 8.288 is a root within the functional error specified, and that the remaining quadratic equation is

$$\lambda^2 + 2.288\lambda + 0.963 = 0 \qquad (3.3\text{-}2a)$$

$$\left(a\lambda^2 + b\lambda + c = 0\right) \qquad (3.3\text{-}2b)$$

with solutions obtained by the quadratic formula

$$\lambda = \frac{-b \pm \sqrt{b^2 - 4ac}}{2a} = \frac{(-2.288) \pm \sqrt{(2.288)^2 - (4)(1)(0.963)}}{(2)(1)}$$

$$\lambda_2, \lambda_3 = -0.55, -1.73, \quad \text{so}$$

$$\sigma_1 = 8.29, \quad \sigma_2 = -0.55, \quad \sigma_3 = -1.73\,(\text{roots})$$

We find the principal stress directions (the eigenvectors) following the procedure in Exercise 2.8:

$$\sigma_1 = 8.288, \quad \begin{bmatrix} -7.29 & 2.00 & 3.00 \\ 2.00 & -6.29 & 4.00 \\ 3.00 & 4.00 & -5.29 \end{bmatrix} \begin{bmatrix} n_1 \\ n_2 \\ n_3 \end{bmatrix} = \begin{bmatrix} 0 \\ 0 \\ 0 \end{bmatrix} \rightarrow \begin{bmatrix} n_1 \\ n_2 \\ n_3 \end{bmatrix} = \begin{bmatrix} 0.441 \\ 0.577 \\ 0.687 \end{bmatrix} \qquad (3.3\text{-}3a)$$

$$\sigma_2 = -0.550, \quad \begin{bmatrix} 1.55 & 2.00 & 3.00 \\ 2.00 & 2.55 & 4.00 \\ 3.00 & 4.00 & 3.55 \end{bmatrix} \begin{bmatrix} m_1 \\ m_2 \\ m_3 \end{bmatrix} = \begin{bmatrix} 0 \\ 0 \\ 0 \end{bmatrix} \rightarrow \begin{bmatrix} m_1 \\ m_2 \\ m_3 \end{bmatrix} = \begin{bmatrix} 0.816 \\ -0.577 \\ -0.039 \end{bmatrix} \qquad (3.3\text{-}3b)$$

$$\sigma_3 = -1.730, \quad \begin{bmatrix} 2.73 & 2.00 & 3.00 \\ 2.00 & 3.73 & 4.00 \\ 3.00 & 4.00 & 4.73 \end{bmatrix} \begin{bmatrix} p_1 \\ p_2 \\ p_3 \end{bmatrix} = \begin{bmatrix} 0 \\ 0 \\ 0 \end{bmatrix} \rightarrow \begin{bmatrix} p_1 \\ p_2 \\ p_3 \end{bmatrix} = \begin{bmatrix} -0.374 \\ -0.578 \\ 0.726 \end{bmatrix} \qquad (3.3\text{-}3c)$$

where n_i, m_i, and p_i are the components of unit vectors $\hat{\mathbf{n}}$, $\hat{\mathbf{m}}$, and $\hat{\mathbf{p}}$lying along the principal directions, which have principal normal stresses $\hat{\sigma}_1$, $\hat{\sigma}_2$, and $\hat{\sigma}_3$, respectively.

Finally, we follow Exercise 2.5 in finding [R], the rotation matrix to transform components to a coordinate system based on the principal axes. We are generally free to choose the order of the axes, except that we require a right-handed system. In this case, the order has been specified in the problem, so we seek [R] such that the new coordinate axes lie parallel as follows:

$$\hat{x}_1' /\!/ \hat{n}$$
$$\hat{x}_2' /\!/ \hat{m}$$
$$\hat{x}_3' /\!/ \hat{p}$$

However, we must be certain that we choose $\hat{x}_1', \hat{x}_2', \hat{x}_3'$ in a right-handed sense. Therefore, we choose

$$\hat{x}_1 = \hat{n} \leftrightarrow (0.441,\ 0.577,\ 0.687) \qquad (3.3\text{-}4a)$$

$$\hat{x}_2 = \hat{m} \leftrightarrow (0.816,\ -0.577,\ -0.039) \qquad (3.3\text{-}4b)$$

$$\hat{x}_3 = \pm\hat{p} \leftrightarrow \pm(-0.374,\ -0.577,\ 0.726) \qquad (3.3\text{-}4c)$$
(depending on the sense needed)

We check $\hat{x}_1 \times \hat{x}_2$ and find that $\hat{x}_3$ = 0.374, 0.577, -0.726 for a right-handed orthogonal system. Therefore, we choose the minus sign in Eq. 3.3-4c. Then, as in Exercise 2.5, the required rotation matrix is

$$[R] = \begin{bmatrix} n_1\ n_2\ n_3 \\ m_1\ m_2\ m_3 \\ p_1\ p_2\ p_3 \end{bmatrix} = \begin{bmatrix} 0.441 & 0.577 & 0.687 \\ 0.816 & -0.577 & -0.039 \\ 0.374 & 0.577 & -0.726 \end{bmatrix} \qquad (3.3\text{-}5)$$

3.6 SPECIAL AND DEVIATORIC STRESS COMPONENTS

Some materials are virtually insensitive to pressure relative to their response to stresses that tend to promote shape distortion.[11] Examples include nearly all liquids and most solids undergoing plastic deformation. The distinction between the two kinds of stress is not of great interest for elastic solids, while gases are virtually insensitive (statically) to deviatoric stresses.

In cases where the hydrostatic pressure is not of great interest, we would like to define a different stress tensor that is unique for all states of stress that differ only by the applied pressure. To do this, we define a mean normal stress from the stress tensor:

---

[11] This qualitative statement applies strictly only to isotropic materials, because anisotropic materials may (or may not) undergo distortion as a response to applied pressure.

$$\sigma_p = -p = \frac{(\sigma_{11} + \sigma_{22} + \sigma_{33})}{3} = \frac{J_1}{3} \tag{3.28}$$

where p has the opposite sign from σ_{ij} because tensile stresses are conventionally positive. We write the stress tensor as the sum of two tensors, a **hydrostatic pressure** corresponding to p, and the remaining part corresponding to distortion (in the isotropic case):

$$\underset{\text{stress}}{\begin{bmatrix} \sigma_{11} & \sigma_{12} & \sigma_{13} \\ \sigma_{21} & \sigma_{22} & \sigma_{23} \\ \sigma_{31} & \sigma_{32} & \sigma_{33} \end{bmatrix}} = \underset{\substack{\text{hydrostatic stress}\\ \text{or}\\ \text{spherical stress}\\ \text{tensor}}}{\begin{bmatrix} \sigma_p & 0 & 0 \\ 0 & \sigma_p & 0 \\ 0 & 0 & \sigma_p \end{bmatrix}} + \underset{\substack{\text{deviatoric stress}\\ \text{tensor}}}{\begin{bmatrix} \sigma_{11} - \sigma_p & \sigma_{12} & \sigma_{13} \\ \sigma_{21} & \sigma_{22} - \sigma_p & \sigma_{13} \\ \sigma_{31} & \sigma_{12} & \sigma_{33} - \sigma_p \end{bmatrix}} \leftrightarrow \tag{3.29}$$

$$[\sigma] = \sigma_p[I] + [\sigma^d] = [\sigma^s] + [\sigma^d] \leftrightarrow \boldsymbol{\sigma}^s + \boldsymbol{\sigma}^d,$$

where $\sigma_i{}^d$ are the principal values of the **deviatoric stress tensor,** σ_d[12]. Since the hydrostatic stress tensor is the same after a rotation of coordinate axes,[13] and, therefore, can never contribute to a shear component, the existence of σ_p does not alter the principal directions. That is, the principal directions of σ are the same as σ^d. The principal values are easily found by

$$\sigma_i = \sigma_i{}^d + \sigma_p \tag{3.30}$$

In fact, it is often numerically easier to find the roots of the cubic characteristic equation for σ^d than for σ because the squared term disappears when the determinant $|\boldsymbol{\sigma}^d - \mathbf{I}\lambda|$ is expanded:

$$\lambda^3 - J^d{}_2\lambda - J^d{}_3 = 0 \tag{3.31}$$

---

[12]We will frequently denote deviatoric stress components by σ'_{ij}, when there is no confusion with the similar notation for transformed components.

[13]This can easily be seen in a variety of ways. Physically, a hydrostatic pressure has no identifiable directionality. Mathematically, a rotation of coordinate system shows that σ^s becomes $\sigma_{ii}' = (R^2{}_{i1} + R^2{}_{i2} + R^2{}_{i3})\sigma_p = \sigma_p$, and $\sigma_{ij}' = 0$ because [R] is orthogonal.

where, here, J^d_2 and J^d_3 are the invariants of σ^d. This equation can be solved[14] without numerical trial and error:

$$\lambda_i = 2\left(\frac{J^d{}_2}{3}\right)^{\frac{1}{2}} \cos \alpha_i \tag{3.32}$$

$$\text{where:} \quad \alpha_1 = \frac{1}{3}\cos^{-1}\left[\frac{J^d{}_3}{2}\left(\frac{3}{J_2^d}\right)^{\frac{3}{2}}\right], \quad 0 \le \alpha_1 < \tfrac{\pi}{3}$$

$$\alpha_2 = \alpha_1 + \frac{2\pi}{3} \tag{3.32a}$$

$$\alpha_3 = \alpha_1 - \frac{2\pi}{3}$$

Thus the principal stresses (i.e., the roots of the cubic equation) can be found by a simple evaluation of trigonometric functions, and the corresponding principal stresses of the stress tensor are given by Eq. 3.30.

For application of plasticity equations, it is often convenient to write the deviatoric invariants explicitly in terms of the principal total stress components:

$$J_1^d = \sigma_1^d + \sigma_2^d + \sigma_3^d \equiv 0 \tag{3.33}$$

$$\begin{aligned} J_2^d &= \left(\sigma_1^d\sigma_2^d + \sigma_1^d\sigma_3^d + \sigma_2^d\sigma_3^d\right) \\ &= \frac{1}{6}\left[(\sigma_1 - \sigma_2)^2 + (\sigma_1 - \sigma_3)^2 + (\sigma_2 - \sigma_3)^2\right] \\ &= \frac{1}{3}\left(\sigma_1^2 + \sigma_2^2 + \sigma_3^2\right)^2 - (\sigma_1\sigma_2 + \sigma_1\sigma_3 + \sigma_2\sigma_3) \end{aligned} \tag{3.34}$$

$$\begin{aligned} &J_3^d + \sigma_1^d\sigma_2^d\sigma_3^d + \frac{1}{27}(2\sigma_1 - \sigma_2 - \sigma_3)(2\sigma_2 - \sigma_1 - \sigma_3)(2\sigma_3 - \sigma_1 - \sigma_2) \\ &= \frac{2}{27}\left(\sigma_1^2 + \sigma_2^2 + \sigma_3^2\right) - \frac{1}{9}\left(\sigma_1\sigma_2^2 + \sigma_2\sigma_1^2 + \sigma_1\sigma_3^2 + \sigma_3\sigma_1^2 + \sigma_2\sigma_3^2 + \sigma_3\sigma_2^2\right) + \frac{4}{9}\sigma_1\sigma_2\sigma_3 \end{aligned} \tag{3.35}$$

---

[14] See, for example, L. E. Malvern, *Introduction to the Mechanics of a Continuous Medium* (New York Prentice-Hall, 1969), 91-93.

Note: We have used the superscript "d" to denote deviatoric representation, but will generally follow the more common notation of a prime for this purpose, whenever there is no confusion with alternate coordinate system notation.

Exercise 3.4 **Find the spherical and deviatoric parts of σ given below, and find the principal components of σ_d and σ.**

$$\boldsymbol{\sigma} \leftrightarrow \begin{bmatrix} 1 & 2 & 3 \\ 2 & 2 & 4 \\ 3 & 4 & 3 \end{bmatrix}$$

$$\sigma_p = \frac{1+2+3}{3} = 2 \qquad (3.41)$$

So:

$$\left[\sigma_p\right] = \begin{bmatrix} 2 & 0 & 0 \\ 0 & 2 & 0 \\ 0 & 0 & 2 \end{bmatrix} \quad \left[\sigma^d\right] = \begin{bmatrix} -1 & 2 & 3 \\ 2 & 0 & 4 \\ 3 & 4 & 1 \end{bmatrix} \qquad (3.4\text{-}2)$$

The principal stresses are found by substitution in Eq. 3.32:

$$J^d{}_2 = -(-1)(0) - (0)(1) - (-1)(1) + 2^2 + 3^2 + 4^2 = 30 \qquad (3.4\text{-}3)$$

$$J^d{}_3 = -1[(0)(1) - (4)(4)] - 2[(2)(1) - (3)(4)] + 3[(2)(4) - (3)(0)] = 60 \qquad (3.4\text{-}4)$$

$$\alpha_1 = \frac{1}{3}\cos^{-1}\left[\frac{60}{2}\left(\frac{3}{30}\right)^{\frac{3}{2}}\right] = 0.10725$$

$$\alpha_2 = 0.10725 + 2.0944 = 2.20165$$

$$\alpha_3 = 0.10725 - 2.0944 = -1.9872$$

$$\sigma_1{}^d = \lambda_1 = 2\left(\frac{30}{3}\right)^{\frac{1}{2}} \cos\alpha_1 = 6.29 \rightarrow \sigma_1 = 6.29 + 2 = 8.29 \qquad (3.4\text{-}5)$$

$$\sigma_2{}^d = \lambda_2 = 2\left(\frac{30}{3}\right)^{\frac{1}{2}} \cos\alpha_2 = -3.73 \rightarrow \sigma_2 = -3.73 + 2 = -1.73$$

$$\sigma_3{}^d = \lambda_3 = 2\left(\frac{30}{3}\right)^{\frac{1}{2}} \cos\alpha_3 = -2.56 \rightarrow \sigma_3 = -2.56 + 2 = -0.56 \tag{3.4-6}$$

Compare the ease with which the principal stresses were found in this case with the procedure followed in Exercise 3.3, for the same stress tensor.

For application of plasticity equations, it is often convenient to write the **deviatoric invariants** explicitly in terms of the principal total stress components:

$$J^d{}_1 = \sigma_1{}^d + \sigma_2{}^d + \sigma_3{}^d = 0$$

$$J^d{}_2 = -(\sigma_1{}^d\sigma_2{}^d + \sigma_2{}^d\sigma_3{}^d + \sigma_1{}^d\sigma_3{}^d) = \frac{1}{6}\left[(\sigma_1 - \sigma_2)^2 + (\sigma_1 - \sigma_3)^2 + (\sigma_2 - \sigma_3)^2\right]$$

$$\begin{aligned} J^d{}_3 &= \sigma_1{}^d\sigma_2{}^d\sigma_3{}^d \\ &= \frac{1}{27}(2\sigma_1 - \sigma_2 - \sigma_3)(2\sigma_2 - \sigma_1 - \sigma_3)(2\sigma_3 - \sigma_1 - \sigma_2) \\ &= \frac{2}{27}(\sigma_1{}^3 + \sigma_2{}^3 + \sigma_3{}^3) - \frac{1}{9}(\sigma_1\sigma_2{}^2 + \sigma_2\sigma_1{}^2 + \sigma_1\sigma_3{}^2 + \sigma_3\sigma_1{}^2 + \sigma_2\sigma_3{}^2 + \sigma_3\sigma_2{}^2) \\ &\quad + \frac{4}{9}\sigma_1\sigma_2\sigma_3 \end{aligned}$$

3.7 A FINAL NOTE

We have followed precisely the approach taken by Cauchy in defining the stress tensor and assigning a physical meaning to it. In fact, the stress tensor so defined is often called the **Cauchy stress tensor**. There are variations on the definition presented here, with slightly different and more abstract physical interpretations. For material considerations, the Cauchy stress tensor always has a better physical interpretation whereas other forms may provide more mathematical convenience in solving certain boundary-value problems. The precise alternate definitions do not make a great deal of sense until the concept of deformation is introduced. Therefore, we shall defer alternate stress-tensor definitions until the end of Chapter 4.

CHAPTER 3 - PROBLEMS

A. Proficiency Problems

1. Calculate the 3-D stress-tensor components for the rectangular material shown in the figure, first in the coordinate system $\hat{\mathbf{x}}_1$, $\hat{\mathbf{x}}_2$, and $\hat{\mathbf{x}}_3$, and then in the coordinate system $\hat{\mathbf{x}}_1'$, $\hat{\mathbf{x}}_2'$, and $\hat{\mathbf{x}}_3'$.

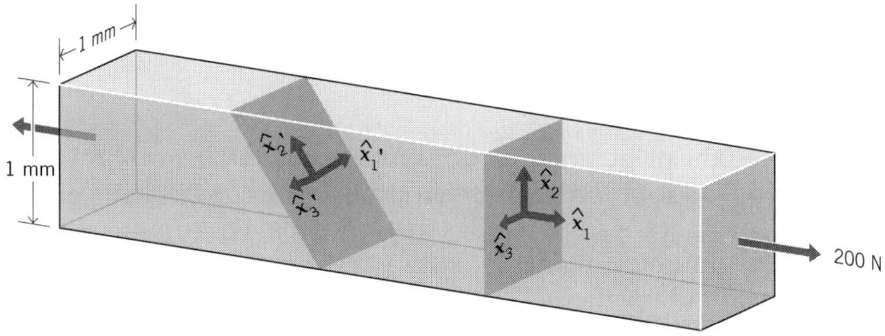

The angle between $\hat{\mathbf{x}}_1'$ and $\hat{\mathbf{x}}_1$ is 30°, and the $\hat{\mathbf{x}}_3'$ and $\hat{\mathbf{x}}_3$ axes are parallel.

2. Given the following stress tensor, find the stress vector acting on planes normal to the unit vectors $\hat{\mathbf{n}}$, $\hat{\mathbf{m}}$, and $\hat{\mathbf{p}}$, also given.

$$\boldsymbol{\sigma} \leftrightarrow \begin{bmatrix} 1 & 2 & 3 \\ 2 & 2 & 4 \\ 3 & 4 & 3 \end{bmatrix} \qquad \begin{aligned} \hat{\mathbf{n}} &\leftrightarrow \frac{1}{\sqrt{3}}(1,1,1) \\ \hat{\mathbf{m}} &\leftrightarrow \frac{1}{\sqrt{6}}(1,2,1) \\ \hat{\mathbf{p}} &\leftrightarrow \frac{1}{\sqrt{2}}(1,1,0) \end{aligned}$$

3. Find the principal stresses, the principal directions, and the rotation matrix for transforming coordinates to the principal coordinate system ($\hat{\mathbf{x}}_1'$ corresponds to σ_{max}, $\hat{\mathbf{x}}_3'$ corresponds to σ_{min}) for the stress tensors given.

$$\text{a.}\ \begin{bmatrix} 3 & -1 & 0 \\ -1 & 3 & 0 \\ 0 & 0 & 1 \end{bmatrix}, \quad \text{b.}\ \begin{bmatrix} 3 & 0 & 0 \\ 0 & 3 & -1 \\ 0 & -1 & 1 \end{bmatrix}, \quad \text{c.}\ \begin{bmatrix} 10 & -5 & 5 \\ -5 & 0 & 5 \\ 5 & 5 & 10 \end{bmatrix}$$

4. Find the invariants for the following stress tensors:

$$\text{a.}\begin{bmatrix} 1.44 & 0.22 & -0.76 \\ 0.22 & 2.25 & -0.38 \\ -0.76 & -0.38 & 2.31 \end{bmatrix}, \text{b.}\begin{bmatrix} 1.75 & 0.35 & -0.75 \\ 0.35 & 2.50 & -0.35 \\ -0.75 & -0.35 & 1.75 \end{bmatrix}, \text{c.}\begin{bmatrix} 1.94 & 0.38 & -0.54 \\ 0.38 & 2.75 & 0.22 \\ -0.54 & -0.22 & 1.31 \end{bmatrix}$$

5. Find and solve the characteristic equations for the stress tensors shown in Problem 4. Use the method followed in Exercise 3.3. (A numerical procedure is required.)

6. Find the principal directions for the stress tensors shown in Problem 4, and find the rotation matrix that transforms components given in the original coordinate system to ones in a principal coordinate system. (Assume that the minimum principal stress acts on a plane with normal $\hat{x}_1'$, and the maximum principal stress acts on a plane with normal $\hat{x}_3'$.)

7. Find the spherical and deviatoric components of the stress tensors given in Problem 4. Find the principal stresses and directions of the deviatoric tensors following the method outlined in Section 3.6. How do these compare with the values for the stress tensor obtained in Problems 5 and 6?

8. Find the spherical, deviatoric, and principal deviatoric components, and principal directions of stress for the following cases:

   a. Uniaxial tension: $\sigma_{11} = \sigma$, other $\sigma_{ij} = 0$

   b. Simple shear: $\sigma_{21} = \sigma_{12} = \sigma$, other $\sigma_{ij} = 0$

   c. Balanced biaxial tension: $\sigma_{11} = \sigma_{22} = \sigma$, other $\sigma_{ij} = 0$

   d. Biaxial shear: $\sigma_{13} = \sigma_{31} = \sigma_A$, $\sigma_{21} = \sigma_{12} = \sigma_B$, other $\sigma_{ij} = 0$

   e. Tension and shear: $\sigma_{11} = \sigma_t$, $\sigma_{13} = \sigma_{31} = \sigma_s$, other $\sigma_{ij} = 0$

B. Depth Problems

9. The reciprocal theorem of Cauchy states that the stress vectors acting on two intersecting planes have the following property:

$$\mathbf{s}_1 \cdot \hat{\mathbf{n}}_2 = \mathbf{s}_2 \cdot \hat{\mathbf{n}}_1$$

where $\mathbf{s}_i$ is the stress vector acting on a plane with normal $\hat{\mathbf{n}}_i$. Show that this principle follows from the symmetry of the stress tensor, or from the equilibrium condition directly.

10. Octahedral planes are ones that have normals forming equal angles with the three principal axes. Show that the normal stress vector is minimized on these planes. (Hint: Consider only the deviatoric stress tensor, **s**, such that $s_{11}+ s_{22}+ s_{33} = 0$. Then show that the normal component of the deviatoric vector disappears on these planes.)

11. Show that the tangential component of the stress vector on the octahedral planes (i.e., the shear component) is equal to

$$\left(\frac{2}{3}J_2'\right)^{1/2}$$

where J_2' is the second invariant of the deviatoric stress tensor.

12. Problems involving cylindrical symmetry often use cylindrical coordinates r, θ, z where

$$r = \left(x_1^2 + x_2^2\right)^{\frac{1}{2}} \quad \text{and} \quad \theta = \tan^{-1}\frac{x}{y}.$$

(Conversely, $x_1 = r\cos\theta$, $x_2 = r\sin\theta$, $x_3 = z$.) For example, consider pure torsion of an elastic, long bar with axis parallel to $\hat{z}$, where $\sigma_{\theta z}$ is a constant on the outside of the bar and other $\sigma_{ij}= 0$. Find the stress tensor in two alternate Cartesian coordinate systems:

a. One that has $\hat{\mathbf{x}}_1$ normal to the cylinder axis and tangent to the cylinder surface, $\hat{\mathbf{x}}_2$ normal to the cylinder surface, and $\hat{\mathbf{x}}_3$ the cylinder axis, such that $\hat{\mathbf{x}}$, is parallel to $\hat{\mathbf{r}}$.

b. One that is fixed in space (i.e., independent of θ), with $\hat{\mathbf{x}}_3 \,//\, \hat{\mathbf{z}}$; $\hat{\mathbf{x}}_1$ lying in the $\theta = 0, z = 0$ direction; and $\hat{\mathbf{x}}_2$ lying in the $\theta = \pi/2,\ z = 0$ direction.

13. Show that:

a. If two roots of the characteristic equation are identical (i.e., degenerate), then any direction normal to the other principal direction (i.e., the one corresponding to the unique root) is a principal direction.

b. If all three roots are identical, all directions are principal.

14. It is often convenient to replace one set of forces with another, statically equivalent set. For example, consider a triangular element of material (assume unit depth normal to the triangle) which is assumed to be a small enough piece of a body to feel only a homogeneous stress, σ_{ij} $(i,j=1,2$, assuming that $\sigma_{i3}=0$, where $\hat{x}_3$ is normal to the triangle). Use a simple, physically motivated procedure to replace σ_{ij} by three forces, $\mathbf{f}_1$, $\mathbf{f}_2$, $\mathbf{f}_3$, acting at the three corners of the triangle. Consider the force transmitted by each face.

15. Physically, why can the entire material loading at a point be reduced to three orthogonal force intensities passing through the point? Why do the shear components disappear along these directions?

16. The following two sets of components correspond to the identical stress tensor, as measured in two coordinate systems, $\hat{x}_1$, $\hat{x}_2$, $\hat{x}_3$, and $\hat{x}_1'$, $\hat{x}_2'$, $\hat{x}_3'$. Find the rotation matrix to transform components from the $\hat{x}_i$ system to the $\hat{x}_i'$ system, and vice versa. (Hint: First find the rotations to the common, principal coordinate systems.)

$$[\sigma]=\begin{bmatrix}1.000 & 1.730 & 1.000\\ 1.730 & 0.750 & 0.433\\ 1.000 & 0.433 & 0.250\end{bmatrix} \quad [\sigma']=\begin{bmatrix}0.500 & 1.414 & 0.500\\ 1.414 & 1.000 & 1.414\\ 0.500 & 1.414 & 0.500\end{bmatrix}$$

17. a. What would happen if we defined a *stress matrix* [15] as a linear operator relating *components* of the area vector defined in one coordinate system (say, $\hat{x}_i$) to the force passing through this area in another coordinate system (say, $\hat{x}_i'$)? (Assume that the two coordinate systems are linked by another linear operator.) How would the components of the stress matrix transform under these conditions?

[15] We will defer until the end of the next chapter whether such a definition is a tensor. However, note that the distinction is very similar to the argument presented in connection with Exercise 2.9.

b. As an alternative definition to part a, define a new stress measure as that relating the force vector to a different area vector: for example, one which is a rotation of the real area vector by a fixed amount.

C. Numerical Problems

18. Write a program to find the roots of a cubic equation based on the procedure presented in Exercise 3.3. Note that the general result from the long division gives the quadratic coefficients in terms of the first root found numerically ($a = 1$, $b = \lambda_1 - J_1$, $c = \lambda_1^2 - J_1\lambda_1 - J_2$) Make sure your program identifies degenerate roots.

19. Write a program to find the principal directions and the rotation matrix to change components to a standard principal coordinate system. (For example, choose $\hat{x}_1{}' =$ the maximum principal stress direction, and $\hat{x}_3{}' =$ the minimum principal stress direction. Make sure the degenerate root case is handled consistently, without stoppage. (Follows Problem 18, Chapter 2.)

20. Write a program to find the components of tensors with ranks 1 to 4 in the principal coordinate system using the rotation matrix found in Problem 2.

CHAPTER 4

Strain

When a real material is loaded, it changes shape, either permanently or temporarily (while the load is present). In a one-dimensional loading, like that pictured in Figure 4.1, the material is elongated along the load axis, and the other two orthogonal directions generally undergo a contraction. In order to understand and model material behavior of this kind, it is necessary to quantify and represent precisely the shape change, or **deformation**, of the body as a response to a load. Then one can proceed to perform experiments and represent the mathematical relationships between material loading (i.e., *stress*) and its shape change. These **constitutive equations** embody the mathematical form of the material response needed to solve mechanical-deformation problems.

In this chapter, we will proceed very much as in the previous one. First, we shall try to perform an intuitive analysis of deformation in one or two dimensions, and then will proceed to define terms more generally and precisely.

Perhaps one clarification should be made. Except for the necessary idea of a continuum (such that geometrical qualities approach definite limits at a point), **kinematics**, the study of material motion, does not rely on material properties. Instead, the subject is closely related to mathematics and geometry. Therefore, although we shall use material ideas to introduce deformation, they are formally unnecessary. When material behavior and other material laws are involved, the more general field is known as **mechanics** or **continuum mechanics** (for a continuum).

4.1 PHYSICAL IDEAS OF DEFORMATION

Let's again consider a long, slender rod, such as a wire or tensile specimen, that is loaded axially (Figure 4.1). This is a convenient geometry because we can safely assume that for a homogeneous material, the loading and deformation throughout the body is uniform and easily calculated. Furthermore, it is easy to identify the material directions that extend without rotation relative to one another. The rod's response to the applied load will generally be to elongate axially and contract in cross section. If the material is isotropic (direction-independent), we expect the contractions to be the same in all directions perpendicular to the tensile axis. Thus the deformation pattern is three-dimensional even though the loading is one-dimensional.

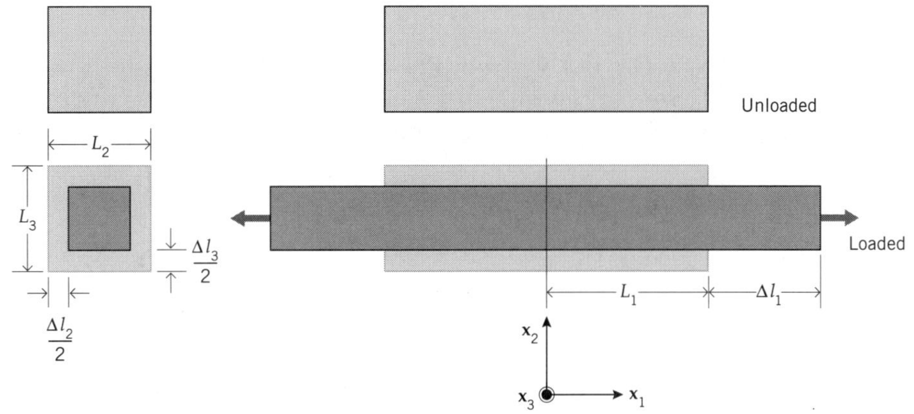

Figure 4.1 Elongation of a long rod under tensile loading.

Specifically, the rod elongates by an amount Δl, as shown in Figure 4.1. Since a bar of twice the length would elongate twice as much (2Δl), we normalize the elongation by the original length in order to obtain the **engineering strain** in the axial direction:

$$e_1 = \frac{\Delta l_1}{L_1} \quad \text{or} \quad \Delta l_1 = L_1 e_1 \tag{4.1a}$$

and similarly in the x_2 and x_3 direction:

$$e_2 = \frac{\Delta l_2}{L_2} \tag{4.1b}$$

and

$$e_3 = \frac{\Delta l_3}{L_3} \tag{4.1c}$$

where, for an isotropic material, $e_2 = e_3$. The symbol e_i is usually reserved for the so-called "engineering strains" in a tensile test as defined by Eqs. 4.1. We can see immediately that the idea of *strain relates the change of length of a* ***material line***[1] *to its original direction and length.*

In order to generalize our idea of strain, we consider an arbitrary material line segment (i.e., one not specifically oriented parallel to $\hat{x}_1$, $\hat{x}_2$, or $\hat{x}_3$),or,

[1] The idea of a material line is conceptually simple. Imagine embedding a straight-line segment in a material and then watching it as the material deforms.

equivalently, a material vector with arbitrary components:[2] $\Delta\mathbf{X} = \Delta X_1\,\hat{x}_1 + \Delta X_2\,\hat{x}_2 + \Delta X_3\,\hat{x}_3$. Since we must find the length of initial and final vectors, the change of length is most simply investigated in squared form:

$$l^2 - L^2 = (\Delta x_1^2 + \Delta x_2^2 + \Delta x_3^2) - (\Delta X_1^2 + \Delta X_2^2 + \Delta X_3^2) \tag{4.2}$$

Because $Dx_i = DX_i\,(1 + e_i)$ from Eq. 4.1, ΔX_i corresponds to L_i, and Δx_i corresponds to l_i, we can rewrite Eq. 4.2 in matrix form in terms of original components and the three values e_1, e_2, and e_3:

$$l^2 - L^2 = [\Delta X_1\ \Delta X_2\ \Delta X_3]\begin{bmatrix}(1+e_1)^2 & 0 & 0\\ 0 & (1+e_2)^2 & 0\\ 0 & 0 & (1+e_3)^2\end{bmatrix}\begin{bmatrix}\Delta X_1\\ \Delta X_2\\ \Delta X_3\end{bmatrix} - [\Delta X_1\ \Delta X_2\ \Delta X_3]\begin{bmatrix}\Delta X_1\\ \Delta X_2\\ \Delta X_3\end{bmatrix} \tag{4.3a}$$

where we have written the matrix form so as to keep the vector components separate. Eq. 4.3a can be written in simpler form by noting that second term on the right side may be considered the product of the identity matrix, which is then combined with the [e] matrix:

$$l^2 - L^2 = [\Delta X_1\ \Delta X_2\ \Delta X_3]\begin{bmatrix}(1+e_1)^2 & 0 & 0\\ 0 & (1+e_2)^2 & 0\\ 0 & 0 & (1+e_3)^2\end{bmatrix}\begin{bmatrix}\Delta X_1\\ \Delta X_2\\ \Delta X_3\end{bmatrix}$$

$$- [\Delta X_1\ \Delta X_2\ \Delta X_3]\begin{bmatrix}1 & 0 & 0\\ 0 & 1 & 0\\ 0 & 0 & 1\end{bmatrix}\begin{bmatrix}\Delta X_1\\ \Delta X_2\\ \Delta X_3\end{bmatrix} \tag{4.3b}$$

$$l^2 - L^2 = [\Delta X_1\ \Delta X_2\ \Delta X_3]\begin{bmatrix}2e_1 + e_1^2 & 0 & 0\\ 0 & 2e_2 + e_2^2 & 0\\ 0 & 0 & 2e_3 + e_3^2\end{bmatrix}\begin{bmatrix}\Delta X_1\\ \Delta X_2\\ \Delta X_3\end{bmatrix} \tag{4.4a}$$

[2] We have broken our rule of using only lowercase letters for vectors in order to distinguish between a material vector before deformation ($\Delta\mathbf{X}$) and the same material vector after deformation ($\Delta\mathbf{x}$). That is, $\Delta\mathbf{X}$ and $\Delta\mathbf{x}$ are spatial vectors, with components ΔX_i and Δx_i expressed in regular Cartesian coordinates, for the same line segment of material before and after deformation. (This segment remains linear because of our imposed homogeneous deformation and by the geometry of the long, slender bar and the homogeneous material.)

$$l^2 - L^2 \approx [\Delta X_1\ \Delta X_2\ \Delta X_3]\begin{bmatrix} 2e_1 & 0 & 0 \\ 0 & 2e_2 & 0 \\ 0 & 0 & 2e_3 \end{bmatrix}\begin{bmatrix} \Delta X_1 \\ \Delta X_2 \\ \Delta X_3 \end{bmatrix} \tag{4.4b}$$

Equation 4.4a relates the squared-length changes of a material vector with original spatial coordinates DX_1, DX_2, DX_3, when the coordinates are given in a Cartesian coordinate system with axes in the directions of the engineering strains e_1, e_2, and e_3. Equation 4.4b is an approximate version, accurate when $e_i << 1$.

In order to make the expression general for any Cartesian coordinate system, we consider that the arbitrary material vector $\mathbf{\Delta X}$ is expressed in any nonspecial coordinate system $\Delta X_i'$ (i.e., $\mathbf{\Delta X} = \Delta X_1'\hat{x}_1' + \Delta X_2'\hat{x}_2' + \Delta X_3'\hat{x}_3'$) where the components of $\mathbf{\Delta X}$ in the new coordinate system ($\Delta X_i'$) are related in the usual way by a standard rotation operation to the components in the special coordinate system (see Section 2.5):

$$\begin{bmatrix} \Delta X_1' \\ \Delta X_2' \\ \Delta X_3' \end{bmatrix} = \begin{bmatrix} R_{11} & R_{12} & R_{13} \\ R_{21} & R_{22} & R_{23} \\ R_{31} & R_{32} & R_{33} \end{bmatrix}\begin{bmatrix} \Delta X_1 \\ \Delta X_2 \\ \Delta X_3 \end{bmatrix} \text{ and}$$

$$[\Delta X_1'\ \Delta X_2'\ \Delta X_3'] = [\Delta X_1\ \Delta X_2\ \Delta X_3]\begin{bmatrix} R_{11} & R_{12} & R_{13} \\ R_{21} & R_{22} & R_{23} \\ R_{31} & R_{32} & R_{33} \end{bmatrix}^T \tag{4.5}$$

Now, Eq. 4.4 can be written in terms of the nonspecial coordinate system by substituting for the special vector components:

$$l^2 - L^2 = \underbrace{[\Delta X_1'\ \Delta X_2'\ \Delta X_3']\begin{bmatrix} R_{11} & R_{12} & R_{13} \\ R_{21} & R_{22} & R_{23} \\ R_{31} & R_{32} & R_{33} \end{bmatrix}}_{[\Delta X_1\ \Delta X_2\ \Delta X_3]}\begin{bmatrix} 2e_1+e_1^2 & 0 & 0 \\ 0 & 2e_2+e_2^2 & 0 \\ 0 & 0 & 2e_3+e_3^2 \end{bmatrix}\underbrace{\begin{bmatrix} R_{11} & R_{12} & R_{13} \\ R_{21} & R_{22} & R_{23} \\ R_{31} & R_{32} & R_{33} \end{bmatrix}\begin{bmatrix} \Delta X_1' \\ \Delta X_2' \\ \Delta X_3' \end{bmatrix}}_{\begin{bmatrix} \Delta X_1 \\ \Delta X_2 \\ \Delta X_3 \end{bmatrix}} \tag{4.6}$$

By grouping the three 3×3 matrices, we arrive at the general Cartesian expression that is the squared-length change that we sought:

$$l^2 - L^2 = [\Delta X_1'\ \Delta X_2'\ \Delta X_3']\begin{bmatrix} \theta_{11} & \theta_{12} & \theta_{13} \\ \theta_{21} & \theta_{22} & \theta_{23} \\ \theta_{31} & \theta_{32} & \theta_{33} \end{bmatrix}\begin{bmatrix} \Delta X_1' \\ \Delta X_2' \\ \Delta X_3' \end{bmatrix} \tag{4.7}$$

$$\text{where:}\quad [\theta] = [R]^T\left[2e_i + e_i^2\right][R],\ \text{or}$$

$$\theta_{ij} = R_{ik}\,R_{jl}\left\{2e_{kr} + (e_{kr})^2\right\}\delta_{rl}$$

$$\left[2e_i + e_i^2\right] = \text{the matrix shown in Eqs. 4.4 and4.6}$$

Thus θ_{ij} are the components of a symmetric tensor, θ, since θ_{ij} clearly transform according to the tensor transformation rule that serves to define tensors (see Section 2.8). Furthermore, $[\theta]$ and $[e^2-1]$ are shown to be alternate components of the same tensor, with the latter taking the simpler form because the components are expressed in a special coordinate system where the axes lie along directions that extend without rotation relative to each other. In the subsequent sections we shall see that θ is, in fact, just two times the conventional definition for **the large strain tensor, E.**

There is one last generalization that must be made in our physical picture of strain. So far, we have focused on a finite material line (which we represented by a vector). While it is always possible to define such a line in the undeformed state, for example, the line will in general be curved after deformation except for the case of homogeneous deformation already considered. In such cases, $(\Delta x_1^2 + \Delta x_2^2 + \Delta x_3^2)^{1/2}$ does not represent the real length of the deformed line. To handle this difficulty, we resort once again to our idea of a continuum and presume the existence of continuous line segments in the deformed and undeformed states, such that $\Delta\mathbf{X} \to d\mathbf{X}$, $\Delta\mathbf{x} \to d\mathbf{x}$ in the infinitesimal limit. With this continuum picture, which we shall mention again in the next section, the strain is defined at a point, exactly as the stress was defined at a point. Thus Eq. 4.7 becomes, in the non-homogeneous (i.e., general) case,

$$(dl)^2 - (dL)^2 = \left[dX_1'\ dX_2'\ dX_3'\right]\begin{bmatrix}\theta_{11} & \theta_{12} & \theta_{13}\\ \theta_{21} & \theta_{22} & \theta_{23}\\ \theta_{31} & \theta_{32} & \theta_{33}\end{bmatrix}\begin{bmatrix}dX_1'\\ dX_2'\\ dX_3'\end{bmatrix} \qquad (4.8)$$

where dX_i are understood to be the components of a vector of infinitesimal length dL before deformation. The corresponding length in the final configuration is dl.

4.2 GENERAL INTRODUCTION TO KINEMATICS

Evolving from the physical picture presented in the previous section, there are many alternate formulations for the geometry of deformation. In this section, we shall attempt to introduce the basic ideas without amplification. In subsequent sections, we shall restrict our attention to the most important formulations and quantities used to represent metal deformation. Since metals usually undergo only nearly infinitesimal elastic deformation (strains of the order of 1/1000), and since the material properties of an element depend on its deformation history, certain forms of kinematic variables are preferred. Quite different choices are made for representing fluids and large-strain elastic materials. Also, metals are usually subject to strain hardening and, to a lesser extent (at low temperatures), strain-rate sensitivity. Therefore, we will concentrate on kinematic variables that focus on these aspects.

The two basic approaches usually followed in solid mechanics are called the **material description** (or **Lagrangian description**) and **spatial description** (or **Eulerian description)**. For nuances of the differences in these approaches and the names associated with them, see standard continuum mechanics texts on the subject.[3]

In the material description, one supposes that the current position of a *material point* is related to the original (or some other reference) position of the *same material point* through smooth, differentiable functions of the form

$$x = \chi\,(\mathbf{X}, t). \tag{4.9}$$

The original position (or reference position of the particle) is given in **material coordinates,** while the final position is given in the particle's **spatial coordinates**. The material coordinates may be considered simply as a mechanism for labeling the same material point for all time.

In fact, in addition to differentiability, our picture of a continuum also requires that Eq. 4.9 is invertible; that is, that there is a precise one-to-one correspondence between positions of each material element at any two times[4] (one can be called the reference time and one the current time without loss of generality). Then Eq. 4.9 implies the existence of the inverse function:

$$\mathbf{X} = \mathbf{X}(x, t) \tag{4.10}$$

where the original position of the material point is a function of its current position.

In fact, Eq. 4.10 can be considered the basis of the spatial or Eulerian description, where one considers the current time and spatial positions as independent

[3] For example: C. Truesdell, *Elements of Continuum Mechanics*, (New York: Springer-Verlag, 1965) or L. E. Malvern, *Introduction to the Mechanics of a Continuous Medium*, (New York: Prentice-Hall, 1969).

[4] If, for example, χ in Eq. 4.9 is a function but $\boldsymbol{X}$ in Eq. 4.10 is not, then one value of **x** could map to multiple values of **X**. That is, two material points labeled originally $\mathbf{X}_1$ and $\mathbf{X}_2$ could occupy one spatial point (**x**) at some later time *t*. Conversely, if $\boldsymbol{X}$ in Eq. 4.10 is a function but χ in Eq. 4.9 is not, then one material point labeled **X** can occupy two distinct spatial positions ($\mathbf{x}_1$, $\mathbf{x}_2$) at some later time *t*. In the first case, material overlaps into itself and interpenetrates; in the second case, a crack is allowed to open in the material. Both possibilities are outlawed by our picture of a continuum. This discussion applies to interior points of a continuum. It may be permissible to have two surface points join (overlap) or separate (forming new surface) in general metal forming and machining problems.

variables. Such a picture is useful in fluid mechanics, where the material properties do not depend on the material history, and in steady-state problems, where each spatial point is associated with a constant material velocity.

Although a function such as **X** (Eq. 4.10) exists in principle, and defines the smooth functions that represent the continuum motion, actually less information is usually available. Consider the flow of a fluid (air, water, etc.) in a vessel, around an airplane, or near other moving boundaries.[5] We can imagine that it might be possible to measure the fluid velocity (i.e., the rate of material motion at a spatial point) at each point in space:

$$v = \upsilon\,(x, t)$$

$$\text{where:} \quad \mathbf{v} = \left(\frac{\partial \mathbf{x}}{\partial t}\right)_{\mathbf{X}} \quad \text{or} \quad \boldsymbol{\upsilon}\,(\mathbf{x}, t) = \frac{\partial}{\partial t}\left[\boldsymbol{\chi}(\mathbf{X}, t)\right]\left(\frac{\partial \mathbf{x}}{\partial t}\right)_{\mathbf{X}} \tag{4.11}$$

The constraint **X** indicates that we must follow a single material particle's motion in order to define the **material velocity**. For example, one can see the difference in the two kinds of the derivatives available from the inverse functions in Eqs. 4.9 and 4.10: $(\partial \mathbf{x}/\partial t)_{\mathbf{X}}$ ignifies the *change of position per unit time of a material particle (labeled* **X***) that happens to appear currently at spatial point* **x**, while $(\partial \mathbf{X}/\partial t)_{\mathbf{x}}$ signifies the rate of change of the material label corresponding to current spatial position **x**. (This latter concept has little or no application in considering material deformation.) Similarly,

$$\left(\frac{\partial v}{\partial t}\right)_{X} = \textbf{material acceleration} \text{ of } X \tag{4.12}$$

$$\left(\frac{\partial \mathbf{v}}{\partial t}\right)_{\mathbf{x}} = \text{local rate of change of } \mathbf{v} \text{ at the spatial point } \mathbf{x} \tag{4.13}$$

where **v** is the material velocity defined in Eq. 4.11.

In summary, these are two basic approaches that we will employ in metal deformation. In the Lagrangian approach, we concentrate on functions relating the spatial positions of a material point labeled by its position at some standard time, perhaps t = 0. In the Eulerian approach, we will assume the existence of exactly the same functions, but we will rely on a current knowledge of material velocities throughout a spatial region at the current time. Both approaches have

[5] In these examples, it should be obvious that we do not usually need to know where each particle of air or water came from in the problem. On the other hand, we need to know that, in principle, one *could* know this, and that one particle of fluid can not split into two, nor can two particles occupy the same spatial point at some other time. Therefore, the existence of differentiable functions such as χ or X can be important while the exact form is not.

advantages, depending on the application.

There are many variations on these two approaches depending on the choice of coordinate system(s) and reference states. All of these are for mathematical convenience and add no new fundamental understanding of the problem. For example, it is possible to hold the relative coordinates of each pair of neighboring material points constant by allowing the local coordinate system to deform along with the material. Then, the material deformation is explored by considering the changing coordinate system for a particle having constant coordinates at each time. Since we wish to avoid the use of curvilinear coordinates as much as possible, we shall defer discussion of these mathematical variations.

4.3 DEFORMATION: FORMAL DESCRIPTION

We shall now proceed to introduce, in concise fashion, the standard measures of material deformation, without regard to the size of the deformation or the time scale. Therefore, we will not make a distinction about the reference and current configurations, how close they are together, which occurred first in time, and so forth. Under these conditions, the quantities derived are applicable to either Lagrangian or Eulerian descriptions of deformation with only a simple change of reference state.[6] In subsequent sections, we will consider the additional topics of infinitesimal strain specializations, rate of material deformation measures, and compatibility requirements for the continuum approach.

Consider an infinitesimal line[7] or vector embedded in a material element undergoing deformation between two states, A and B. Figure 4.2 shows the graphical situation. As the material deforms, the infinitesimal element of material containing our tagged line is carried from a spatial point in space **X** to a spatial point **x**, by a displacement of **u**. The continuum equations of motion for the particle at **X** are

$$\mathbf{x} = \boldsymbol{\chi}(\mathbf{X}, t) \quad \text{or} \quad \mathbf{u} = \boldsymbol{\upsilon}(\mathbf{X}) = \mathbf{x}(\mathbf{X}) - \mathbf{X} \tag{4.14}$$

The infinitesimal vector is transformed by defining the deformation gradient, **F**

$$d\mathbf{x} = \mathbf{F}\, d\mathbf{X} \leftrightarrow \begin{bmatrix} dX_1 \\ dX_2 \\ dX_3 \end{bmatrix} = \begin{bmatrix} F_{11} & F_{12} & F_{13} \\ F_{21} & F_{22} & F_{23} \\ F_{31} & F_{32} & F_{33} \end{bmatrix} \begin{bmatrix} dX_1 \\ dX_2 \\ dX_3 \end{bmatrix} \leftrightarrow dx_i = \left(\frac{\partial x_i}{\partial X_j} \right) dX_j \leftrightarrow d\mathbf{x} = \left(\frac{\partial \mathbf{x}}{\partial \mathbf{X}} \right) d\mathbf{X} \tag{4.15}$$

while a similar transformation is made for the **relative displacements** of the

[6] The important distinctions between kinds of kinematic description become clear only upon introducing constitutive equations and energy principles, which must be complementary.

[7] Conversely, one could consider a finite line in a region of homogeneous deformation with the same result.

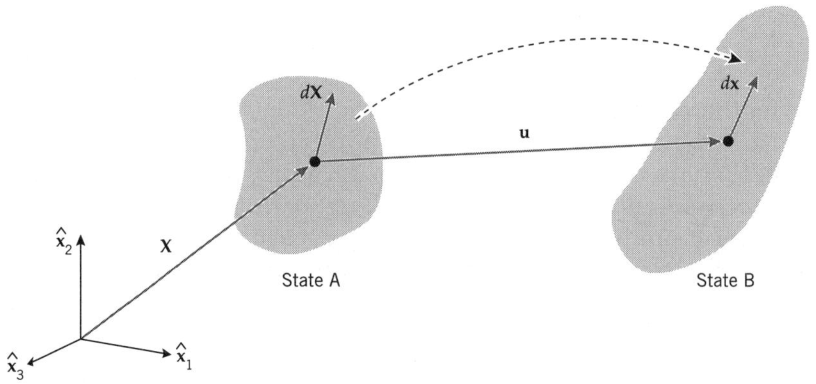

Figure 4.2 Motion of a material vector during deformation from state A to state B.

head and tail of material vector d**X**:

$$\mathbf{du} = \mathbf{J}\mathbf{dX} \leftrightarrow \begin{bmatrix} du_1 \\ du_2 \\ du_3 \end{bmatrix} = \begin{bmatrix} J_{11} & J_{12} & J_{13} \\ J_{21} & J_{22} & J_{23} \\ J_{31} & J_{32} & J_{33} \end{bmatrix} \begin{bmatrix} dX_1 \\ dX_2 \\ dX_3 \end{bmatrix} \leftrightarrow du_i = \left(\frac{\partial u_i}{\partial X_j} \right) dX_j \tag{4.16}$$

where d**u** = d**x** – d**X**. Therefore **J**, the **Jacobian matrix** or **displacement gradient tensor**, can be defined in terms of **F**, or vice versa:

$$\begin{gathered} \mathbf{du} = \mathbf{dx} - \mathbf{dX} = \mathbf{J}\ \mathbf{dX} \\ \mathbf{dx} = \mathbf{J}\mathbf{dX} + \mathbf{dX} = \mathbf{J}\ \mathbf{dX} + \mathbf{I}\ \mathbf{dX} \\ \mathbf{dx} = (\mathbf{J} + \mathbf{I})\ \mathbf{dX} \\ \text{Thus: } \mathbf{F} = \mathbf{J} + \mathbf{I},\ \ \mathbf{J} = \mathbf{F} - \mathbf{I} \end{gathered} \tag{4.17}$$

If **F** or **J** is known throughout a body for all times, a complete picture of the local deformation is available for each material point. Excluded from this description are translations of points and relationships of nonadjacent material points, which we assume do not affect the material response.[8] On the other hand, **F** and **J** do include rigid-body rotations that do not contribute to shape change and internal stress. Therefore, for formulating material constitutive equations, it is better to define a measure of deformation that ignores such non-shape-changing rotations.

In order to ignore rotations, we focus on the stretch of a material element[9] and in the process define deformation and strain tensors. Starting from the definition of **F**, we find the required representation:

$$ds^2 = dx^T\,dx = d\mathbf{X}^T\,\mathbf{F}^T\,\mathbf{F}\,d\mathbf{X} \leftrightarrow \left[dX_1\;dX_2\;dX_3\right]\begin{bmatrix} F_{11} & F_{21} & F_{31} \\ F_{12} & F_{22} & F_{32} \\ F_{13} & F_{23} & F_{33}\end{bmatrix}\begin{bmatrix} F_{11} & F_{12} & F_{13} \\ F_{21} & F_{22} & F_{23} \\ F_{31} & F_{32} & F_{33}\end{bmatrix}\begin{bmatrix} dX_1 \\ dX_2 \\ dX_3\end{bmatrix} \leftrightarrow$$

$$ds^2 = \frac{\partial x_i}{\partial X_j}\frac{\partial x_i}{\partial X_k}dX_j dX_k \tag{4.18}$$

$$\text{or}\quad ds^2 = d\mathbf{X}^T\,\mathbf{C}\,d\mathbf{X}^T \qquad \text{where} \qquad \mathbf{C} = \mathbf{F}^T\mathbf{F} \leftrightarrow C_{ij} = \frac{\partial x_k}{\partial X_i}\frac{\partial x_k}{\partial X_j} \tag{4.19}$$

where ds^2 is the squared final length of the original material vector d**X**, and **C** is called the **deformation tensor** or **Green deformation tensor** or **Cauchy deformation tensor** (depending on how the reference state is chosen).

By referring to Section 4.1 and Eqs. 4.2 through 4.7, we can see that the generalization of strain in a tensile test involves *the change* of length of an arbitrary material line segment. Therefore, we modify the foregoing equations to arrive at a general definition of strain consistent with the definition motivated by the tensile test:

$$ds^2 - dS^2 = d\mathbf{X}^T\mathbf{C}\,d\mathbf{X} - d\mathbf{X}^T\,d\mathbf{X} =$$

$$d\mathbf{X}^T\,\mathbf{C}\,d\mathbf{X} - d\mathbf{X}^T\mathbf{I}\,d\mathbf{X} = d\mathbf{X}^T\,(\mathbf{C}-\mathbf{I})\,d\mathbf{X} \tag{4.20}$$

$$ds^2 - dS^2 = d\mathbf{X}^T\,(2\mathbf{E})\,d\mathbf{X}$$

$$\text{where } \mathbf{E} = \frac{1}{2}(\mathbf{C}-\mathbf{I}) = \frac{1}{2}(\mathbf{F}^T\mathbf{F}-\mathbf{I}) \leftrightarrow E_{ij} = \frac{1}{2}\left[\frac{\partial x_k}{\partial X_i}\frac{\partial x_k}{\partial X_j} - \delta_{ij}\right] \tag{4.21}$$

---

[8] If, for example, material elements could interact at long distances, the idea of stress (force exerted across an interface area) would be far less useful. Instead, we would be required to consider interacting volumes of material, as for the case of long-range internal gravitational, or electrostatic/magnetic interactions. Fortunately, such interactions are very short range or of negligible magnitudes in materials, at roughly the length scale where the continuum assumptions break down.

[9] It is always necessary to assume that we follow the deformation, stretch, etc., of a single material element, rather than some nonphysical distortion involving different material elements. The latter do not involve deformation, but rather some quantity that has only mathematical significance. See, for example, the discussion following Eq. 4.11.

where dS^2 is the original material line length squared, equal to $d\mathbf{X}^T \cdot d\mathbf{X}$. **E** is the **strain tensor**. The factor $\frac{1}{2}$ enters conventionally, such that the conventional **infinitesimal strain components** (next section) are the small limit of **E**. Note that **E** behaves consistently with our view of tensile strain (and all components are zero for no deformation) , whereas the deformation tensor becomes the identity tensor for no deformations. Otherwise, both **C** and **E** remove the rigid-body rotation that appears in **F** or **J** and can be used equally for formulating constitutive equations. **C** and **E** are both symmetric by definition (Eqs. 4.19 and 4.21) and thus will have three real eigenvalues corresponding to orthogonal eigendirections (principal directions of strain). Since **C** and **E** differ mainly by a constant added to the diagonal elements, the principal directions are identical. (That is, when a rotation of coordinate system is found such that the non-diagonal components are zero for one of the tensors, the rotation is identical for the other. This can easily be seen by examining the forms of the invariants J_2 and J_3, Eqs. 2.40 or 3.22).

As shown below, the three orthogonal principal strain directions are material lines, specified in the original spatial directions, which retain the same directions relative to each other before and after the deformation. The principal values correspond to the extension of these material lines. Thus in Figure 4.1 we were "lucky" enough to choose the principal coordinate system for the long rod so that the the nine tensor components were reduced to the principal values:

$$E_{ii} \approx e_i \quad \text{or} \quad C_{ii} \approx 1 + 2e_i \tag{4.22}$$

The interpretation for principal strain directions is a bit more difficult to show than for stress because **E** and **C** are defined quadratically in d**X**. The mathematical basis for the computation is, of course, the same. Any real, symmetric matrix has three orthogonal directions with the desired property: [E] [dX] = λ[dX]. Because a rigid rotation or translation of material produces no strain ([E] = [0], or [C] = [I]), a combined deformation and rotation appears only as a deformation in [E] and [C].

Exercise 4.1 **Show that the principal directions of E or C are material directions that do not change their relative orientations before and after the deformation.**

To show that the principal directions of strain represent fixed relative directions before and after deformation, we simply start with the fixed angle requirement and derive the characteristic equation. Assume that we choose two unit vectors, $d\hat{\mathbf{X}}_1$ and $d\hat{\mathbf{X}}_2$ in the undeformed, reference state which correspond to the same material vectors in the deformed state, $d\mathbf{x}_1 = \lambda_1 d\hat{\mathbf{x}}_1$ and $d\mathbf{x}_2 = \lambda_2 d\hat{\mathbf{x}}_2$, where we have assigned the arbitrary vector magnitudes λ_1 and λ_2. Before the deformation, the cosine of the angle between the unit vectors is given simply by

$$\cos\theta_{\text{before}} = \left[d\hat{X}_1\right]^T \left[d\hat{X}_2\right]. \tag{4.1-1}$$

A similar expression can be written in the deformed configuration:

$$\cos\theta_{after} = \frac{[d\hat{X}_1]^T[d\hat{X}_2]}{\lambda_1\lambda_2} \tag{4.1-2}$$

and can be rewritten using the definition of the deformation gradient, **F** (Eq. 4.15):

$$\cos\theta_{after} = \frac{[d\hat{X}_1]^T}{\lambda_1}\,[F]^T\,[F]\,\frac{[d\hat{X}_2]}{\lambda_2}$$

$$= \frac{1}{\lambda_1\lambda_2}[d\hat{X}_1]^T[C]\,[d\hat{X}_2]. \tag{4.1-3}$$

Next, we enforce the requirement that $\theta_{before} = \theta_{after}$:

$$\cos\theta_{before} = \cos\theta_{after}$$

$$[d\hat{X}_1]^T\,[d\hat{X}_2] = \frac{1}{\lambda}\,[d\hat{X}_1]^T\,[C]\,[d\hat{X}_2] \tag{4.1-4}$$

$$0 = [d\hat{X}_1]^T\,[C]\,[d\hat{X}_2] - \lambda[d\hat{X}_1]^T\,[I]\,[d\hat{X}_2], \quad \text{where } \lambda = \lambda_1\lambda_2 \tag{4.1-5}$$

$$0 = [d\hat{X}_1]^T\,[C - \lambda I]\,[d\hat{X}_2] \tag{4.1-6}$$

Equation 4.1-6 only has nontrivial solutions if

$$|\mathbf{C} - \lambda\mathbf{I}| = 0, \tag{4.1-7}$$

which is exactly the characteristic equation. In terms of **E**, Eq. 4.17 can be rewritten to emphasize that the solutions are the same:

$$|\mathbf{C} - \lambda\mathbf{I}| = |2\mathbf{E} - (\lambda - 1)\mathbf{I}| = 2^3\left|\mathbf{E} - \left(\frac{\lambda - 1}{2}\right)\mathbf{I}\right| \tag{4.1-8}$$

Thus the principal values of **E** and **C** differ: $l_c = 2l_E + 1$, while the principal directions are the same.

Exercise 4.2 **Consider three modes of in-plane deformation: balanced biaxial stretch (BB), plane strain deformation (PS), and simple shear (SS). Find F, R, u, C, and E for a given δ specified over a unit square, as shown:**

We can write the deformation gradient for each case by noting that $d\mathbf{x}_1 = \mathbf{F}d\hat{\mathbf{X}}_1$ and $dx_2 = dF\hat{X}_2$ in each case and by expressing the orientation of these same material vectors after the deformation. For example, for the balanced biaxial case,

$$\begin{bmatrix} 0 \\ 1+\delta \end{bmatrix} = [F]\begin{bmatrix} 0 \\ 1 \end{bmatrix}, \quad \text{and} \quad \begin{bmatrix} 1+\delta \\ 0 \end{bmatrix} = [F]\begin{bmatrix} 1 \\ 0 \end{bmatrix} \tag{4.2-1}$$

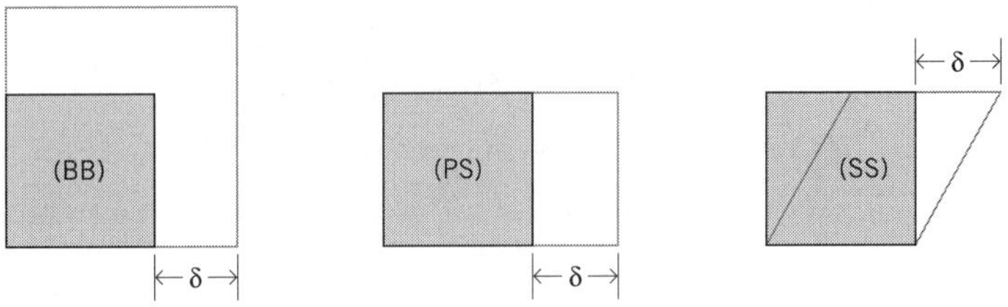

$$\text{So, } \begin{bmatrix} 1+\delta & 0 \\ 0 & 1+\delta \end{bmatrix} = [F]\begin{bmatrix} 1 & 0 \\ 0 & 1 \end{bmatrix}, \ [F_{BB}] = \begin{bmatrix} 1+\delta & 0 \\ 0 & -1+\delta \end{bmatrix}, \ [C_{BB}] = \begin{bmatrix} (1+\delta)^2 & 0 \\ 0 & (1+\delta)^2 \end{bmatrix} \quad (4.2\text{-}2)$$

Similarly, for the other two cases,

$$[F_{PS}] = \begin{bmatrix} 1+\delta & 0 \\ 0 & 1 \end{bmatrix}, \quad [F_{SS}] = \begin{bmatrix} 1 & \delta \\ 0 & 1 \end{bmatrix} \quad (4.2\text{-}3)$$

$$[C_{PS}] = \begin{bmatrix} (1+\delta)^2 & 0 \\ 0 & 1 \end{bmatrix}, \ [C_{SS}] = \begin{bmatrix} 1 & 0 \\ \delta & 1 \end{bmatrix}\begin{bmatrix} 1 & \delta \\ 0 & 1 \end{bmatrix} = \begin{bmatrix} 1 & \delta \\ \delta & 1+\delta^2 \end{bmatrix}$$

The stretch and rotation matrices are very simple for the first two cases because **F** = **u** and **Q** = **I.** That is, **F** is symmetric and positive-definite, and thus there are no rigid-body rotations.

Simple shear, however, includes a systematic rotation of the material lines corresponding to the principal directions. **F** in this case is no longer symmetric or positive-definite. In order to find **R u**, we simply solve for $\mathbf{C}^{1/2}$, substitute for **u,** and then solve for **R**:

$$[C_{SS}] = [F]^T[F] = \begin{bmatrix} 1 & \delta \\ \delta & 1+\delta^2 \end{bmatrix}. \quad (4.2\text{-}4)$$

The principal values of $\mathbf{C}_{ss}$ are

$$\lambda_1 = C_1 = 1 + \frac{1}{2}\delta^2 + \frac{1}{2}\delta\frac{1}{2}\sqrt{4+9^2} \quad (4.2\text{-}5a)$$

$$\lambda_2 = C_2 = 1 + \frac{1}{2}\delta^2 - \frac{1}{2}\delta\sqrt{4+9^2} \quad (4.2\text{-}5b)$$

or, in terms of $b = \left[\sqrt{\left(\delta^2 + 2\delta + 3\right)^2 - 4}\right]$,

$$C_1 = \frac{1}{2}\left[b + \sqrt{(b+2)(b-2)}\right] \quad (4.2\text{-}6a)$$

$$C_2 = \frac{1}{2}\left[b - \sqrt{(b+2)(b-2)}\right] \tag{4.2-6b}$$

So that in the principal axes, **U** takes the form

$$[u] = \begin{bmatrix} \sqrt{\lambda_1} & 0 \\ 0 & \sqrt{\lambda_2} \end{bmatrix} \tag{4.2-6c}$$

4.4 SMALL STRAIN

In the previous section, we introduced the basic concepts used often in solid kinematics. Within the framework of a continuum, there were no approximations made, and we were able to relate infinitesimal material vectors, their orientations and lengths, before and after deformation. When the material deformation is very small (infinitesimal), one can use a limiting procedure to arrive at a mathematically simple (but perhaps conceptually more difficult) picture of strain. Historically, such a measure has been used successfully for the elastic response of metals. Furthermore, such measures are also useful as infinitesimal steps in a large deformation, always with a need to be careful to account for additive errors and reference configurations.

For comparison with the previous section, it is most convenient to start with the kinematic variables based on relative displacements of a material line or vector, because these displacements may be considered vanishingly small relative to the material line consideration. Thus, when $|\mathbf{du}| << |\mathbf{dx}|$, the displacement gradient [J] approaches zero (see Eq. 4.16, for example) and allows the simplifications in strain.

Recall the general definitions of strain, and the alternate forms

$$2[E] = [F]^T[F] - I = [J+I]^T[J+I] - I \tag{4.23}$$

$$2[E] = \left[J^T J + J^T + J + I\right] - I$$

or

$$[E] = \frac{\left[J^T + J\right]}{2} + \frac{J^T J}{2} \tag{4.24}$$

and introduce a new tensor,

$$[\varepsilon] = \frac{\left[J^T + J\right]}{2} \tag{4.25}$$

where we have dropped the terms $\mathbf{J}^T\mathbf{J}$ in Eq. 4.25 because it is second order relative to J^T and J whenever $|J| << 1$ (equivalent to $|du| << |dx|$). We use the common notation of ε to represent the **small-strain tensor** and **E** for the **large-strain tensor**.

[ε] is clearly the symmetric part of [J] formed by linearly averaging the off-diagonal elements across the diagonal. It is convenient to relate this back to the original picture of [J] (Eq. 4.16) and to define the antisymmetric part of [J] similarly:

$$[\omega] = \frac{\left[J - J^T\right]}{2} \tag{4.26}$$

such that

$$[J] = [\varepsilon] + [\omega] \tag{4.27}$$

where [ε] is identified as the strain (or shape-change part) of the small displacement deformation, and [ω] is the **rigid-body rotational part**. Thus the relative displacements of each material line consist of a strain part and a rotational part:

$$|du| = [\varepsilon]\,|dx| + [\omega]\,|dx| \tag{4.28}$$

Exercise 4.3 **Show that [Ω] does in fact represent a rigid rotation through a small angle.**

Using the rotation matrix [R] to rotate a vector about $\hat{X}_3$, find the new coordinate of the tip of material vectors originally lying along $\hat{X}_1$ and $\hat{X}_2$.

$$dx = \mathbf{R} \cdot dX \tag{4.3-1}$$

$$\begin{bmatrix} dx_1 \\ 0 \\ 0 \end{bmatrix} = \begin{bmatrix} \cos\theta & -\sin\theta & 0 \\ \sin\theta & \cos\theta & 0 \\ 0 & 0 & 1 \end{bmatrix} \begin{bmatrix} dX_1 \\ 0 \\ 0 \end{bmatrix} = \begin{bmatrix} dX_1 \cos\theta \\ dX_1 \sin\theta \\ 0 \end{bmatrix} \tag{4.3-1a}$$

$$\begin{bmatrix} 0 \\ dx_2 \\ 0 \end{bmatrix} \leftrightarrow \begin{bmatrix} \cos\theta & -\sin\theta & 0 \\ \sin\theta & \cos\theta & 0 \\ 0 & 0 & 1 \end{bmatrix} \begin{bmatrix} 0 \\ dX_2 \\ 0 \end{bmatrix} = \begin{bmatrix} -dX_2 \sin\theta \\ dX_2 \cos\theta \\ 0 \end{bmatrix} \tag{4.3-1b}$$

The relative displacement of the tips of these material vectors (**dx** - **dX**) is related to their initial coordinates via the displacement gradient:

$$\mathbf{du} = (\mathbf{R}-\mathbf{I}) \cdot dX \tag{4.3-2}$$

$$\begin{bmatrix} du_1 \\ du_2 \\ du_3 \end{bmatrix} = \begin{bmatrix} \cos\theta - 1 & -\sin\theta & 0 \\ \sin\theta & \cos\theta - 1 & 0 \\ 0 & 0 & 0 \end{bmatrix} \begin{bmatrix} dX_1 \\ dX_2 \\ dX_3 \end{bmatrix} \tag{4.3-2a}$$

We make the usual substitutions for small angles, namely that $\cos\theta \approx 1$ and $\sin\theta \approx \theta$. Now the displacement gradient matrix becomes

$$J = \begin{bmatrix} 0 & -\theta & 0 \\ \theta & 0 & 0 \\ 0 & 0 & 1 \end{bmatrix} \tag{4.3-3}$$

which is the form (antisymmetric, with zero diagonal elements) of $[\omega]$, the small deformation rotation.

Exercise 4.4 **Derive the definition of the small-strain tensor directly in 2-D by considering the infinitesimal displacements of a square material element.**

Although the rigid-body displacements might be large,[10] we ignore them by choosing a translated coordinate system at the same *material* origin. At this site, we picture the undeformed element of material and compare with the deformed shape to obtain the deformation. Since the relative displacements are second order, we make no distinction between the O and O′ coordinate systems. Then, using our intuitive understanding of strain (the change of relative position per unit length of material

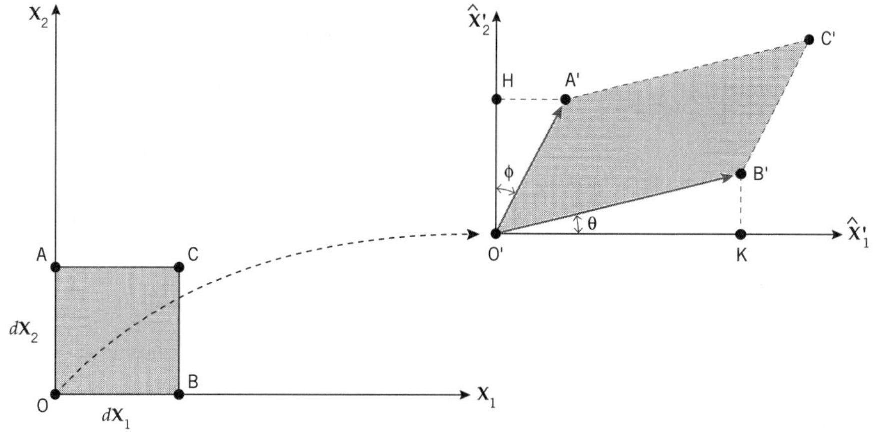

line), we can evaluate the quantities.

[10] But note that the rigid-body displacements cannot have a large rotation because this would violate the requirement that $|\mathbf{du}| \ll |\mathbf{dx}|$. We have drawn large displacements in the figure for clarity only.

ε_{11} : OB $\rightarrow$ O′B′, with an extensional strain of $\frac{|O'B'|-|OB|}{|OB|}$ (4.4-1)

$(\Delta u_1/\Delta l_1)$

$$OB = dx_1 \tag{4.4-2}$$

$$O'B' = dx_1 + \frac{\partial u_1}{\partial x_1} dx_1 + (\text{higher order terms in } x_1 \text{ and } x_2) \tag{4.4-3}$$

$$\varepsilon_{11} = \frac{dx_1 + \frac{\partial u_1}{\partial x_1} dx_1 - dx_1}{dx_1} = \frac{\partial u_1}{\partial x_1} \tag{4.4-4}$$

ε_{22} : OA $\rightarrow$ OA′, with an extensional strain of $\frac{|O'A'|-|OA|}{|OA|}$ (4.4-5)

$(\Delta u_2/\Delta l_2)$

$$OC = dx_2 \tag{4.4-6}$$

$$O'A' = dx_2 + \frac{\partial u_2}{\partial x_2} dx_2 + (\text{higher order terms}) \tag{4.4-7}$$

$$\varepsilon_{22} = \frac{dx_2 + \frac{\partial u_2}{\partial x_2} dx_2 - dx_2}{dx_2} = \frac{\partial U_2}{\partial x_2} \tag{4.4-8}$$

ε_{12} : OB $\rightarrow$ OB′, with a rotation of θ
$(\Delta u_1/\Delta l_2)$ or (4.4-9)
$(\Delta u_2/\Delta l_1)$

$$\text{where } \tan\theta \approx \theta = \frac{KB'}{KO'} = \frac{\frac{\partial u_2}{\partial x_1} dx_1}{dx_1} = \frac{\partial u_2}{\partial x_1} + (\text{higher order terms}) \tag{4.4-10}$$

and OA $\rightarrow$ OA′ with a rotation of Φ (4.4-11)

$$\text{where } \tan\phi \approx \phi = \frac{HA'}{HO'} = \frac{\frac{\partial u_1}{\partial X_2} dx_2}{dx_2} = \frac{\partial u_1}{\partial x_2} \quad (4.4\text{-}12)$$

To remove the possibility of rigid-body rotation, we are interested only in

$$\theta + \phi = \frac{\partial u_1}{\partial x_2} + \frac{\partial u_2}{\partial x_1}, \text{ so}$$

$$\gamma_{12} = \frac{\partial u_1}{\partial x_2} + \frac{\partial u_2}{\partial x_1} \quad (4.4\text{-}13)$$

$$\text{or } \varepsilon_{12} = \frac{1}{2}\left(\frac{\partial u_1}{\partial x_2} + \frac{\partial u_2}{\partial x_1}\right) \quad (4.4\text{-}14)$$

The factor of $\frac{1}{2}$ is employed so that the strains ε_{11}, ε_{22}, and ε_{12} are the components of a tensor. In non-tensor engineering mechanics, one often encounters ε_{12}, which is the total angle change between the sides of a square element.
Because

$$\frac{\partial u_1}{\partial x_1} = \frac{1}{2}\left(\frac{\partial u_1}{\partial x_1} + \frac{\partial u_1}{\partial x_1}\right)$$

it is easily seen that the components derived from the small-strain geometry are exactly the same as the ones taken from the limit of the large-strain tensor:

$$\varepsilon_{ij} = \frac{1}{2}\left(\frac{\partial u_i}{\partial x_j} + \frac{\partial u_j}{\partial x_i}\right) \leftrightarrow [\varepsilon] = \frac{1}{2}[J]^T + [J] \quad (4.4\text{-}15)$$

4.5 RATE OF DEFORMATION

In the foregoing, we used the basic Lagrangian method of following a material element from one time to another. This is the usual approach for solid mechanics when the material properties depend on the material history, and a steady state is not established. However, when considering rates of deformation, it is usually more convenient to adopt the Eulerian approach, where a velocity field is considered the principal information available. Then, using proper geometrical constructions, it is possible to compute the rates from the nearby velocity field.

Consider a material vector **dx** at time t_o. If **v**(**x**) is the material velocity field specified in terms of spatial position, then the relative velocity of the head to the tail is

$$dv_i = \left(\frac{\partial v_i}{\partial x_j}\right) dx_j \quad (4.29)$$

or

$$[dv] = [L]\,[dx] \leftrightarrow \mathbf{dv} = \mathbf{L} \cdot \mathbf{dx} \tag{4.30}$$

where

$$L_{ij} = \left(\frac{\partial v_i}{\partial x_j} \right) \tag{4.31}$$

and **L** is called the **velocity gradient** or the **spatial gradient of velocity**.

> ***Note:*** *The connection with **J**, the deformation gradient, is clear if one considers two times separated infinitesimally, say t and t + dt. Then,* $du_i / dt = dv_i$*, and* $d\mathbf{J} / dt = \mathbf{L}$*, as long as the reference state for **J** is chosen to be the state at t. This is necessary because **L** is based on the current spatial positions of a material element, that is, in the usual Eulerian way.*

In an infinitesimal time, the relative displacements of material points are infinitesimal relative to their separation ($|\mathbf{du}| << |\mathbf{dx}|$) so that the small-strain formulation is exactly satisfied. Thus we follow the procedure of linear separation of **L** into symmetric parts, exactly corresponding to the deconvolution of [J] into [E] and [ε]:

$$[L] = [D]+[W] \tag{4.32}$$

$$\text{where } 2[D] = [L] + [L]^T \leftrightarrow D_{ij} = \frac{1}{2}\left(\frac{\partial v_i}{\partial x_j} + \frac{\partial v_j}{\partial x_i} \right) \leftrightarrow \text{rate of deformation tensor} \tag{4.33}$$

$$2[W] = [L] - [L]^T \leftrightarrow W_{ij} = \frac{1}{2}\left(\frac{\partial v_i}{\partial x_j} - \frac{\partial v_j}{\partial x_i} \right) \leftrightarrow \text{spin tensor or vorticity tensor} \tag{4.34}$$

The role of [D] and [W] can be seen by considering the small displacement counterparts [ε] and [ω], which have already been discussed:[11]

$$[\varepsilon] = [D]dt \text{ or } [\dot{\varepsilon}] = [D] \tag{4.35}$$

$$[\omega] = [W]dt \text{ or } [\dot{\omega}] = [W] \tag{4.36}$$

---

[11] Some authors are careful to point out that $[\dot{\varepsilon}_{ij}] \neq [D_{ij}]$ because $[\dot{\varepsilon}_{ij}]$ is based on material derivatives while $[D_{ij}]$ relies on spatial derivatives. In fact, there is no distinction when the displacements are truly infinitesimal, because $\hat{x}_i = \hat{X}_i$. In fact, this is a matter of convention: we denote $\dot{\boldsymbol{\varepsilon}} = \mathbf{D}$ in all cases and must remember that $\dot{\boldsymbol{\varepsilon}} \neq d\boldsymbol{\varepsilon}/dt$.

(infinitesimal displacements only)

Thus [D] is shown to be the rate of change of [ε] (or [E] with the reference state at the curved state) with time, whereas [W] is the rate of change of [**w**] with time. It should be noted well that these correspondences are precise only for *vanishingly small displacements.* When the small-displacement quantities are considered to be approximations of the large-strain values, then these equivalences do not hold.

Exercise 4.5 **Show how the Eulerian forms L and D are related to the rate of change of the Lagrangian quantities F and E.**

The basic difference between the Eulerian quantities and the Lagrangian quantities lies in the current basis for the former and the original (undeformed) basis for the latter. For example, compare:

$$L_{ij} = \frac{\partial v_i}{\partial x_j} \tag{4.5-1}$$

$$\dot{F}_{ij} = \frac{\partial v_i}{\partial X_j} \tag{4.5-2}$$

where the definition for **F** is taken from a time derivative of Eq. 4.15, and noting that $v_i = \frac{\partial x_i}{\partial t}$, the current velocity of a material particle currently at **x** and originally at **X**. The only difference in the equations is apparent: the denominator of **L** uses a current spatial increment dx whereas **F** is based on the variation with respect to the original increment corresponding to this current increment. We can relate the two simply:

$$\mathrm{dx} = \mathbf{F}\mathrm{d}\mathbf{X} \tag{4.5-3}$$

$$\mathbf{dv} = \left(\frac{\partial \mathbf{x}}{\partial \mathbf{t}}\right)_{\mathbf{X}} = \dot{\mathbf{F}}\mathrm{d}\mathbf{X} \tag{4.5-4}$$

$$\left(\frac{\partial \mathbf{v}}{\partial \mathbf{X}}\right)_{\mathbf{X}} = \dot{\mathbf{F}} \tag{4.5-5}$$

But

$$\mathbf{L} = \frac{\partial \mathbf{v}}{\partial \mathbf{x}} = \frac{\partial \mathbf{v}}{\partial \mathbf{X}}\frac{\partial \mathbf{X}}{\partial \mathbf{x}} = \frac{\partial \mathbf{v}}{\partial \mathbf{X}}\left(\frac{\partial \mathbf{x}}{\partial \mathbf{X}}\right)^{-1} \tag{4.5-6}$$

so

$$\mathbf{L} = \dot{\mathbf{F}}\,\mathbf{F}^{-1}, \text{ or } \dot{\mathbf{F}} = \mathbf{L}\,\mathbf{F}$$

and, since

$$\mathbf{E} = \frac{1}{2}\left(\mathbf{F}^T\ \mathbf{F} - \mathbf{I}\right) \leftrightarrow E_{ij} = \frac{1}{2}\left[\frac{\partial X_k}{\partial X_i}\frac{\partial X_k}{\partial X_j} - \delta_{ij}\right] \tag{4.5-7}$$

the Lagrangian rate of strain, $\dot{\mathbf{E}}$, may be found:

$$\dot{\mathbf{E}} = \frac{1}{2}\left[\dot{\mathbf{F}}^T\ \mathbf{F} + \mathbf{F}^T\ \dot{\mathbf{F}}\right] \tag{4.5-8}$$

But, from above $\dot{\mathbf{F}}^T = \mathbf{F}^T\mathbf{L}^T$, and $\dot{\mathbf{F}} = \mathbf{LF}$, so

$$\dot{\mathbf{E}} = \frac{1}{2}\left[\mathbf{F}^T\ \mathbf{L}^T\mathbf{F} + \mathbf{F}^T\ \mathbf{L}\ \mathbf{F}\right] = \mathbf{F}^T\left[\frac{1}{2}\left(\mathbf{L}^T + \mathbf{L}\right)\right]\mathbf{F} \tag{4.5-9}$$

and, since **L** = **D** + **W** where **D** is symmetric and **W** is anti-symmetric,

$$\dot{\mathbf{E}} = \mathbf{F}^T\mathbf{D}\ \mathbf{F} \tag{4.5-10}$$

Following the same procedure as in Exercise 4.4, other useful equations relating various measures of deformation can be derived:

$$\dot{\mathbf{C}} = 2\dot{\mathbf{E}} \tag{4.5-11}$$

$$\mathbf{L} = \dot{\mathbf{F}}\ \mathbf{F}^{-1} \tag{4.5-12}$$

$$\dot{\mathbf{E}}^* = \mathbf{D} - \left(\mathbf{E}^*\mathbf{L} - \mathbf{L}^T\mathbf{E}^*\right) \tag{4.5-13}$$

where **E*** is the Eulerian strain tensor, referred to the current configuration.

When **F** is small (infinitesimal), all of the strain rates become the same:

$$\mathbf{D} = \dot{\mathbf{E}} = \dot{\mathbf{E}}^* \quad \text{(infinitesimal deformation)}$$

4.6 SOME PRACTICAL ASPECTS OF KINEMATIC FORMULATIONS

In the foregoing sections, we have attempted to outline the basic kinematics needed for metal-forming analysis without providing a great deal of theory. Many of the esoteric concepts in general theory are not critical to practical analysis. However, several abstract concepts do frequently arise in the application of metal plasticity and elasto-plasticity that are potential sources of trouble. In this section, we comment briefly on some of these points.

One-Dimensional Plasticity: The Tensile Test

An ideal tensile test is one in which the stresses and strains are evenly distributed throughout the specimen and for which the stress state is well known: $s_1 = s$, other $s_{ij} = 0$. Therefore, the principal stress axes are well known (and are fixed in space) and for an isotropic material we assume that the principal axes are identical for strain. Under these conditions, we usually restrict our attention to the tensile axis, recording e and s in this direction, which does not rotate.

In the time-honored way, we define engineering quantities in terms of the original specimen:

$$\sigma_{eng} = \frac{f_{axial}}{a_{original}} \tag{4.37a}$$

$$e_{eng} = \frac{\Delta l}{l_{original}} \tag{4.37b}$$

However, both quantities have little material significance when the deformation is large. For large deformation, the **true stress** is defined in terms of the current, material, cross-sectional area:

$$\sigma_{true} = \frac{f_{axial}}{a_{current}} \tag{4.38}$$

while the **true strain** is defined infinitesimally because the current length is always changing:

$$d\varepsilon_{true} = \frac{dl}{l_{current}} \tag{4.39}$$

Fortunately, because the tensile axis does not rotate and therefore the spatial and material directions coincide for all times, it is easy to integrate Eq. 4.39 without careful distinctions between these directions:

$$\varepsilon_{true} = \int_{S_1}^{S_2} d\varepsilon_{true} = \int_{l_1}^{l_2} \frac{dl}{l} = \ln \frac{l_2}{l_1} \tag{4.40}$$

We have used S_1 and S_2 to refer to "state 1" and "state 2" so as not to confuse the

We have used S_1 and S_2 to refer to "state 1" and "state 2" so as not to confuse the extent of deformation with change of time.[12]

Thus the true strain relates the length of the same material element between two states of deformation. It provides a measure of the material lines' stretch and has the advantage of being additive, such that it serves as a correct indicator of the material's stretch throughout the deformation, rather than having an arbitrary reference (as the engineering strain has):

$$\varepsilon_{total} = \int_{S_o}^{S_B} d\varepsilon = \int_{S_o}^{S_A} d\varepsilon + \int_{S_A}^{S_B} d\varepsilon = \ln\frac{l_a}{l_o} + \ln\frac{l_B}{l_A} = \varepsilon_A + \varepsilon_B \tag{4.41}$$

The integrals from S_0 to S_A and from S_A to S_B are independent of one another; that is, they do not require knowledge about the initial specimen's length for evaluation. (When the initial specimen's length is used as a reference—for example, to compute engineering strains—the strains in the various paths are not additive and cannot be defined purely internal to that path segment.)

Plasticity laws based on the flow theory involve the use of incremental, infinitesimal strain measures ($d\varepsilon$ or $\dot{\varepsilon}$), similar to fluid flow laws. At any instant in time, these present no particular conceptual problem because, as shown in the previous section, Eulerian and Lagrangian descriptions are identical for infinitesimal deformation when the current state is adopted as the reference state. (This is equivalent to equating Eqs. 4.37b and 4.40 by substituting dl for Δl, and $l_{current}$ for $l_{original}$.) However, the current material properties can depend in a complex way on the entire deformation history of a material element. For this reason, it is necessary to follow the deformation history of each element along large-deformation paths. For the tensile test, this is accomplished by the integration shown in Eq. 4.40, which relates a total strain variable, ε, to an incrementally defined one, $d\varepsilon$. Such a correspondence becomes much more difficult along general deformation paths.

"True" Strain in General Deformation

It is very attractive to seek a materially relevant large-strain quantity equivalent to the true strain, e, in the tensile test. However, two conceptual problems enter when general deformation is considered. First, there is an unknown rigid-body material rotation that must be considered, such that material directions may not remain identical to fixed spatial directions. Second, large strain is a two-point tensor, and the path between the two

[12] Since a tensile test is usually carried out at constant boundary velocity, the distinction between time and amount of deformation is unimportant. However, there are some important conceptual advantages to separating time from extent of deformation as variables characterizing the extent or progress of the deformation.

points is unknown. However, this path is critical to the material view of the deformation. (In the tensile test, the path was specified for all time, so this uncertainty did not arise.)

One tempting approach is based on the Eulerian view of material flowing past a fixed spatial point. The rate of deformation, **D**, has already been equated with the material strain rate, ε, Eq. 4.35. By introducing an infinitesimal time increment, we can look at our infinitesimal increment of true strain:

$$d\varepsilon = \mathbf{D}dt \tag{4.42}$$

Then, purely formally, we can define the result of an integration as the true strain:

$$\Delta\boldsymbol{\varepsilon} = \int_{l_1}^{l_2} d\boldsymbol{\varepsilon} = \int_{t_1}^{t_2} \mathbf{D}dt \tag{4.43}$$

Unfortunately, for Eq. 4.43 to have any physical significance, the integration must be carried out for the same material element, not at a fixed point in space. The integration of Eq. 4.43 is therefore a material time integral and its evaluation may be very difficult. In fact, integrals similar to Eq. 4.43 must always be carried out (usually in some approximate form) in order to evaluate the plastic strain hardening of a material, but the loss of the two-point nature of ε means that it is seldom used in this form.

A second approach to approximating the material strains (as opposed to the geometric strains) for a large path starts with the Lagrangian approach. The total extensions and rotations of each infinitesimal material line between two times are completely specified by **F**, the deformation gradient. However, since the path between the two states is unknown, this measure has no direct relevance to the material view of deformation, only the change in geometry between the two states. The extensions and relative rotations along the path are completely unknown but may be critical to the state of material.

However, at least some improvement can be made in the kinematic variables by separating rigid rotation (**R**), which doesn't contribute to strain, and stretch (**U**) as follows:

$$\mathbf{F} = \mathbf{R} \cdot \mathbf{U} \tag{4.44}$$

This so-called polar decomposition[13] (as opposed to the linear decomposition of **J** into ε and ω, or **L** into **D** and **W**) is required because the **R** must be orthogonal to be a true large-rotation tensor. (ω + **I** and **W**Δt + **I** are orthogonal only for vanishingly small angles of rotation.) **R** is called the **rotational tensor**. It rotates a vector aligned to the principal material line in the initial configuration into a

[13] Because the order of polar decomposition is important, one can equally define **F** = **V** · **R**, where **V** is called the left stretch tensor. The order of operation corresponds to the decomposition of the deformation path into steps taken in a different order.

vector aligned to this same material line in the deformed configuration. However, all other material lines (except the ones corresponding to the initial principal directions of **C** or **E**) rotate by other amounts that can be found only from **F** (or, equivalently, **R** and **U**).

U is a symmetric, positive definite tensor called the **right stretch tensor** (because of its placement to the right side of **R** in Eq. 4.44). **U** operates on any material vector in the initial configuration to give the length change (stretch) of all vectors and also gives additional amounts of rotation for all nonprincipal material vectors.

The decomposition of **F** into **R** and **U** corresponds to a sequential three-stage deformation, Figure 4.3.

1. Stretch by **U** to **x**′
2. Rigid rotation by **R**
3. A possible translation to **X** (for curvilinear coordinate applications only)

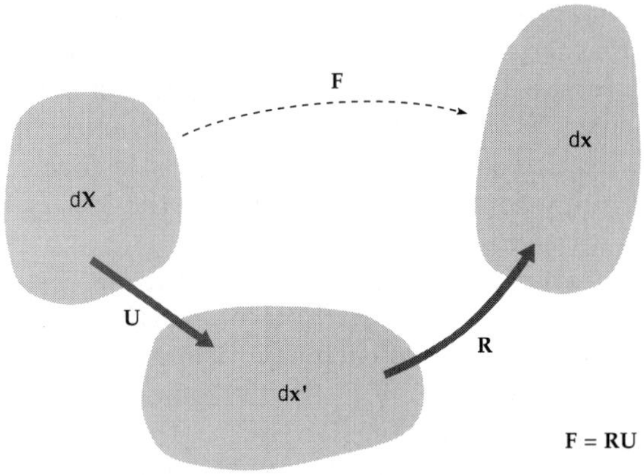

Figure 4.3 Conceptual decomposition of a general deformation into an initial stretch step followed by a rigid-body rotation.

The decomposition to obtain **U** is useful from a material point of view because the stretch of the material lines in the principal directions is available from the first step of the deformation (**X** to **x**′, Figure 4.3):

$$d\mathbf{x}' = \mathbf{U} \cdot d\mathbf{X} \tag{4.45}$$

or

$$ds^2 = d\mathbf{x}^T\, d\mathbf{x} = d\mathbf{X}^T\, \mathbf{U}^T\mathbf{U}\; d\mathbf{X} = d\mathbf{X}^T\, \mathbf{U}^2\, d\mathbf{X} \tag{4.46}$$

or

$$\frac{ds^2}{dS^2} = \left(\frac{dl}{dL}\right)^2 = \frac{d\mathbf{X}^T}{\left|d\mathbf{X}^T\right|}\mathbf{U}^2\frac{d\mathbf{X}}{\left|d\mathbf{X}\right|} = \hat{\mathbf{n}}^T\mathbf{U}^2\,\hat{\mathbf{n}} \tag{4.46a}$$

where $\frac{dl}{dL}$ is the **stretch ratio** for a material line with original length dL and final length dl, and u^2 means $u^T \cdot u$ (= u^2 because u is symmetric). When $\hat{\mathbf{n}}_i$ are principal directions, it is possible to take the square root of U^2 and thus define the principal stretch ratios:

$$\lambda_i = \frac{dl_i}{dL_i} = U_i \qquad \text{(principal components of } \mathbf{U}\text{)}, \tag{4.47}$$

where dl_i and dL_i are the final and original lengths of a material line corresponding to principal direction $\hat{\mathbf{n}}_i$ **of U**.

Comparison of Eq. 4.47 with Eq. 4.40 illustrates the correspondence with 1-D ideas of true strain. In fact, the true extensional strain of a material line lying in principal direction $\mathbf{n}_i$ is simply given by[14]

$$\varepsilon_i\,(\text{true}) = \ln\,\frac{l_i}{L_i} = \ln\,\lambda_i \tag{4.48}$$

However, it should be clearly noted that while the decomposition illustrated in Figure 4.2 solves the conceptual difficulty of following the extension of material lines, the decomposition may be carried out in alternate ways. It therefore is an unverified assumption to allow a two-point representation of strain to serve as an approximation to a possibly path-dependent material strain measure. Conversely, the proper integration of $d\varepsilon$, Eq. 4.43, along a material trajectory is proper, but is too cumbersome to be used except in numerical treatments. These limitations should be kept in mind whenever attempting to apply large-strain measures to material laws written in rate-of-strain quantities.

While the fact that **F** can be decomposed into a pure rotation (**R**) and a positive definite symmetric matrix (**U**) is theoretically interesting, **U** is seldom used because it involves fractional powers of the usual kinematic quantities. In fact, it may be shown that

$$\mathbf{U} = \mathbf{C}^{1/2} = [2\mathbf{E} + \mathbf{I}]^{\frac{1}{2}} \tag{4.48a}$$

[14] Some authors use the notation $\varepsilon_{true} = \ln \mathbf{U}$, which can be defined properly in terms of principal values.

where the fractional power of **C** signifies a tensor with the same principal directions as **C**, but with principal values equal to the square roots of C_i. (Thus, note that **U**, **C**, and **E** have identical principal directions and differ only in the way the stretch along those three directions is expressed.)

With the aid of the relationship above, Eq. 4.48a, the principal stretch ratios that correspond to these orthogonal material directions in the initial and final states, and that play the 3-D role of tensile stretch ratio dl/dL except for the unknown path between initial and final states, are given by

$$\lambda_i = U_i = C_i^{\frac{1}{2}} = (2E_i + 1)^{\frac{1}{2}} \tag{4.48b}$$

where U_i, C_i, and E_i are the principal values corresponding to the principal material direction in the X_i reference system.

4.7 COMPATIBILITY CONDITIONS

Compatibility refers to the continuum requirement for deformation that we have discussed in this and the previous chapter. It requires that the kinematic deformation variables (strain, stretch, deformation gradient, etc.) be compatible with continuous, single-valued deformation functions as assumed in Section 4.2: $\mathbf{x} = \chi(\mathbf{X}, t)$ (Eq. 4.9). Physically, this requires that one material particle cannot occupy two spatial positions at once (by opening a hole, for example), and that two material particles cannot occupy one position at an instant (by folding into itself, for example). But exceptions to these rules can be readily introduced locally to treat crack opening or folding.

Compatibility is automatically satisfied when the mechanical problems are formulated and solved in terms of displacement functions that are required to be continuous. Smooth-displacement functions automatically ensure that Eq. 4.9 is continuous, and physically that no voids or incursions occur. However, it is usually more convenient to formulate problems for closed-form solution (classical elasticity problems, for example) in terms of strains.[15] In this case, continuous strain functions *do not* ensure that the assumption of material continuity is preserved, and auxiliary conditions must be applied.

As a simple illustration of separate compatibility conditions, we consider the well-known application to 2-D, small-strain deformation. (It should be noted that the idea of compatibility, while completely general for 3-D and large strains, seldom arises in other cases because closed-form solutions are seldom possible to find.) Considering $\mathbf{X}_1$ and $\mathbf{X}_2$ directions separately, two smooth-displacement functions are required that satisfy a given material law and equilibrium condition:

---

[15] In formulating mechanical problems for numerical solution, compatibility is either relaxed or it is automatically satisfied by the formulation. Seldom is it convenient to consider both equilibrium and compatibility conditions directly in a numerical framework.

$$u_1 = u_1 (X_1, X_2) = u_1(x_1, x_2) \tag{4.49a}$$

$$u_2 = u_2 (X_1, X_2) = u_2(x_1, x_2) \tag{4.49b}$$

where U_i are continuous and single-valued, and the distinction between X_i and x_i is lost for small displacement.

The material law, however, is written in terms of strains, and thus the equilibrium equation is also. Assuming infinitesimal components, the strains are related to spatial variations of the displacement function via

$$\varepsilon_{ij} = \frac{1}{2}\left(\frac{\partial u_i}{\partial x_j} + \frac{\partial uj}{\partial x_i}\right) \tag{4.50a}$$

or

$$\varepsilon_{11} = \frac{\partial u_1}{\partial x_1} \qquad \varepsilon_{22} = \frac{\partial u_2}{\partial x_2} \qquad \varepsilon_{12} = \frac{1}{2}\left(\frac{\partial u_1}{\partial x_2} + \frac{\partial u_2}{\partial x_1}\right). \tag{4.50b}$$

Thus, when solving the mechanical problem in terms of three strain functions, ε_{11} (x_1, x_2), ε_{22} (x_1, x_2), and ε_{12} (x_1, x_2), there is no assurance that the two displacement functions, Eqs. 4.49, are smooth.

We can find a supplementary condition on the **variations of strains** (the three strains at any point remain independent) throughout the continuum by using the well-known property of continuous, smooth functions, namely that they are differentiable. Thus U_1 and U_2 are required to be smooth and continuous, such that

$$\frac{\partial^2 u_1}{\partial x_1 \partial x_2} = \frac{\partial^2 u_1}{\partial x_2 \partial x_1} \quad \text{and} \quad \frac{\partial^2 u_2}{\partial x_1 \partial x_2} = \frac{\partial^2 u_2}{\partial x_2 \partial x_1} \tag{4.51}$$

Also, we take one more derivative to find a useful expression in terms of strain variations:

$$\frac{\partial^2 \varepsilon_{11}}{\partial x_2^2} = \frac{\partial^3 u_1}{\partial x_1 \partial x_2{}^2} = \frac{\partial^3 u_1}{\partial x_2{}^2 \partial x_1} \tag{4.51a}$$

$$\frac{\partial^2 \varepsilon_{22}}{\partial x_1^2} = \frac{\partial^3 u_2}{\partial x_2 \partial x_1{}^2} = \frac{\partial^3 u_2}{\partial x_1{}^2 \partial x_2} \tag{4.51b}$$

Equations 4.51 a and b show how the variation of the normal strain components in the orthogonal directions are related to the variation of the shear components. Usually, Eqs. 4.51 are summed and the continuity of ε_{12} (x_1, x_2) is used to arrive at St.-Venant's compatibility equations:

$$\frac{\partial^2 \varepsilon_{11}}{\partial x_2^2} = \frac{\partial^2 \varepsilon_{22}}{\partial x_1^2} = 2\frac{\partial^2 \varepsilon_{12}}{\partial x_1 \partial x_2} \tag{4.52}$$

Equations 4.51 and 4.52 ensure that the three smooth-strain functions are derivable from the two smooth-displacement functions. This can be extended to 3-D calculation; by similar handling it leads to three compatibility equations.

4.8 ALTERNATE STRESS MEASURES

Just as there are alternate ways of looking at deformation (Lagrangian vs. Eulerian, or engineering strain vs. true strain), it is possible to define alternate measures of stress. From a material point of view, the **Cauchy stress**, already defined in detail, is the physically relevant quantity. The other measures can be viewed as mathematical permutations of this measure.

The important principle in choosing mechanical variables is that they are **complementary**. In particular, the stress and strain measures used must be **work-conjugate**; that is, an increment of strain multiplied by the current stress must be a proper infinitesimal incremental work quantity per volume of material. This means that the material volumes, areas, and lengths must be well-defined and consistent between the two variables.

The Cauchy stress tensor σ (introduced in detail in Chapter 3) arises from a Eulerian view of the deformation, where all variables are defined at an instant in time. (In one dimension, it corresponds to the true stress.) Thus σ relates the current force acting on a current material area:

$$\mathbf{f}_{\text{current}} = \sigma \cdot \mathbf{a}_{\text{current}} \tag{4.53}$$

The work-conjugate measure of deformation is **D**, the deformation rate tensor, in rate form (i.e., when multiplied $\boldsymbol{\sigma} \cdot \mathrm{D} = \text{power}$). In incremental form, **D**dt (when interpreted in the material sense) becomes the incremental variable:

$$\text{Power} = \sigma \cdot \mathbf{D} \qquad \delta\,\text{work} = \sigma \cdot \mathbf{D}dt = \sigma \cdot d\varepsilon \tag{4.54}$$

In the Lagrangian point of view, deformation is considered between an initial state (perhaps undeformed) and a final, deformed one. In one dimension, the corresponding variables are engineering stress and engineering strain. There are two stress quantities (sometimes called pseudo-stress tensors) that are typically used. In addition, the Cauchy stress is sometimes altered by a scalar factor such that the mechanical work is expressed per volume of original material:

$$\sigma^{(1)} = \sigma \frac{\rho}{\rho_o} \tag{4.55}$$

$$\sigma^{(1)} d\varepsilon = \frac{\rho}{\rho_o} \sigma \, d\varepsilon = \frac{\delta \text{ work}}{\text{original material volume}} \tag{4.56}$$

The **first Piola-Kirchoff stress tensor** ,$\Gamma^{(1)}$, gives the actual force $\mathbf{df}$ acting on a deformed area, but it is arranged to operate on the corresponding undeformed area:

$$\mathbf{df} = \Gamma^{(1)} \mathbf{da}_o \tag{4.57}$$

where $\mathbf{da}_o$ is the vector quantity representing the infinitesimal area of material in the original configuration that becomes $\mathbf{da}$ in the deformed configuration. It is not symmetric.

The **second Piola-Kirchoff stress tensor** yields the current force as transformed back to the original configuration, again related to an area expressed in the original configuration:

$$\mathbf{df}_o = \Gamma^{(2)} \mathbf{da}_o \tag{4.58}$$

where $\mathbf{df}_o$ is related to the current force acting on an actual element of area in the deformed configuration:

$$\mathbf{df} = \mathbf{F}\mathbf{df}_o. \tag{4.59}$$

In order to find the relationships among σ, $\Gamma^{(1)}$, and $\Gamma^{(2)}$, we present two intermediate results without derivation.[16] The volume change of a deformed element is given by the density change ($\rho_o \rightarrow \rho$) or the determinant of $\mathbf{F}$ (usually called the Jacobian):

$$\frac{d(\text{vol})}{d(\text{vol})_o} = \frac{\rho}{\rho_o} = J = |\mathbf{F}| \tag{4.60}$$

J can also be shown equal to the square root of the third invariant of $\mathbf{C}$, the deformation tensor:

$$J = \frac{d(\text{vol})}{d(\text{vol})_o} = \sqrt{C_1 \, C_2 \, C_3} \tag{4.61}$$

[16] See any mechanics textbook for the derivations—for example, L.E. Malvern, *Introduction to the Mechanics of a Continuous Medium*, (Englewood Cliffs, N.J.: Prentice-Hall, 1969), 167-170. However, note that vectors are often expressed in row form in this reference, whereas we show vectors in column form whenever possible.

The other intermediate result relates the size and normal of an element of area before and after deformation. It may be shown that $\mathbf{da}_o$ transforms to $\mathbf{da}$ (these are area vectors defined in the usual way, normal to the area element, with lengths equal to the areas) as follows:

$$\mathbf{da} = \frac{\rho_o}{\rho}\left[F^{-1}\right]^T \mathbf{da}_o \tag{4.62}$$

With these purely geometric results, we can find the relationships among the Cauchy stress and Piola-Kirchoff stresses. To do so, we simply equate the real force acting on a current element of material and then compare the descriptions.

$$\mathbf{df} = \boldsymbol{\sigma} \cdot \mathbf{da} = \boldsymbol{\Gamma}^{(1)}\mathbf{da}_o = \mathbf{F}\boldsymbol{\Gamma}^{(2)}\mathbf{da}_o \tag{4.63}$$

where the last tensor arises from a minor rearrangement of Eq. 4.59. From this equality (Eq. 4.63) and with the use of Eq. 4.62, all of the required relationships can be formed by simple matrix multiplication:

$$\boldsymbol{\Gamma}^{(2)} = \frac{\rho_o}{\rho}\mathbf{F}^{-1}\left[\mathbf{F}^{-1}\right]\boldsymbol{\sigma}^T \quad \text{or} \quad \boldsymbol{\sigma} = \frac{\rho}{\rho_o}\boldsymbol{\Gamma}^{(1)}\mathbf{F}^T \tag{4.64}$$

$$\boldsymbol{\Gamma}^{(1)} = \mathbf{F}\boldsymbol{\Gamma}^{(2)} \quad \text{or} \quad \boldsymbol{\Gamma}^{(2)} = \mathbf{F}^{-1}\boldsymbol{\Gamma}^{(1)} \tag{4.65}$$

$$\boldsymbol{\Gamma}^{(2)} = \frac{\rho_o}{\rho}\mathbf{F}^{-1}\boldsymbol{\sigma}\left[\mathbf{F}^{-1}\right]^T \quad \text{or} \quad \boldsymbol{\sigma} = \frac{\rho}{\rho_o}\mathbf{F}\,\boldsymbol{\Gamma}^{(2)}\mathbf{F}^T \tag{4.66}$$

These equations show that the second Piola-Kirchoff stress tensor is symmetric, whereas the first Piola-Kirchoff stress tensor is not. The lack of symmetry of the latter often makes it inconvenient for manipulation.

Any of the various stress quantities can be used in formulating a mechanical problem. The only caution that must be followed is that the proper stress- and strain-rate quantities must yield a proper mechanical power per unit volume of material. Such quantities are called **work-conjugate variables** and are correctly defined for use with energy principles (see Chapter 7).

As noted earlier, the Cauchy stress, σ, and the deformation velocity tensor, **D**, are work-conjugate. This is easily demonstrated by considering an infinitesimal right parallelepiped of material in the current, deformed configuration as in Figures 3.3 and 3.5. The forces acting on each face are $\mathbf{f} = \sigma\, \mathbf{da}$ ($f_i = \sigma_{ij} da_j$). Without limiting the generality of the illustration, we can assume that the origin of the cube does not translate (a rigid translation does not do internal work in any case). Then the internal power done by the cube of material is the sum

TABLE 4.1

Common Work-Conjugate Measures of Stress and Deformation Rate

Stress	*Strain Rate*	*Type of Motion*	*Power*	*Incremental Work Form*
σ Cauchy stress $\mathbf{df} = \sigma\, \mathbf{da}$	$\mathbf{D}$ rate of deformation $\mathbf{D} = \frac{\partial \mathbf{v}}{\partial \mathbf{x}}$	Eulerian	$\sigma \cdot \mathbf{D} = \frac{\text{power}}{\text{current volume}}$	$\sigma \cdot d\varepsilon$ $(d\varepsilon = \mathbf{D}dt)$
$\sigma^{(1)}$ volume = corrected Cauchy stress $\mathbf{df} = \frac{\rho}{\rho_o}\sigma^{(1)}\mathbf{da}$	$\mathbf{D}$ rate of deformation $\mathbf{D} = \frac{\partial \mathbf{v}}{\partial \mathbf{x}}$	Eulerian, per original material volume	$\sigma^{(1)} \cdot \mathbf{D} = \frac{\text{power}}{\text{original volume}}$	$\sigma^{(1)} \cdot d\varepsilon$ $(d\varepsilon = \mathbf{D}dt)$
$\Gamma^{(1)}$ first Piola = Kirchoff stress $\mathbf{df}_o = \Gamma^{(1)}\mathbf{da}$	$\dot{\mathbf{F}}$ rate of deformation gradient $\dot{\mathbf{F}} = \frac{d\mathbf{F}}{dt} = \frac{\partial^2 \mathbf{x}}{\partial \mathbf{X} \partial t}$	mixed: Eulerian area, Lagrangian force	$\Gamma^{(1)} \cdot \dot{\mathbf{F}} = \frac{\text{power}}{\text{original volume}}$	$\Gamma^{(1)}d\mathbf{F}$
$\Gamma^{(2)}$ second Piola = Kirchoff stress $\mathbf{df}_o = \Gamma^{(2)}\mathbf{da}$	$\dot{\mathbf{E}}$ or $\frac{1}{2}\dot{\mathbf{C}}$ rate of Lagrangian strain $\dot{\mathbf{E}} = \frac{1}{2}\dot{\mathbf{C}} = \mathbf{F}^T\mathbf{D}\mathbf{F}$	Lagrangian	$\Gamma^{(2)} \cdot \dot{\mathbf{E}} = \frac{\text{power}}{\text{original volume}}$	$\Gamma^{(2)}d\mathbf{E}$ or $\frac{1}{2}\Gamma^{(2)}d\mathbf{C}$

Notes: $\mathbf{df}$ = force vector acting on a current element of area $\mathbf{da}$
$\mathbf{df}_o$ = $\mathbf{df}$ transformed back to the original configuration: $\mathbf{df} = \mathbf{F}\mathbf{df}_o$
$\mathbf{da}_o$ = the original area vector corresponding to the current area:

$$\mathbf{da} = \frac{\rho_o}{\rho}\left[\mathbf{F}^{-1}\right]^T \mathbf{da}_o$$

of **f** • d**v** for the three moving faces of the cube. (The velocities are differential because the joint at the origin is stationary and the element is infinitesimal.) The velocities themselves are d**v** = **L**d**x**, so the total power is the sum over the three (moving) faces:

$$\text{power} = \sigma \cdot \mathbf{L} \tag{4.67}$$

However, because σ is symmetric, only the symmetric part of **L** contributes to the inner product, so Eq. 4.67 reduces to

$$\text{power} = \sigma \cdot \mathbf{D} \tag{4.68}$$

where **D** is the symmetric part of **L**. Physically, it can easily be seen that the velocity of each plane caused by the shear-symmetric part of **L** (ω) lies in the plane and therefore does no work in conjunction with the force acting on the plane. For the alternate stress measures, it can be readily shown that the work-conjugate quantities are those shown in Table 4.1 on page 149. Any of the power-conjugate quantities can be viewed in incremental work form by multiplying the rate term by dt.

CHAPTER 4 - PROBLEMS

A. Proficiency Problems

1. Given: $\mathbf{F} \leftrightarrow \begin{bmatrix} 1 & 2 \\ 3 & 4 \end{bmatrix}$

   a. Find E_{ij} and ε_{ij}, the components of the large- and small-strain tensors, respectively.

   b. Using **E** and ε directly, find the new length of the vectors OA and AB shown here. Note that the original vectors are of unit length.

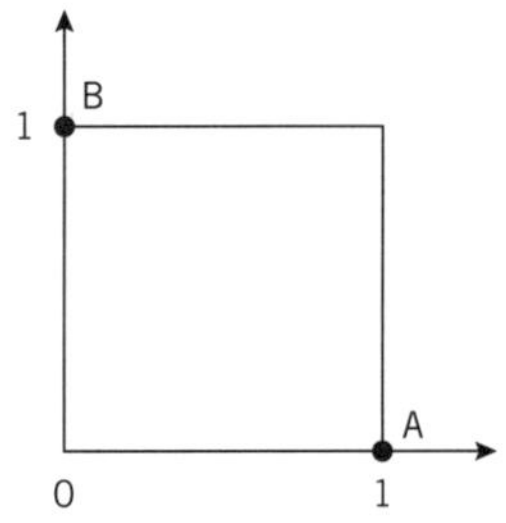

c. Why are the deformed lengths of O′A′ and O′B′ different when calculated using the two different measures of deformation?

2. Given this diagram for an assumed homogeneous deformation, write down the deformation gradient, **F**:

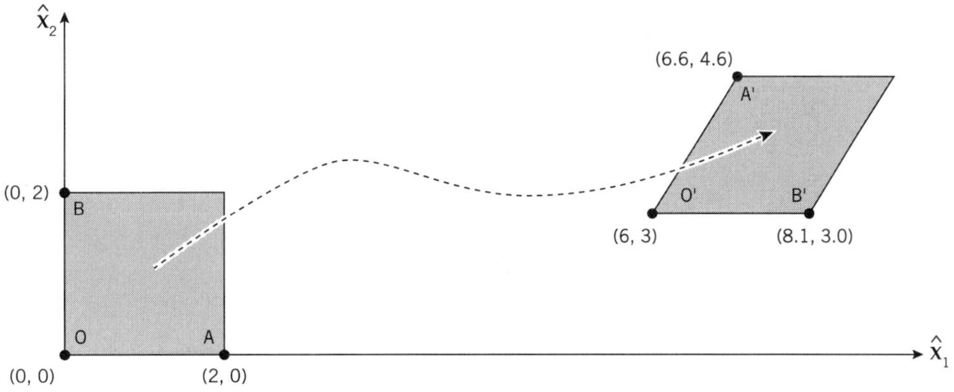

3. Given: $\mathbf{F} = \begin{bmatrix} 0.1 & 0.2 & 0.5 \\ 0.3 & 0.4 & 0.6 \\ 0.7 & 0.8 & 0.9 \end{bmatrix}$

Find: J, E, and ε.

4. In the following diagram, a point in a continuum (O) moves to a new point (O′) as shown.

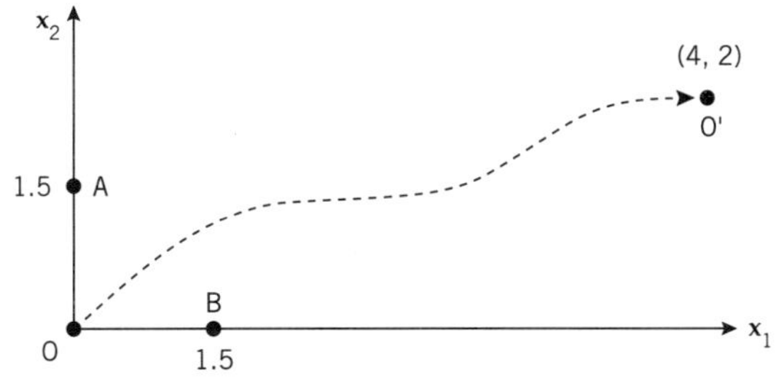

a. Find the new points A′ and B′, assuming homogeneous deformation for the following two cases:

$$\mathbf{F} = \begin{bmatrix} 1 & 2 \\ 3 & 4 \end{bmatrix} \qquad \mathbf{J} = \begin{bmatrix} 2 & 1 \\ 4 & 3 \end{bmatrix}$$

b. For each deformation, find **E,** the large strain tensor.

5. Imagine that a line segment OP is embedded in a material that is deformed to a new state. The line segment becomes O′P′ after deformation, as shown:

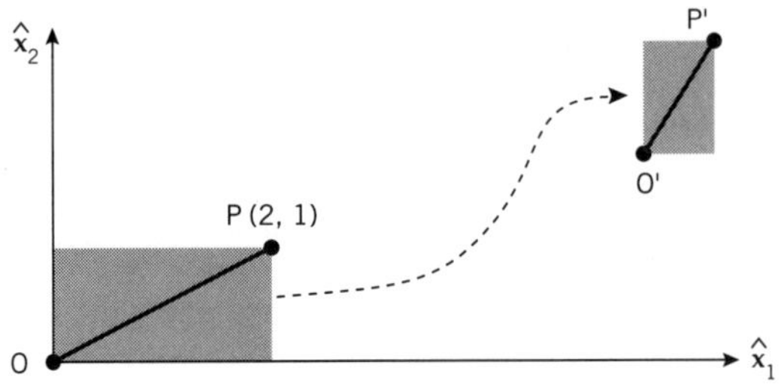

a. Find the vector components of $\overline{O'P'}$:

$$\mathbf{F} = \begin{bmatrix} 1 & 2 \\ 3 & 4 \end{bmatrix}$$

b. Find the length of O′P′ if

$$\mathbf{C} = \begin{bmatrix} 1 & 3 \\ 3 & 2 \end{bmatrix}$$

c. Find the components of $\overline{OP}$ if

$$\mathbf{J} = \begin{bmatrix} 0 & 2 \\ 3 & 3 \end{bmatrix}$$

6. A homogeneous deformation is imposed in the plane of the sheet. Two lines painted on the surface move as shown, with coordinates measured as shown:

a. Find **F**, the deformation gradient.

b. Starting with **F**, find **C**, **E**, **I**, and ε.

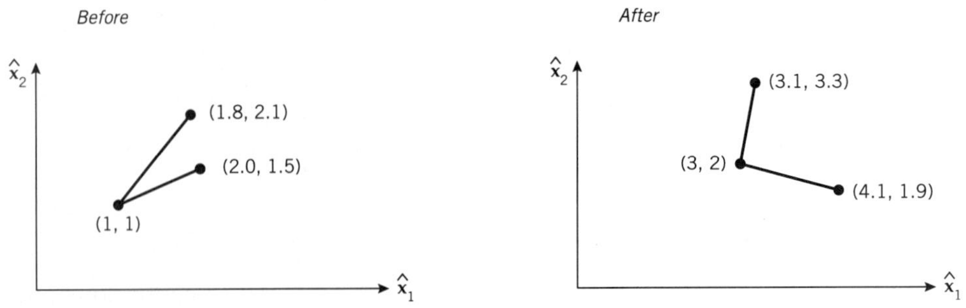

c. Find the principal strains and axes of **E**.

7. At time t, the position of a material particle initially at (X_1, X_2, X_3) is

$$x_1 = X_1 + a\,X_2$$

$$x_2 = X_2 + a\,X_1$$

$$x_3 = X_3$$

Obtain the **unit elongation** (i.e. change in length per unit initial length) of an element initially in the direction of $\hat{\mathbf{x}}_1 + \hat{\mathbf{x}}_2$.

8. Take fixed right-handed axes $\mathbf{x}_1$, $\mathbf{x}_2$, $\mathbf{x}_3$. Write down the deformation gradient matrix, $\frac{\partial X_i}{\partial x_j}$ for the deformation of a body from $\mathbf{x}$ to $\mathbf{X}$ for the following:

a. Right-handed rotation of 45° about $\hat{\mathbf{x}}_1$.

b. Left-handed rotation of 45° about $\hat{\mathbf{x}}_2$.

c. Stretch by a stretch ratio of 2 in the $\hat{\mathbf{x}}_3$ direction.

d. Stretch by a stretch ratio of $\frac{1}{2}$ in the $\hat{\mathbf{x}}_2$ direction.

e. Right-handed rotation of 90° about $\hat{\mathbf{x}}_3$.

Find the total deformation matrix for these motions carried out sequentially. Using this result, check the final volume ratio.

9. From the following mapping, find **C**, **U**, and **R**:

$$\begin{bmatrix} x_1 \\ x_2 \\ x_3 \end{bmatrix} = \begin{bmatrix} 2 & 0 & 0 \\ 0 & 3 & 4 \\ 0 & 4 & -3 \end{bmatrix} \begin{bmatrix} X_1 \\ X_2 \\ X_3 \end{bmatrix}$$

Check whether this is a permissible deformation in a continuous body.

10. Check the compatibility of the following strain components:

$$\boldsymbol{\varepsilon} \leftrightarrow \begin{bmatrix} x_1+x_2 & x_1+2x_2 & 0 \\ x_1+2x_2 & x_1+x_2 & 0 \\ 0 & 0 & x_3 \end{bmatrix} \qquad \boldsymbol{\varepsilon} \leftrightarrow \begin{bmatrix} x_1^2 & x_1^2+x_2^2 & x_2x_3 \\ x_2^2+x_3^2 & x_2^2 & 0 \\ x_3 & 0 & 0 \end{bmatrix}$$

B. Depth Problems

11. Consider the extension of an arbitrary small-line element AB. Start by examining how $(A'B')^2$ is related to $(AB)^2$ using the small extensional strain along that direction, e_n. Show that for small strains and displacements, rotations do not cause extension; that is, if extensions are zero, strains are zero.

12. In sheet forming, one often measures strains from a grid on the sheet surface. Then one can plot these strains as a function of the original length along an originally straight line:

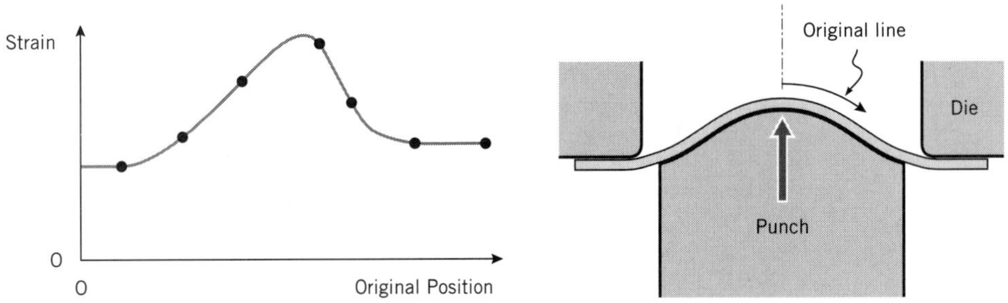

As shown in diagram (*b*) (for a simple forming operation), this originally straight line is curved and stretched. If the edges of the sheet do not move (stretch boundary conditions), develop a rule that the measured strain distribution must follow. Consider that the original sheet length l_o becomes l at some later time.

13. Consider a 1-inch square of material deformed in the following ways:

For each case:

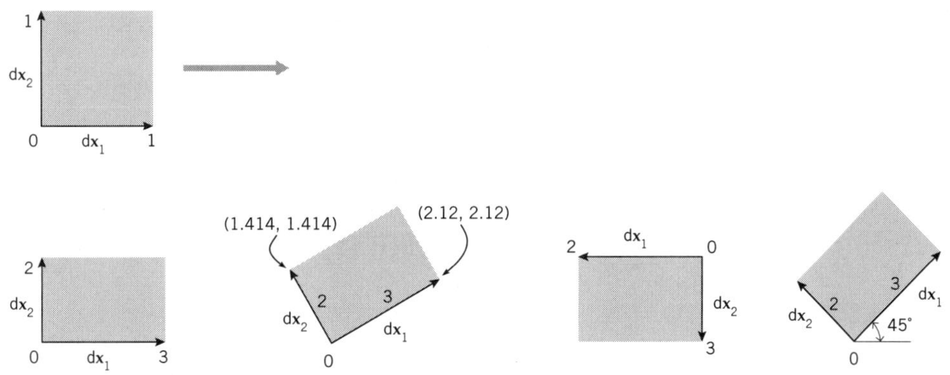

(a) (b) (c) (d)

a. Find **F**.

b. Find **C** and **E**.

c. Find the principal values and directions of **E**.

d. Find the material principal directions after deformation.

e. Which of these cases (a-d) are mechanically the same under isotropic conditions? Under general anisotropy?

14. Given that $[\mathbf{C}] = \begin{bmatrix} 1 & 0 \\ 0 & 2 \end{bmatrix}$, find **F** when

a. the principal material axes do not rotate, and

b. the principal material axes rotate by 30° counterclockwise. Express your answers in the original coordinate system.

15. Derive a set of compatibility equations corresponding to Eq. 4.52 for the three-dimensional case.

C. Computational Problems

16. Consider a triangle of material defined by three points: A, B, and C, with coordinates before and after an in-plane, homogeneous deformation as follows:

before: (A_1, A_2) (B_1, B_2) (C_1, C_2)

after: (a_1, a_2) (b_1, b_2) (c_1, c_2)

a. Find **F**, **E**, **U**, **R**, and **C** in terms of the original and final coordinates.

b. Find the principal stretch ratios and true principal strains.

c. Assuming that strains are small (but displacements and rigid rotations may be large), find simplified expressions for principal strains and small-strain components.

d. Assume that all displacements are small, and find simplified expressions for principal strains and small-strain components.

17. Repeat Problem 16 for the 3-D deformation where point $A(A_1, A_2, A_3)$ moves to $a(a_1, a_2, a_3)$, and so on.

CHAPTER 5

Standard Mechanical Principles[1]

Many alternate forms of equation can be derived to describe the condition of mechanical equilibrium, both for continuous bodies or for discrete assemblages of finite elements. In the latter case, it is possible to consider local forms for isolated elements, and global forms describing the body that has been modeled discretely.

Special mathematical tools are required for transforming one description of mechanical equilibrium to another. These are described briefly in this chapter and they are demonstrated by classically establishing the most important laws of deformable bodies. Integral forms are derived, and, finally, variational formulations are presented for the most standard classes of mechanical behavior: elasticity, plasticity, and viscoplasticity.

5.1 MATHEMATICAL THEOREMS

Throughout this chapter, frequent use will be made of the **Green theorem**, or one of its various forms. The most simple expression is analyzed, and the mathematical proof is given in a simple case, which can be generalized sufficiently to cover cases of practical interest for applications. As we are not concerned with pure mathematical issues, we shall make the assumption that the domains of integration (volume, surface, or curve) and the functions (scalar, vector, or tensor) are regular enough for all the integrals to be defined.[2] A theorem for the necessary and sufficient condition for an integral to be null is given.

[1] This chapter is more mathematical than the others. It develops principles essential for finite element analysis of forming operations, but does not contain material needed to understand the remaining content of this book and thus may be skipped at a first reading if desired.

[2] dV is the volume element and dS is the surface element.

The Green Theorem

Let Ω be a finite domain with boundary $\partial\Omega$, where f and g are two scalar functions defined on Ω. Then for any set of coordinates x_k ($k = 1, 2$, or 3) corresponding to a position vector **x**, we have

$$\int_\Omega f\frac{\partial g}{\partial x_k} dV = \int_{\partial\Omega} f\, g\, n_k\, dS - \int_\Omega \frac{\partial f}{\partial x_k} g\, dV \tag{5.1}$$

(Green's theorem)

when **n** is the unit outward normal to $\partial\Omega$ with components (n_1, n_2, n_3). Equation 5.1 is a generalization of the rule of integration by parts of simple integrals. The proof is relatively simple if we suppose that a straight line parallel to the $\hat{x}$ axis intersects the boundary $\partial\Omega$ of the domain Ω in, at most, two points (see Figure 5.1, which is drawn in three dimensions with $k = 3$).

Our hypothesis allows us to define a curve in the $\hat{x}_3$ plane and to decompose the boundary $\partial\Omega$ into an upper part $\partial\Omega_U$ and a lower part $\partial\Omega_L$. More precisely, C is defined so that for any point M inside C, the straight line containing M and parallel to Ox_3 intersects $\partial\Omega_U$ at the point M_U.

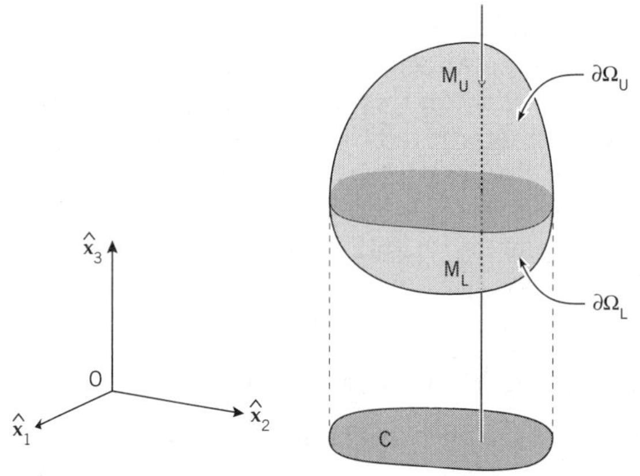

Figure 5.1 Proof of the Green theorem.

With these assumptions, the corresponding x_3 coordinate can be obtained (at least theoretically) by

$$x_3 = h_U(x_1, x_2)$$

and similarly for the lower boundary $\partial\Omega_L$:

$$x_3 = h_L(x_1, x_2)$$

With this notation, the left-hand side of Eq. 5.1 can be rewritten as follows:

$$\int_{\Omega} f \frac{\partial g}{\partial x_3} dV = \int_{C} dx_1\, dx_2 \int_{h_L(x_1,x_2)}^{h_U(x_1,x_2)} f \frac{\partial g}{\partial x_3} dx_3 \tag{5.2}$$

In Eq. 5.2, the one-dimensional integral can be transformed by integrating by parts with respect to x_3 according to

$$\int_{h_L}^{h_U} f \frac{\partial g}{\partial x_3} dx_3 = f_U\, g_U - f_L\, g_L - \int_{h_L}^{h_U} \frac{\partial f}{\partial x_3} g\, dx_3 \tag{5.3}$$

where we introduced the notation

$$f_U = f\left[x_1,\, x_2,\, h_U(x_1,\, x_2)\right]$$

(Similar notation is introduced for g_U, f_L and g_L).

Now Eq. 5.3 allows the transformation of Eq. 5.2 into

$$\int_{\Omega} f \frac{\partial g}{\partial x_3} dx_3 = \int_{C} (f_U\, g_U - f_L\, g_L)\, dx_1\, dx_2 - \int_{C} dx_1\, dx_2 \int_{h_L}^{h_U} \frac{\partial f}{\partial x_3} g\, dx_3 \tag{5.4}$$

We shall focus first on the first integral appearing in the right-hand side of Eq. 5.4. If the normal to $\partial\Omega$ is oriented outward, we can write on $\partial\Omega_U$,

$$dx_1\, dx_2 = \cos\gamma_U\, dS = n_{U3}\, dS \tag{5.5a}$$

which means that the projection of the elementary surface dS onto the $\hat{x}_3$ plane involves the cosine of the normal $\mathbf{n}_U$ with the $\hat{x}_3$ axis, as pictured in Figure 5.2.

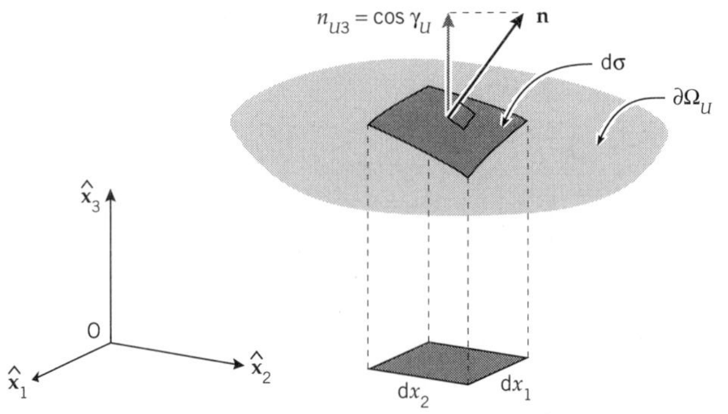

Figure 5.2 Projection of the elementary surface on the ox_1x_2 plane.

Similarly, the corresponding equality for the lower surface may be written

$$dx_1\, dx_2 = -\cos\gamma_L\, dS = -n_{L3}\, dS \tag{5.5b}$$

With the help of Eqs. 5.5a and 5.5b, we can write

$$\begin{aligned}\int_C (f_U\, g_U - f_L\, g_L)\, dx_1\, dx_2 &= \int_{\partial\Omega_U} f_U\, g_U\, n_{U3}\, dS - \int_{\partial\Omega_L} f_U\, g_L\, (-n_{L3})\, dS \\ &= \int_{\partial\Omega} f\, g\, n_3\, dS\end{aligned} \tag{5.6}$$

The second integral in the right-hand side of Eq. 5.4 is rewritten with the definition of 3-D integrals and gives

$$\int_C dx_1\, dx_2 \int_{h_L}^{h_U} \frac{\partial f}{\partial x_3} g\, dx_3 = \int_\Omega \frac{\partial f}{\partial x_3} g\, dV \tag{5.7}$$

Finally, if the right-hand side of Eq. 5.4 is transformed with the help of Eqs. 5.6 and 5.7, we see that it is equivalent to Eq 5.1, where k is chosen equal to 3.

To complete the proof, one has to remark that if a straight line, parallel to $\hat{x}_i$ intersects the boundary in more than two points, the domain can be decomposed into several subdomains for which this property holds. Then, in each subdomain, Eq. 5.1 can be written and one has only to note that for those parts of the boundary inside Ω, and therefore belonging to two adjacent subdomains, the surface integrals vanish as they correspond to opposite normals.

The Ostrogradski (Divergence) Theorem

The theorem may be stated as follows: for any vector field **w** defined in W with boundary $\partial\Omega$,

$$\int_{\Omega} \operatorname{div}(\omega)\, dV = \int_{\partial\Omega} \omega \cdot v\, dS \tag{5.8}$$

The proof is easily obtained by using the Green theorem. If f = 1 and g = w_k are put into Eq. 5.1, we get

$$\int_{\Omega} \frac{\partial w_k}{\partial x_k}\, dV = \int_{\partial\Omega} w_k n_k\, d\, S \tag{5.9}$$

In Equation 5.9, no summation is made on the k index, but if we sum over it, we obtain

$$\int_{\Omega} \sum_{k} \frac{\partial w_k}{\partial x_k}\, dV = \int_{\partial\Omega} \sum_{k} w_k\, n_k\, d\, S. \tag{5.10}$$

(divergence theorem)

Equation 5.10 is the component form of Eq. 5.8, presented with the vector notation. This is also called the **divergence theorem**, and it is often used when the material flow is described by a specified velocity field **v**.

Time Derivative of an Integral with a Variable Domain

We suppose that the domain is time dependent, denoted by Ω_t, and moves with a velocity field **v**.[3] A time-dependent scalar function f is considered; that is, $f = f(\mathbf{x}_0, t)$.[4] With these hypotheses it will be shown that

$$\frac{d}{dt}\left(\int_{\Omega_t} f\, dV\right) = \int_{\Omega_t} \frac{\partial f}{\partial t}\, dV + \int_{\partial\Omega_t} f\mathbf{v}\cdot\mathbf{n}\, dS \tag{5.11}$$

---

[3] The velocity vector need not be defined in the whole domain Ω here; it can be defined only on the boundary $\partial\Omega$.

[4] The notation $\mathbf{x}_0$ on Ω_t (and $\mathbf{x}_1$ on $\Omega_{t+\Delta t}$) is chosen in order to indicate clearly that we are not concerned with a mechanical problem here, where the coordinates refer to material points which move as the body is deformed. In other words, the theorem can be established if the velocity field is defined only on the boundary.

To give an idea of the proof, the left-hand side of Eq. 5.11 is written according to the definition of the time derivative:

$$\lim_{\Delta t \to 0} \frac{1}{\Delta t}\left[\int_{\Omega_{t+\Delta t}} f(\mathbf{x}_1, t+\Delta t)\, dV - \int_{\Omega_t} f(\mathbf{x}_0, t)\, dV\right] \tag{5.11a}$$

or, in a more convenient but obviously equivalent form,

$$\lim_{\Delta t \to 0} \int_{\Omega_{t+\Delta t}} \frac{f(\mathbf{x}_1, t+\Delta t) - f(\mathbf{x}_1, t)}{\Delta t}\, dV$$

$$\lim_{\Delta t \to 0} \frac{1}{\Delta t}\left[\int_{\Omega_{t+\Delta t}} f(\mathbf{x}_1, t)\, dV - \int_{\Omega_t} f(\mathbf{x}_0, t)\, dV\right] \tag{5.12}$$

When Δt tends to zero, the first integral in Eq. 5.12 tends to the limit

$$\int_{\Omega_t} \frac{\partial f(\mathbf{x}_0, t)}{\partial t}\, dv \tag{5.13}$$

The second term in Eq. 5.12 can be transformed if one recalls that, up to the first order in Δt, $\partial\Omega_{t+\Delta t}$ is obtained from $\partial\Omega_t$ with the velocity $\mathbf{v}$, according to

$$\mathbf{x}_1 = \mathbf{x}_0 + \Delta t\, \mathbf{v} \tag{5.14a}$$

If Δt is small enough, the displacement can be viewed as oriented along the normal to Ω_t. If $\mathbf{x}_0$ is on $\partial\Omega_t$, the corresponding point on $\partial\Omega_{t+\Delta t}$ will be $\mathbf{x}'$, with

$$\mathbf{x}' = \mathbf{x} + (\mathbf{v} \cdot \mathbf{n})\, \Delta t\, \mathbf{n} \tag{5.14b}$$

The situation is represented in Figure 5.3, where we introduce

$$\Omega' = \Omega_t \cap \Omega_{t+\Delta t} \qquad \Omega_{t+\Delta t} = \Omega' + \Delta\Omega_+ \qquad \Omega_t = \Omega' = \Delta\Omega_- \tag{5.15}$$

In other words, W_+ is the domain between ∂W_t and ∂W_{t+Dt} when $\mathbf{v} \cdot \mathbf{n} > 0$; and W_- is the domain between ∂W_t and ∂W_{t+Dt} for $\mathbf{v} \cdot \mathbf{n} < 0$. With this notation, the second term of Eq. 5.12 is also equal to

$$\frac{1}{\Delta t}\left[\int_{\Omega_+} f(\mathbf{x}_0,t)\,dV - \int_{\Omega_-} f(\mathbf{x}_1,t)\,dV\right] \tag{5.16}$$

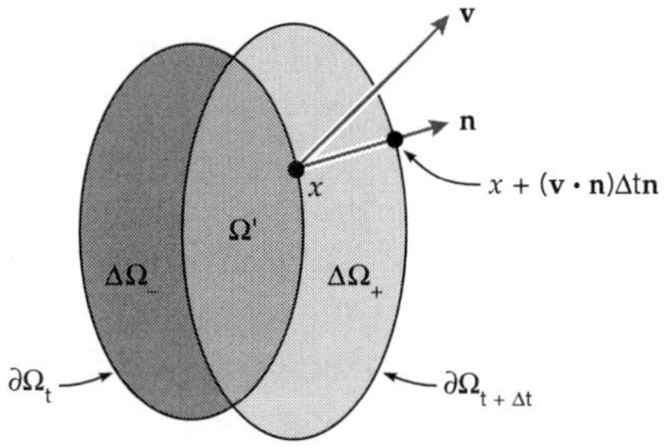

Figure 5.3 Transformation of the difference of volume integrals (2-D representation).

Now, when Δt is small, $\Delta\Omega_\pm$ can be described by

$$\mathbf{x}' = \mathbf{x}_0 + \xi\,(\pm\,\mathbf{n}) \quad \text{with} \quad \mathbf{x}_0 \in \partial\Omega_{t\pm} \quad \text{and} \quad 0 \le \xi \le \pm\,(\mathbf{v}\cdot\mathbf{n})\,\Delta t$$

where $\partial\Omega_{t+}$ corresponds to the boundary $\partial\Omega_t$ with $\mathbf{v}\cdot\mathbf{n} \ge 0$, and $\partial\Omega_{t-}$ to the boundary $\partial\Omega_t$ with $\mathbf{v}\cdot\mathbf{n} \le 0$. This notation allow us to rewrite Eq. 5.16 as

$$\frac{1}{\Delta t}\left[\int_{\partial\Omega_{t+}} dS \int_0^{\mathbf{v}\cdot\mathbf{n}\Delta t} f(\mathbf{x}_0 + \xi\mathbf{n}, t)\,d\xi - \int_{\partial\Omega_{t-}} dS \int_0^{-\mathbf{v}\cdot\mathbf{n}\Delta t} f(\mathbf{x}_0 - \xi\mathbf{n}, t)\,d\xi\right] \tag{5.17}$$

If Δt tends to zero, we have

$$\lim_{\Delta t \to 0} \frac{1}{\Delta t}\int_0^{\pm\mathbf{v}\cdot\mathbf{n}\Delta t} f(\mathbf{x}_0 \pm \xi\mathbf{n}, t)\,d\xi = \pm\mathbf{v}\cdot\mathbf{n}\,f(\mathbf{x}_0,\,t) \tag{5.18}$$

so that Eq. 5.17 gives

$$\int_{\partial\Omega_{t+}} \mathbf{v}\cdot\mathbf{n}\,f(\mathbf{x},t)\;dS - \int_{\partial\Omega_{t-}} (-\mathbf{v}\cdot\mathbf{n})\,f(\mathbf{x},t)\,dS = \int_{\partial\Omega} \mathbf{v}\cdot\mathbf{n}\,f(\mathbf{x},t)\,dS \tag{5.19}$$

The proof is completed when the limit values given by Eq. 5.14 and Eq. 5.19 are substituted for the first and second terms respectively in Eq. 5.13.

Theorem: A Necessary and Sufficient Condition for an Integral to be Null

If a continuous function f is defined in a domain Ω, and if for all possible subdomains w of Ω we have

$$\int_{\omega} f\, dV = 0 \tag{5.20}$$

then the function f must be equal to zero on the whole domain ω.

The proof makes use of the continuity of f. For example, if for a given geometrical point, with coordinate vector $\mathbf{x}'$, we have $f(\mathbf{x}') > 0$, then there exists a subdomain ω' of Ω, containing $\mathbf{x}'$, and verifying the condition[5]

$$\forall\, \mathbf{x} \in \omega' \qquad f(\mathbf{x}) \geq \frac{1}{2}\, f(\mathbf{x}') \tag{5.21}$$

By integration of both sides of the previous inequality (Eq. 5.21), we obtain

$$\int_{\omega} f(\mathbf{x})\, dV \geq \int_{\omega'} f(\mathbf{x})\, dV \geq \int_{\omega'} \frac{1}{2} f(\mathbf{x}')\, dV = \frac{1}{2} f(\mathbf{x}') \int_{\omega'} dV > 0 \tag{5.22}$$

It is clear that Eq. 5.22 contradicts the initial hypothesis given by Eq. 5.20 so that the function f cannot be different from zero on any part of the domain Ω. Finally, it is obvious that if $f(\mathbf{x}) = 0$ in the domain Ω, then the integral is null.

5.2 THE MATERIAL DERIVATIVE

The **material derivative** is defined as the "total" time derivative when the particle is followed in the material flow. It can be applied to a function, a vector, or a tensor.

Material Derivative of a Scalar Function

If the scalar function F depends on the initial coordinates $\mathbf{X}$ and the time t - that is, when a Lagrangian description is used (see Section 4.2)-the material derivative is defined by

$$\frac{dF}{dt}(\mathbf{X}, t) = \frac{\partial F}{\partial t}(\mathbf{X}, t) \tag{5.23}$$

---

[5] This property of continuous functions will be used without proof. Standard mathematical texts may be consulted for proofs.

When a function f is defined in terms of the present value of the coordinate **x**, which corresponds to the Euler description (see also Section 4.2), and the time t, it is written f(**x**, t). In order to define properly the material derivative, the dependence of the material coordinate vector in terms of the initial coordinates **X** must be recalled by the continuum mapping function **x**(**X**, t). Thus the definition of the material derivative is

$$\frac{df}{dt}(\mathbf{x},t) = \lim_{\Delta t \to 0} \frac{1}{\Delta t}\{f[\mathbf{x}(\mathbf{X},t+\Delta t),t+\Delta t] - f[\mathbf{x}(\mathbf{X},t),t]\} \tag{5.24}$$

Equation 5.24 can be transformed according to

$$\begin{aligned}\frac{df}{dt}(\mathbf{x},t) = {} & \lim_{\Delta t \to 0}\left(\frac{1}{\Delta t}\{f[\mathbf{x}(\mathbf{X},t+\Delta t),t+\Delta t] - f[\mathbf{x}(\mathbf{X},t),t+\Delta t]\}\right) \\ & + \lim_{\Delta t \to 0}\left(\frac{1}{\Delta t}\{f[\mathbf{x}(\mathbf{X},t),t+\Delta t] - f[\mathbf{x}(\mathbf{X},t),t]\}\right)\end{aligned} \tag{5.25}$$

which gives the result

$$\frac{df}{dt}(\mathbf{x},t) = \frac{\partial f}{\partial t}(\mathbf{x},t) + \sum_i \frac{\partial f}{\partial x_i}(\mathbf{x},t)\frac{\partial x_i(\mathbf{X},t)}{\partial t} \tag{5.26}$$

Finally, if the definition of the velocity of the material point with coordinate vector **x** is introduced into Eq. 5.26, we obtain the desired result:

$$\frac{df}{dt}(\mathbf{x},t) = \frac{\partial f}{\partial t}(\mathbf{x},t) + \sum_i \frac{\partial f}{\partial x_i}(\mathbf{x},t)\, v_i(\mathbf{x},t) \tag{5.27}$$

This is often denoted by the more compact form:

$$\frac{df}{dt} = \frac{\partial f}{\partial t} + \mathbf{grad}(f)\cdot\mathbf{v} \tag{5.28}$$

Material Derivative of a Vector or a Tensor Field

The same approach can be carried out for each component separately. The case of a vector $\mathbf{U}(\mathbf{X}, t)$ as a function of the Lagrange variables will give

$$\frac{d\mathbf{U}}{dt}(\mathbf{X}, t) = \frac{\partial \mathbf{U}}{\partial t}(\mathbf{X}, t) \tag{5.29}$$

with components

$$\frac{dU_i}{dt}(\mathbf{X}, t) = \frac{\partial U_i}{\partial t}(\mathbf{X}, t) \tag{5.30}$$

With the Euler description, the vector field is denoted by $\mathbf{u}(\mathbf{x}, t)$, and its material derivative is

$$\frac{d\mathbf{u}}{dt} = \frac{\partial \mathbf{u}}{\partial t} + \mathbf{grad}(\mathbf{u}) \cdot \mathbf{v} \tag{5.31}$$

where the components are given by

$$\frac{du_i}{dt}(\mathbf{x}, t) = \frac{\partial u_i}{\partial t}(\mathbf{x}, t) + \sum_j \frac{\partial u_i}{\partial x_j}(\mathbf{x}, t)\, v_j(\mathbf{x}, t) \tag{5.32}$$

An important material derivative, the acceleration, is obtained from the velocity field:

$$\mathbf{a} = \frac{d\mathbf{v}}{dt} = \frac{\partial \mathbf{v}}{\partial t} + \mathbf{grad}(\mathbf{v}) \cdot \mathbf{v} \tag{5.33}$$

Note that the gradient of a vector is an operator with two indices. Three indices are necessary for the gradient of a tensor, the analogous of Eq. 5.32 being for the tensor σ:[6]

[6] Note that the material derivative of a tensor is not generally frame invariant—that is, if the basis function for the spatial coordinates is changed, the new material derivative of the *same tensor* σ will be different.

$$\frac{d\sigma_{ij}}{dt}(\mathbf{x}, t) = \frac{\partial \sigma_{ij}}{\partial t}(\mathbf{x}, t) + \sum_k \frac{\partial \sigma_{ij}}{\partial x_k}(\mathbf{x}, t)\, v_k(\mathbf{x}, t) \tag{5.34a}$$

which can also be written

$$\frac{d\boldsymbol{\sigma}}{dt} = \frac{\partial \boldsymbol{\sigma}}{\partial t} + \mathbf{grad}(\boldsymbol{\sigma}) \cdot \mathbf{v} \tag{5.34b}$$

Derivative of an Integral

This topic was already addressed in the previous section (see Eq. 5.11). Another useful form is obtained by first applying the Green theorem (Eq. 5.1) with f and g = $\mathbf{v}_i$, and summing the contributions for i = 1, 2, 3, which gives

$$\int_{\Omega_t} f \sum_i \frac{\partial v_i}{\partial x_i}\, dV = \int_{\partial\Omega_t} f \sum_i v_i\, n_i\, dS - \int_{\Omega_t} \sum_i \frac{\partial f}{\partial x_i} v_i\, dV \tag{5.35}$$

The surface integral is eliminated between Eqs. 5.11 and 5.35, leaving

$$\frac{d}{dt}\int_{\Omega_t} f\, dV = \int_{\Omega_t} \left[\frac{\partial f}{\partial t} + \sum_i \frac{\partial f}{\partial x_i} v_i + f \sum_i \frac{\partial v_i}{\partial x_i} \right] dV \tag{5.36}$$

Introducing the material derivative of f, given by Eq. 5.27, and the divergence of the velocity field,

$$\mathrm{div}(\mathbf{v}) = \sum_i \frac{\partial v_i}{\partial x_i}$$

we get

$$\frac{d}{dt}\int_{\Omega_t} f\, dV = \int_{\Omega_t} \left[\frac{df}{dt} + f\, \mathrm{div}(\mathbf{v}) \right] dV \tag{5.37}$$

which is also often used in the form:

$$\frac{d}{dt}\int_{\Omega_t} f\, dV = \int_{\Omega_t} \left[\frac{\partial f}{\partial t} + \mathrm{div}(f\, \mathbf{v}) \right] dV \tag{5.38}$$

5.3 SUMMARY OF IMPORTANT MECHANICAL EQUATIONS

In the following section, the general physical principles of conservation will be applied to any (small) domain, in order to obtain the local laws for conservation in the mechanical field theory.

Mass Conservation: The Equation of Continuity

We assume that the mass density r is a function of position **x** and time t. The total mass (M) of the system in the domain Ω_t is evaluated by

$$M = \int_{\Omega_t} \rho\,(\mathbf{x}, t)\, dV \tag{5.39}$$

If we consider any subdomain w_t embedded in Ω_t, the corresponding mass m_ω should remain constant during all deformation of the material. This mass conservation principle is written:

$$\frac{dm_\omega}{dt} = \frac{d}{dt}\int_{\omega_t} \rho(\mathbf{x}, t)\, dV = 0 \tag{5.40}$$

According to Eq. 5.37, Eq. 5.40 can be rewritten as follows:

$$\int_{\omega_t} \left(\frac{d\rho}{dt} + \rho\, \mathrm{div}(\mathbf{v})\right) dV = 0 \tag{5.41}$$

Because the equality stated by Eq. 5.41 must hold for any subdomain w_t of Ω_t, the general mathematical theorem of Section 5.1 states that the integrand must be equal to zero everywhere in the domain Ω; that is,

$$\frac{d\rho}{dt} + \rho\, \mathrm{div}(\mathbf{v}) = 0 \tag{5.42}$$

which is equivalent to

$$\frac{\partial\rho}{\partial t} + \mathrm{div}(\rho\mathbf{v}) = 0 \tag{5.43}$$

This is the general form of the mass conservation principle, also called the **continuity equation**. If the material is incompressible, then its density is a constant and Eq. 5.42 reduces to

$$\mathrm{div}(\mathbf{v}) = 0 \tag{5.44}$$

Exercise 5.1 **Provide an elementary proof of Eq. 5.42 by calculating the density evolution of a small cube during a time increment, and the material flux across the faces of the cube.**

The small cube is represented in the following figure:

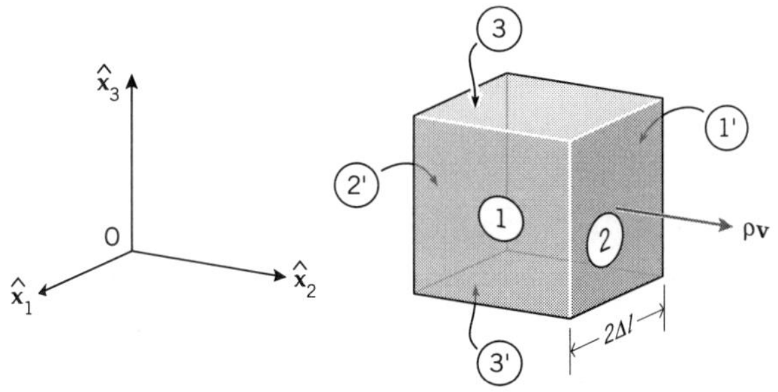

During the time increment Δt, the density increase $\Delta\rho$ comes from the sum of material fluxes across the six faces of the cube with side $2\Delta l$, and with center of coordinates (x_1, x_2, x_3).

The mass flux through face number 1 with outward normal is

$$\Delta m_1 = \rho(x_1 + \Delta l,\ x_2, x_3)\ v_1(x_1 + \Delta l,\ x_2, x_3)\ \Delta l^2\ \Delta t \tag{5.1-1}$$

Similarly, face $1'$, with opposite normal, gives

$$\Delta m_1{}' = -\rho(x_1 - \Delta l,\ x_2, x_3)\ v_1(x_1 - \Delta l,\ x_2, x_3)\ \Delta l^2\ \Delta t \tag{5.1-2}$$

and we observe that by combining Eqs. 5.1-1 and 5.1-2, we get

$$\Delta m_1 + \Delta m_1{}' = \frac{\partial}{\partial x_1}\ (\rho v_1)\ \Delta l^3\ \Delta t \tag{5.1-3}$$

Two other expressions similar to Eq. 5.1-3 can be written by considering the fluxes across faces 2 and $2'$, and across faces 3 and $3'$, such that the total mass increment is

$$\Delta m = \Delta m_1 + \Delta m_1{}' + \Delta m_2 + \Delta m_2{}' + \Delta m_3 + \Delta m_3{}' = \left(\sum_i \frac{\partial}{\partial x_i}(\rho v_i) \right) \Delta l^3\ \Delta t \tag{5.1-4}$$

We remark that the density change is linked to the mass transfer by

$$\Delta\rho\ \Delta l^3 + \Delta m = 0 \tag{5.1-5}$$

The proof is completed by substituting Δm between Eqs. 5.1-4 and 5.1-5, dividing the resulting equation by $\Delta t\ \Delta l^3$, and letting Δt tend to zero, which obtains

$$\frac{\partial \rho}{\partial t} + \mathrm{div}(\rho \mathbf{v}) = \frac{\partial \rho}{\partial t} + \mathrm{grad}(\rho) \cdot \mathbf{v} + \rho\, \mathrm{div}(\mathbf{v}) = 0 \tag{5.1-6}$$

which is clearly equal to Eq. 5.42.

The following exercise allows us to give a useful expression for the time derivative of an integral containing the material density ρ:

$$\frac{d}{dt} \int_{\omega_t} \rho\, f\, dV = \int_{\omega_t} \rho\, \frac{df}{dt}\, dV \tag{5.45}$$

Exercise 5.2 **Show that for any function f, Eq. 5.45 is verified.**

From Eq. 5.37 we have

$$\frac{d}{dt} \int_{\omega_t} \rho\, f\, dV = \int_{\omega_t} \left[\frac{d(\rho\, f)}{dt} + \rho\, f\, \mathrm{div}(\mathbf{v}) \right] dV \tag{5.2-1}$$

but it is easy from the definition of the material derivative to convince oneself that

$$\frac{d(\rho\, f)}{dt} = \frac{d\rho}{dt} f + \rho \frac{df}{dt} \tag{5.2-2}$$

The substitution of Eq. 5.2-2 into Eq. 5.2-1 gives

$$\frac{d}{dt} \int_{\omega_t} \rho\, f\, dV = \int_{\omega_t} \left[\frac{d\rho}{dt} f + \rho \frac{df}{dt} + \rho\, f\, \mathrm{div}(\mathbf{v}) \right] dV \tag{5.2-3}$$

Taking into account Eq. 5.42 allows the transformation of Eq. 5.2-3 into Eq. 5.45.

Linear Momentum Conservation: The Equation of Motion

At time t, to any subdomain w_t, by definition we can associate the linear momentum:

$$\int_{\omega_t} \rho\, \mathbf{v}\, dV \tag{5.46}$$

The fundamental law of mechanics states that the time derivative of the linear momentum is equal to the sum of the applied forces:

$$\frac{d}{dt}\int_{\omega_t} \rho\, \mathbf{v}\, dV = \int_{\partial\omega_t} \boldsymbol{\sigma}\cdot\mathbf{n}\, dS + \int_{\omega_t} \rho\, \mathbf{g}\, dV \tag{5.47}$$

where the contact forces result from the stress tensor σ, and the volumic forces from the vector field **g** (generally the gravity field). The left-hand side of Eq. 5.47 is transformed with the help of Eq. 5.45: if we put $f = v_i$, we get

$$\frac{d}{dt}\int_{\omega_t} \rho\, v_i\, dV = \int_{\omega_t} \rho\, \frac{dv_i}{dt}\, dV \tag{5.48}$$

The acceleration $\mathbf{a} = d\mathbf{v}/dt$ is introduced in Eq. 5.48, and the fundamental law of mechanics (Eq. 5.47) is now written

$$\int_{\omega_t} \rho\, \mathbf{a}\, dV = \int_{\partial\omega_t} \boldsymbol{\sigma}\cdot\mathbf{n}\, dS + \int_{\omega_t} \rho\, \mathbf{g}\, dV \tag{5.49}$$

A further transformation can be performed on the first integral of the right hand side of Eq. 5.49. The component number i is considered, the Ostrogradski theorem (Eq. 5.11) is applied with $k = j$ and $w_j = \sigma_{ij}$, so that we obtain

$$\int_{\partial\omega_t} \sum_j \sigma_{ij} n_j\, dS = \int_{\omega_t} \sum_j \frac{\partial \sigma_{ij}}{\partial x_j}\, dV \tag{5.50}$$

Finally, Eq. 5.49 is modified with the help of Eq. 5.50, leading to

$$\int_{\omega_t} \rho\, \mathbf{a}\, dV = \int_{\omega_t} \mathrm{div}(\boldsymbol{\sigma})\, dV + \int_{\omega_t} \rho\, \mathbf{g}\, dV \tag{5.51}$$

which must be true for any subdomain ω_t and which gives the **equation of motion**:

$$\rho\, \mathbf{a} = \mathrm{div}(\boldsymbol{\sigma}) + \rho\, \mathbf{g} \tag{5.52a}$$

or the component form,

$$\rho\, a_i = \sum_j \frac{\partial \sigma_{ij}}{\partial x_j} + \rho\, g_i \tag{5.52b}$$

An important particular case deserves special attention: equilibrium problems where the acceleration is equal to zero, or quasi-static problems where the acceleration is considered as negligible.[7] In this case, the equation of motion reduces to the **equilibrium equation**:

$$\mathrm{div}(\boldsymbol{\sigma}) + \rho \mathbf{g} = 0 \tag{5.52c}$$

Conservation of Angular Momentum and Symmetry of the Stress Tensor

When there is no internal torque, the second equation of motion, involving the equilibrium of momentum, will be written. Again for any subdomain ω_t,

$$\frac{d}{dt}\int_{\omega_t} \mathbf{x} \times \rho\, \mathbf{v}\, dV = \int_{\partial\omega_t} \mathbf{x} \times \boldsymbol{\sigma} \cdot \mathbf{n}\, dS + \int_{\omega_t} \mathbf{x} \times \rho\, \mathbf{g}\, dV \tag{5.53}$$

This simply states that the time rate of change of angular momentum is equal to the sum of moments. The right-hand side of equation of Eq. 5.53 is transformed with the rule for calculating the material derivative of an expression with the density (Eq. 5.45), and using the linearity of the cross product:

$$\frac{d}{dt}\int_{\omega_t} \rho\, \mathbf{x} \times \mathbf{v}\, dV = \int_{\omega_t} \rho \frac{d\mathbf{x}}{dt} \times \mathbf{v}\, dV + \int_{\omega_t} \rho\, \mathbf{x} \times \frac{d\mathbf{v}}{dt}\, dV \tag{5.54}$$

In the first term of the right-hand side we note that

$$\frac{d\mathbf{x}}{dt} \times \mathbf{v} = \mathbf{v} \times \mathbf{v} = 0$$

---

[7] This is true in most forming operations, at least as a first approximation. Most often for solid metals, in cold or hot forming processes, the gravity forces can also be neglected.

which allows us to rewrite 5.53 as

$$\int_{\omega_t} \mathbf{x} \times \rho\, \mathbf{a}\, dV = \int_{\partial\omega_t} \mathbf{x} \times \boldsymbol{\sigma} \cdot \mathbf{n}\, dS + \int_{\omega_t} \mathbf{x} \times \rho\, \mathbf{g}\, dV \tag{5.55}$$

Now we compute the cross product of **x** with each term of Eq. 5.52, and integrate over the domain ω_t, giving

$$\int_{\omega_t} \mathbf{x} \times \rho\, \mathbf{a}\, dV = \int_{\omega_t} \mathbf{x} \times \mathrm{div}(\boldsymbol{\sigma})\, dV + \int_{\omega_t} \mathbf{x} \times \rho\, \mathbf{g}\, dV \tag{5.56}$$

Comparison of Eqs. 5.55 and 5.56 allows us to deduce the equality:

$$\int_{\partial\omega_t} \mathbf{x} \times \boldsymbol{\sigma} \cdot \mathbf{n}\, dS = \int_{\omega_t} \mathbf{x} \times \mathrm{div}(\boldsymbol{\sigma})\, dV \tag{5.57}$$

Transformation of the left-hand side of Eq. 5.57 can be made more easily using the component form presented in Chapter 2:

$$(\mathbf{x} \times \boldsymbol{\sigma} \cdot \mathbf{n})_i = \sum_{jkl} \varepsilon_{ijk}\, x_j\, \sigma_{kl}\, n_l \tag{5.58}$$

so that the component number i of Eq. 5.57 can be written with the help of the Ostrogradski Divergence theorem:

$$\int_{\partial\omega_t} \sum_{jkl} \varepsilon_{ijk}\, x_j\, \sigma_{kl}\, n_l\, dV = \int_{\omega_t} \sum_{jkl} \frac{\partial\left(\varepsilon_{ijk}\, x_j\, \sigma_{kl}\right)}{\partial x_l}\, dV \tag{5.59}$$

The right-hand side integrand of Eq. 5.59 can be written successively:

$$\sum_{jkl}\left(\varepsilon_{ijk}\, x_j \frac{\partial \sigma_{kl}}{\partial x_l} + \varepsilon_{ijk}\, \delta_{lj}\, \sigma_{kl}\right) = \sum_{jkl} \varepsilon_{ijk}\, x_j \frac{\partial \sigma_{kl}}{\partial x_l} + \sum_{jk} \varepsilon_{ijk}\, \sigma_{jk} \tag{5.60}$$

so that we can deduce from Eqs. 5.59 and 5.60:

$$\int_{\partial\omega_t} \sum_{jkl} \varepsilon_{ijk}\, x_j\, \sigma_{kl}\, n_l\, dV = \int_{\omega_t} (\mathbf{x} \times \mathrm{div}(\boldsymbol{\sigma}))_i\, dV + \int_{\omega_t} \sum_{jk} \varepsilon_{ijk}\, \sigma_{jk}\, dV \tag{5.61}$$

where $(a)_i$ is the component number i of the a vector. The conclusion comes from the comparison of Eqs. 5.57 and 5.61:

$$\int_{\omega_t} \sum_{jk} \varepsilon_{ijk}\, \sigma_{jk}\, dV = 0 \tag{5.62}$$

As Eq. 5.62 holds for any subdomain ω_t, a local relation holds in the whole domain Ω_t, which corresponds to the integrand. The three components are:

$$\begin{aligned} \sigma_{23} - \sigma_{32} &= 0 \\ \sigma_{31} - \sigma_{13} &= 0 \\ \sigma_{12} - \sigma_{21} &= 0 \end{aligned} \tag{5.63}$$

which is only the translation of the symmetry of the Cauchy stress tensor σ.

5.4 THE VIRTUAL WORK PRINCIPLE

Definition of the Mechanical Problem

We shall consider a mechanical problem in a domain Ω, with prescribed boundary conditions on $\partial\Omega$. The problem is **well-posed** if it satisfies the following constraints on the boundary conditions. For any point lying on the boundary $\partial\Omega$, and for any component,[8] one and only one condition is enforced, either on the displacement **u** (or **v**, if the problem is formulated in terms of the velocity field), or the stress vector **T**. This is expressed by the equation

$$\text{on } \partial\Omega \quad u_i = u_i^d \qquad \text{or} \qquad T_i = T_i^d \quad i = 1, 2, 3 \tag{5.64}$$

(where the exclusive "or " is used here).

---

[8] A local orthonormal reference system can be defined for each point of the boundary. This definition can be the most convenient one with one axis normal to the boundary, when contact conditions are to be imposed.

A **virtual displacement field, δu,** is defined in the domain Ω. This displacement is said to be **statically admissible** if for any point belonging to the boundary, and for any component, it is null when the velocity component is prescribed:

$$\text{on } \partial\Omega \quad \delta u_i = 0 \quad \text{if} \quad u_i = u_i^d \quad i = 1, 2, 3 \tag{5.65}$$

To the virtual displacement, a virtual strain tensor δε can be associated in the usual way:

$$\delta\varepsilon_{ij} = \frac{1}{2}\left(\frac{\partial \delta u_i}{\partial x_j} + \frac{\partial \delta u_j}{\partial x_i}\right) \tag{5.66}$$

Now a stress field σ is considered, which must be only **statically admissible**; that is, it does not necessarily correspond to the solution of our mechanical problem (which in fact is not completely defined, as the material mechanical behavior is not prescribed yet).σ is statically admissible if

- it verifies the equilibrium equation (Eq. 5.52c) in the domain
- it satisfies the stress vector conditions on the boundary:

$$\text{on } \partial\Omega \quad \sum_j \sigma_{ij} n_j = T_i^d \quad \text{if} \quad T_i = T_i^d \quad i = 1, 2, 3 \tag{5.67}$$

The Virtual Work Principle

With these hypotheses, the virtual work principle states that the virtual work of the external forces is equal to the virtual work of the internal forces:

$$\int_{\partial\Omega} \boldsymbol{\sigma} \cdot \mathbf{n} \cdot \delta\mathbf{u}\, dS + \int_{\Omega} \rho\, \mathbf{g} \cdot \delta\mathbf{u}\, dV = \int_{\Omega} \boldsymbol{\sigma} \cdot \delta\boldsymbol{\varepsilon}\, dV \tag{5.68}$$

To establish the proof of this statement, Eq. 5.52c is multiplied by $\delta\mathbf{u}$ and integrated over the whole domain, yielding

$$\int_{\Omega} \text{div}(\boldsymbol{\sigma}) \cdot \delta\mathbf{u}\, dV + \int_{\Omega} \rho\, \mathbf{g} \cdot \delta\mathbf{u}\, dV = 0 \tag{5.69}$$

The first integral in Eq. 5.69 is transformed using the Green theorem for each component. We put into Eq. 5.1 the following substitutions: $f = \delta u_i$, $g = \sigma_{ij}$, and $k = j$, and we obtain

$$\int_{\Omega} \delta u_i \frac{\partial \sigma_{ij}}{\partial x_j} \, dV = \int_{\partial\Omega} \delta u_i \, \sigma_{ij} \, n_j \, dS - \int_{\Omega} \sigma_{ij} \frac{\partial \delta u_i}{\partial x_j} \, dV \tag{5.70}$$

A summation on the i and j indices is made, and the symmetry of the σ tensor is taken into account, so that Eq. 5.70 leads to

$$\int_{\Omega} \sum_{ij} \frac{\partial \sigma_{ij}}{\partial x_j} \delta u_i \, dV = \int_{\partial\Omega} \sum_{ij} \sigma_{ij} \, n_j \, \delta u_i \, dS - \int_{\Omega} \sum_{ij} \sigma_{ij} \, \delta\varepsilon_{ij} \, dV \tag{5.71}$$

which is rewritten in the more compact equivalent form:

$$\int_{\Omega} \mathrm{div}(\boldsymbol{\sigma}) \cdot \delta\mathbf{u} \, dV = \int_{\partial\Omega} \boldsymbol{\sigma} \, \mathbf{n} \cdot \delta\mathbf{u} \, dV - \int_{\Omega} \boldsymbol{\sigma} \cdot \delta\boldsymbol{\varepsilon} \, dV \tag{5.72}$$

The right-hand side of Eq. 5.72 is substituted into Eq. 5.69 and, after rearrangement, gives Eq. 5.68, which is the desired result. In this presentation, we should introduce the virtual work result as a theorem rather than as a principle, because it can proved. However, if the virtual work principle is stated first as valid for any virtual kinematically admissible displacement field, then the equilibrium equation can be deduced.

Generalization of the Virtual Work Principle

If acceleration is not neglected, an analogous derivation can be made from Eq. 5.52a, which can be viewed as expressing a dynamic equilibrium. The mathematical expression we obtain is simply

$$\int_{\partial\Omega} \boldsymbol{\sigma}\mathbf{n} \cdot \delta\mathbf{u} \, dS + \int_{\Omega} \rho \, \mathbf{g} \cdot \delta\mathbf{u} \, dV = \int_{\Omega} \rho \, \mathbf{a} \cdot \delta\mathbf{u}^* \, dV + \int_{\Omega} \boldsymbol{\sigma} \cdot \delta\boldsymbol{\varepsilon} \, dV \tag{5.73}$$

Exactly the same approach can be followed with a virtual velocity field $\mathbf{v}^*$, satisfying similar conditions on the domain boundary. This may be more convenient if the mechanical problem is formulated in term of the velocity field $\mathbf{v}$.

5.5 VARIATIONAL FORM OF MECHANICAL PRINCIPLES

Introduction to Formal Variational Calculation

Without entering deeply into mathematical issues, for which the interested reader can refer to specialized books, the objective of this section is to provide tools that can be used to verify that a variational formulation corresponds to a given problem expressed in terms of partial differential equations.

We assume first that the unknown is the scalar function u,[8] which is defined in a domain Ω, and is prescribed (u^d) on part of the domain boundary ($\partial\Omega^u$). In order to shorten the notation, we shall make use of the convention

$$u_{,i} = \frac{\partial u}{\partial x_i} \tag{5.74}$$

We suppose here that the variational problem is to find u defined in Ω that:

- is equal to u^d on $\partial\Omega^u$
- minimizes the functional I :

$$I(u) = \int_\Omega F\left(u, u_{,1}, u_{,2}, u_{,3}\right) dV + \int_{\partial\Omega^p} f(u) dS \tag{5.75}$$

Here $\partial\Omega^p$ is the complementary part of $\partial\Omega$ with respect to $\partial\Omega^u$.

We admit that the solution, which minimizes the functional of Eq. 5.75, is such that, for any variation δu of u, we have

$$\delta I \, (u) = 0 \tag{5.76}$$

More generally, if Eq. 5.76 is satisfied, the functional is **stationary** for the current value of u (it is not necessarily a minimum or a maximum). To express this condition, the functional I is differentiated in the same way as a function of four independent arguments:

$$\delta I \, (u) = \int_\Omega \left(\frac{\partial F}{\partial u} \delta u + \sum_{i=1}^{3} \frac{\partial F}{\partial u_{,i}} \delta u_{,i} \right) dV + \int_{\partial\Omega^p} \frac{\partial f(u)}{\partial u} \delta u \, dS = 0 \tag{5.77}$$

---

[8] Generalization to vector unknown functions is straightforward, and is treated in the Problems section.

In Eq. 5.77, each term of the summation with index i can be transformed according to the Green theorem (Eq. 5.1):

$$\int_{\Omega} \frac{\partial F}{\partial u_{,i}} \delta u_{,i} \, dV = \int_{\partial\Omega} \frac{\partial F}{\partial u_{,i}} n_i \, \delta u \, dS - \int_{\Omega} \frac{\partial}{\partial x_i}\left(\frac{\partial F}{\partial u_{,i}}\right) \delta u \, dV \tag{5.78}$$

By putting Eq. 5.78 into Eq. 5.77, we obtain

$$\int_{\Omega} \left\{ \frac{\partial F}{\partial u} - \sum_i \frac{\partial}{\partial x_i}\left(\frac{\partial F}{\partial u_{,i}}\right) \right\} \delta u \, dV + \int_{\partial\Omega^p} \frac{\partial f}{\partial u} \delta u \, dS + \int_{\partial\Omega} \sum_i \frac{\partial F}{\partial u_{,i}} n_i \, \delta u \, dS = 0 \tag{5.79}$$

As this equality is valid for any δu, we can first choose $\delta u = 0$ on $\partial\Omega$, with the consequence that the second integral in Eq. 5.79 vanishes identically. Thus one can see that the first integral must be null for any $\delta u \neq 0$ inside Ω, so that the following equation must hold:

$$\frac{\partial F}{\partial u} - \frac{\partial}{\partial x_i}\left(\frac{\partial F}{\partial u_{,i}}\right) = 0 \tag{5.80}$$

which is the **partial differential equation (PDE)** that the function u satisfies in Ω. Now we decide to choose any δu equal to zero only on the part $\partial\Omega^u$ of $\partial\Omega$ where u is prescribed. With Eqs. 5.79 and 5.80, it can be concluded that the following equation must hold:[9]

$$\frac{\partial f}{\partial u} + \sum_i \frac{\partial F}{\partial u_{,i}} n_i = 0 \tag{5.81}$$

This is the **natural boundary condition (NBC)** that the function u must verify on $\partial\Omega^p$.

Formal Variational Calculation for Mechanical Problems

The mechanical problems are formulated in terms of the displacement vector field **u** (or **v**, the velocity field) and the strain tensor ε (or the strain-rate tensor $\dot{\varepsilon}$). We shall suppose again that the mechanical problem is defined in a

---

[9] This is in fact a generalization of the last mathematical theorem of Section 5.1 to surface integral.

domain Ω, and the displacement (or some components of the displacement) is imposed on the part $\partial\Omega^u$ of the boundary, while $\partial\Omega^s$ is the part of the boundary where at least one component of the displacement is not prescribed. Using analogous notations as in the previous subsection, we define the functional I by

$$I(\mathbf{u}) = \int_{\Omega} F(\mathbf{u}, \boldsymbol{\varepsilon})\, dV + \int_{\partial\Omega^s} f(\mathbf{u})\, dS \tag{5.82}$$

It is worthy to note that we have considered the case where F is a function of the displacement **u** and of the strain tensor $\boldsymbol{\varepsilon}$. After a treatment that is formally equivalent to the previous one, we obtain the PDE as

$$\frac{\partial F(\mathbf{u}, \boldsymbol{\varepsilon})}{\partial u_i} - \sum_j \frac{\partial}{\partial x_j}\left(\frac{\partial F(\mathbf{u}, \boldsymbol{\varepsilon})}{\partial \varepsilon_{ij}}\right) = 0 \tag{5.83}$$

Only for the components i, which do not correspond to a prescribed displacement, we have the NBC on $\partial\Omega^s$:

$$\frac{\partial f(\mathbf{u})}{\partial u_i} + \sum_j \frac{\partial F(\mathbf{u}, \boldsymbol{\varepsilon})}{\partial \varepsilon_{ij}} n_j = 0 \tag{5.84}$$

One can remark that the term involving the partial derivatives of the function F with respect to the components of the strain tensor can be identified with the stress tensor, which is plausible for hypo-elastic cases:

$$\frac{\partial F(\mathbf{u}, \boldsymbol{\varepsilon})}{\partial \varepsilon_{ij}} = \sigma_{ij} \tag{5.85}$$

Equation 5.83 thus is the equilibrium equation with body forces corresponding to the derivative of F with respect to the displacement:

$$\frac{\partial F(\mathbf{u}, \boldsymbol{\varepsilon})}{\partial u_i} = g_i \tag{5.86}$$

The boundary condition in terms of stress vector is obtained from Eqs. 5.84 and 5.85:

$$T_i = \sum_j \sigma_{ij}\, n_j = -\frac{\partial f(\mathbf{u})}{\partial u_i} \tag{5.87}$$

The same approach can be followed with the velocity field **v** as the main unknown.

5.6 THE WEAK FORM

In continuum mechanics, and in physics, it is often difficult or impossible to derive a functional formulation, such that the solution satisfies a maximum, minimum, or stationary principle. However, an integral formulation is highly desirable in building a finite element approximation. Instead of satisfying exactly (i. e., point-wise) the partial differential equations arising from a physical problem, it is often necessary to verify it in an average sense with an integral form, and to introduce a family of functions that can be considered as **weighting functions**. The resulting formulation is called a **weak form**. We do not necessarily have a strict equivalence with the initial equations; they are satisfied only in a weaker sense.

The most frequently used of these formulations is the Galerkin approach, which we will examine briefly for mechanical problems. A family of vector weight functions ($\mathbf{w}_k$; k = 1 to n_w) defined in the domain Ω is considered. The equation of motion (Eq. 5.52a) is satisfied in the weak sense:

$$\int_\Omega \rho\, \boldsymbol{\gamma}\cdot \mathbf{w}_k \, dV = \int_\Omega \mathrm{div}(\boldsymbol{\sigma})\cdot \mathbf{w}_k \, dV + \int_\Omega \rho\, \mathbf{g}\cdot \mathbf{w}_k \, dV \tag{5.88}$$

The first integral of the right-hand side of Eq. 5.88 is transformed in the usual way with the Green theorem:

$$\int_\Omega \mathrm{div}(\boldsymbol{\sigma})\cdot \mathbf{w}_k \, dV = \int_{\partial\Omega} \boldsymbol{\sigma}\cdot \mathbf{n}\cdot \mathbf{w}_k \, dS - \int_\Omega \boldsymbol{\sigma}\cdot \frac{\partial \mathbf{w}_k}{\partial \mathbf{x}} \, dV \tag{5.89}$$

Each vector function $\mathbf{w}_k$ is chosen in such a way that it is equal to zero on $\partial\Omega^u$, where the displacement (or the velocity) is prescribed; and we replace $\boldsymbol{\sigma}\cdot\mathbf{n}$ on $\partial\Omega^s$ where the stress vector is prescribed by its value $\mathbf{T}^d$,[10] so that Eq. 5.88 is rewritten:

$$\int_\Omega \mathrm{div}(\boldsymbol{\sigma})\cdot \mathbf{w}_k \, dV = \int_{\partial\Omega^s} \mathbf{T}^d \cdot \mathbf{w}_k \, dS - \int_\Omega \boldsymbol{\sigma}\cdot \frac{\partial \mathbf{w}_k}{\partial \mathbf{x}} \, dV \tag{5.90}$$

Equation 5.90 is substituted in Eq. 5.88 and the weak form is obtained:

$$\int_\Omega \rho\, \boldsymbol{\gamma}\cdot \mathbf{w}_k \, dV + \int_\Omega \boldsymbol{\sigma}\cdot \frac{\partial \mathbf{w}_k}{\partial \mathbf{x}} \, dV - \int_\Omega \rho\, \mathbf{g}\cdot \mathbf{w}_k \, dV - \int_{\partial\Omega^s} \mathbf{T}^d \cdot \mathbf{w}_k dS = 0 \tag{5.91}$$

---

[10] General conditions for a well-posed mechanical problem can be considered in the same way as in Eqs. 5.64 and 5.65.

If the family of weight functions is large enough, by doing the same transformations in the reverse way, we can guess that the initial PDE, and BC on the stress, will be satisfied approximately. This approach is analogous to the virtual work principle: it is obvious if we select a family of virtual displacements $\delta \mathbf{u}_k = \mathbf{w}_k$. But it is important to note that the same method can be extended to a much wider class of problems, including thermal analysis.

5.7 VARIATIONAL PRINCIPLES FOR SIMPLE CONSTITUTIVE EQUATIONS

Variational Principle for Elastic Material

The problem is defined in terms of the displacement $\mathbf{u}$, and the associated strain tensor ε, in a domain Ω. The material is isotropic and linear-elastic, and obeys Hooke's law:

$$\boldsymbol{\sigma} = \lambda\, \theta\, \mathbf{1} + 2\, \mu\, \boldsymbol{\varepsilon} \tag{5.92}$$

On a part $\partial\Omega^u$ of the boundary, $\partial\Omega$ of Ω, at least one component of the displacement is prescribed:

$$u_i = u_i^d \quad \text{on} \quad \partial\Omega^u \tag{5.93}$$

On $\partial\Omega^s$ at least one component of the stress vector is imposed:

$$T_i = \sum_j \sigma_{ij}\, n_j = T_i^d \quad \text{on} \quad \partial\Omega^s \tag{5.94}$$

For simplicity, we shall assume that no temperature change occurs in the domain Ω, so that the free-energy-state function for an elastic material becomes

$$\rho\, \psi = \frac{1}{2}\, \lambda\, \theta^2 + \mu\, \boldsymbol{\varepsilon} \cdot \boldsymbol{\varepsilon} \tag{5.95}$$

We see immediately that

$$\boldsymbol{\sigma} = \frac{\partial}{\partial \boldsymbol{\varepsilon}}\, (\rho\, \psi) \tag{5.96}$$

This is theoretically sufficient to build the variational form according to Section 5.5, but the following derivation will permit a proof that the functional is minimum for the solution. The quadratic form of Eq. 5.95 allows us to write exactly the second order Taylor expansion for any variation $\delta\mathbf{u}$ of the displacement with the associated strain $\delta\varepsilon$:

$$\rho\,\psi(\boldsymbol{\varepsilon}+\delta\,\boldsymbol{\varepsilon}) = \rho\,\psi(\boldsymbol{\varepsilon}) + \frac{\partial}{\partial\boldsymbol{\varepsilon}}(\rho\,\psi)\cdot\delta\,\boldsymbol{\varepsilon} + \frac{1}{2}\left(\frac{\partial^2}{\partial\boldsymbol{\varepsilon}^2}(\rho\,\psi)\cdot\delta\,\boldsymbol{\varepsilon}\right)\cdot\delta\,\boldsymbol{\varepsilon} \tag{5.97}$$

Looking at Eq. 5.95 it is easy to convince oneself that:

$$\left(\frac{\partial^2}{\partial\boldsymbol{\varepsilon}^2}(\rho\,\psi)\cdot\delta\boldsymbol{\varepsilon}\right)\cdot\delta\boldsymbol{\varepsilon} \geq 0 \tag{5.98}$$

and that the equality can occur only if $\delta\varepsilon = 0$. Equations 5.92 and 5.96 to 5.98 allow the transformation of Eq. 5.97 into

$$\rho\,\psi(\boldsymbol{\varepsilon}+\delta\,\boldsymbol{\varepsilon}) \geq \rho\,\psi(\boldsymbol{\varepsilon}) + \frac{\partial}{\partial\boldsymbol{\varepsilon}}(\rho\,\psi)\cdot\delta\,\boldsymbol{\varepsilon} = \rho\,\psi(\boldsymbol{\varepsilon}) + \boldsymbol{\sigma}\cdot\delta\,\boldsymbol{\varepsilon} \tag{5.99}$$

By integration over the domain Ω, it is also

$$\int_\Omega \rho\,\psi(\boldsymbol{\varepsilon}+\delta\,\boldsymbol{\varepsilon})\,dV \geq \int_\Omega \rho\,\psi(\boldsymbol{\varepsilon})\,dV + \int_\Omega \boldsymbol{\sigma}\cdot\delta\,\boldsymbol{\varepsilon}\,dV \tag{5.100}$$

We suppose that $\mathbf{u}$, ε, and σ correspond to the solution, so that σ satisfies the equilibrium equation, which can be written with the integral form:

$$\int_\Omega \boldsymbol{\sigma}\cdot\delta\,\boldsymbol{\varepsilon}\,dV = \int_{\partial\Omega^s} (\boldsymbol{\sigma}\,\mathbf{n})\cdot\delta\mathbf{u}\,dS \tag{5.101}$$

No variation $\delta\mathbf{u}$ of $\mathbf{u}$ is allowed on the part $\partial\Omega^u$ of the boundary where the displacement is prescribed, and σ satisfies the boundary condition given by Eq. 5.94, allowing Eq. 5.101 to be rewritten:

$$\int_\Omega \sigma\cdot\delta\,\varepsilon\,dV = \int_{\partial\Omega^s} \mathbf{T}^d\cdot\mathbf{S}\mathbf{u}\,dS \tag{5.102}$$

Combining Eqs. 5.100 and 5.102 leads to

$$\int_\Omega \rho\,\psi(\boldsymbol{\varepsilon}+\delta\,\boldsymbol{\varepsilon})\,dV - \int_{\partial\Omega^s} \mathbf{T}^d\cdot(\mathbf{u}+\delta\mathbf{u})\;dS \geq \int_\Omega \rho\,\psi(\boldsymbol{\varepsilon})\,dV - \int_{\partial\Omega^s} \mathbf{T}^d\cdot\mathbf{u}\,dS \tag{5.103}$$

This last equation suggests the introduction of the energy functional Π:

$$\Pi(\boldsymbol{\varepsilon}) = \int_{\Omega} \rho\, \psi(\boldsymbol{\varepsilon})\, dV - \int_{\partial\Omega^s} \mathbf{T}^d \cdot \mathbf{u}\, dS \tag{5.104}$$

with the boundary condition given by Eq. 5.84, in order to rewrite Eq. 5.93 in a more compact way:

$$\Pi(\boldsymbol{\varepsilon} + \delta\, \boldsymbol{\varepsilon}) \geq \Pi(\boldsymbol{\varepsilon}) \tag{5.105}$$

Eq. 5.105 shows that, if no rigid body motion is allowed, the energy functional Π has a strict minimum that corresponds to the solution of the linear elastic problem.

Plastic or Viscoplastic Materials

A general viscoplastic potential φ is introduced as a function of the irreversible strain rate $\dot{\boldsymbol{\varepsilon}}$, with the following properties:

- φ is convex and positive
- $\varphi(0) = 0$
- the stress tensor is derived from φ by

$$\boldsymbol{\sigma} = \frac{\partial \varphi(\dot{\boldsymbol{\varepsilon}})}{\partial \dot{\boldsymbol{\varepsilon}}} \quad \text{or} \quad \sigma_{ij} = \frac{\partial \varphi(\dot{\varepsilon}_{ij})}{\partial \dot{\varepsilon}_{ij}} \tag{5.106}$$

For analysis of a viscoplastic material deformed in a domain Ω, the solution is expressed in terms of a velocity field **v**. The boundary conditions are the following:

- On a part $\partial\Omega^v$ of the boundary, the velocity is prescribed (or some of its components); i. e., $v_i = v_i^d$.
- On another part $\partial\Omega^s$, the stress vector is imposed (or some of its components, with the usual condition of a well-posed mechanical problem): $T_i = T_i^d$.

With these hypotheses, a functional **Φ** can be built by putting

$$\Phi(\mathbf{v}) = \int_{\Omega} \varphi(\dot{\boldsymbol{\varepsilon}})\, dV - \int_{\partial\Omega^s} \mathbf{T}^d \cdot \mathbf{v}\, dS \tag{5.107}$$

The derivation of Section 5.5 (Eqs. 5.82 to 5.87), where one has to replace the displacement by the velocity vector, allows verification that the functional is stationary for the solution of the viscoplastic problem defined by Eq. 5.106.

For an isotropic material, the plastic potential is a function only of the three invariants of the strain rate tensor, namely,

$$I_1 = \mathrm{tr}(\dot{\boldsymbol{\varepsilon}}), \qquad I_2 = \sum_{ij} \dot{\varepsilon}_{ij}^{\,2}, \qquad I_3 = \det(\dot{\boldsymbol{\varepsilon}})$$

The third invariant is generally omitted, and for **incompressible viscoplasticity,** the first one is equal to zero by definition. In this case, we introduce the equivalent strain rate:

$$\dot{\bar{\varepsilon}} = \left(\frac{2}{3} I_2\right)^{1/2} = \left(\frac{2}{3}\sum_{ij} \dot{\varepsilon}_{ij}^{\,2}\right)^{1/2} \tag{5.108}$$

and put (using the same function name for simplicity) $\varphi(\dot{\boldsymbol{\varepsilon}}) = \varphi(\dot{\bar{\varepsilon}})$, so that we can rewrite Eq. 5.106 in terms of the deviatoric stress tensor **s**:

$$\mathbf{s} = \frac{\partial \varphi(\dot{\bar{\varepsilon}})}{\partial \dot{\bar{\varepsilon}}} \frac{\partial \dot{\bar{\varepsilon}}}{\partial \dot{\boldsymbol{\varepsilon}}} = \frac{2}{3} \frac{1}{\dot{\bar{\varepsilon}}} \frac{\partial \varphi(\dot{\bar{\varepsilon}})}{\partial \dot{\bar{\varepsilon}}} \dot{\boldsymbol{\varepsilon}} \tag{5.109}$$

A useful form of the viscoplastic potential is given by the Norton-Hoff law:

$$\varphi(\dot{\bar{\varepsilon}}) = \frac{K}{m+1} \left(\sqrt{3}\, \dot{\bar{\varepsilon}}\right)^{m+1} \tag{5.110}$$

With Eq. 5.109, the Norton-Hoff viscoplastic potential allows formulation of the constitutive equation:

$$\mathbf{s} = 2\,K \left(\sqrt{3}\, \dot{\bar{\varepsilon}}\right)^{m-1} \dot{\boldsymbol{\varepsilon}} \tag{5.111}$$

The functional for the **Norton-Hoff** viscoplastic material is a particular case of Eq. 5.107:

$$\Phi(\mathbf{v}) = \int_{\Omega} \frac{K}{m+1}\left(\sqrt{3}\, \dot{\bar{\varepsilon}}\right)^{m+1} dV - \int_{\partial\Omega^s} \mathbf{T}^d \cdot \mathbf{v}^d\, S \tag{5.112}$$

The velocity field must satisfy the boundary conditions as usual, but also the incompressibility constraint div(**v**) = 0.

Von Mises plasticity can be considered as a limiting case of the Norton-Hoff viscoplasticity with m = 0 and $K = \sigma_0 / \sqrt{3}$, so that the functional given by Eq. 5.112 also applies to plastic materials.

CHAPTER 5 PROBLEMS

A. Proficiency problems

1. Adapt the demonstration of the Green theorem (Eq. 5.1) to the following:

   a. A curve C, and show that

$$\int_C f \frac{dg}{dl} dl = f(B)\, g(B) - f(A)\, g(A) - \int_C \frac{df}{dl}\, g\, dl,$$

   where dl is the curvilinear element and A and B are the extremities of the curve. Examine the case when the curve is closed (A = B).

   b. A surface S defined in the plane (with coordinates x_1 and x_2), so that

$$\int_S f \frac{\partial g}{\partial x_i} dS = \int_{\partial S} f\, g\, n_i\, dl - \int_S \frac{\partial f}{\partial x_i}\, g\, dS$$

   where i = 1 or 2, ∂S is the curve limiting the surface S (with no hole), and **n** is the unit normal vector to ∂S.

2. Apply the 2-D form of the Green theorem (see Problem 1) to the functions $f = x_1 + x_2$ and $g = x_1 - x_2$ in the square domain $[0, 1]^2$, for i = 1 and i = 2. Compute directly the integrals and verify the Green theorem for this specific example.

3. A 2-D velocity field is given by these expressions:

$$v_1 = a\, x_1^2 + 2\, b\, x_1\, x_2 + c\, x_1\, x2^2$$
$$v_2 = a'\, x_1^2 + 2\, b'\, x_1\, x_2 + c'\, x_1\, x2^2$$

   depending on the six parameters a, b, c, a′, b′, c′. Determine the relations the parameters must satisfy in order for the velocity field to be incompressible.

Keeping these relations in mind, show directly that the material flow through the hatched unit triangle is equal to zero.

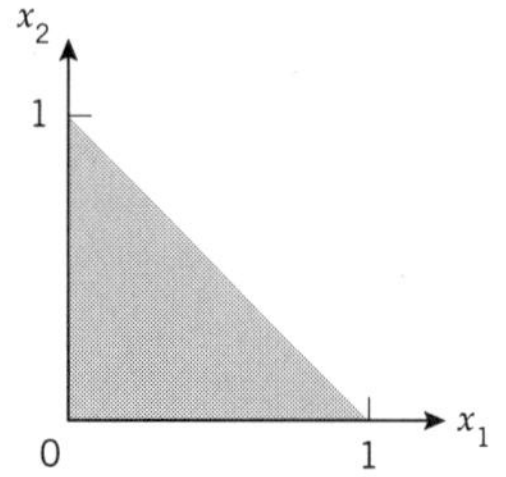

4. A porous material with an initial relative density of $\rho_r^0 < 1$ is densified by a uniform rate of volume change equal to c. Express the law of relative density ρ_r as a function of time t, the beginning of the process corresponding to t = 0.

5. The plane strain upsetting of a rectangular section with height 2h and width 2a is shown in the figure.

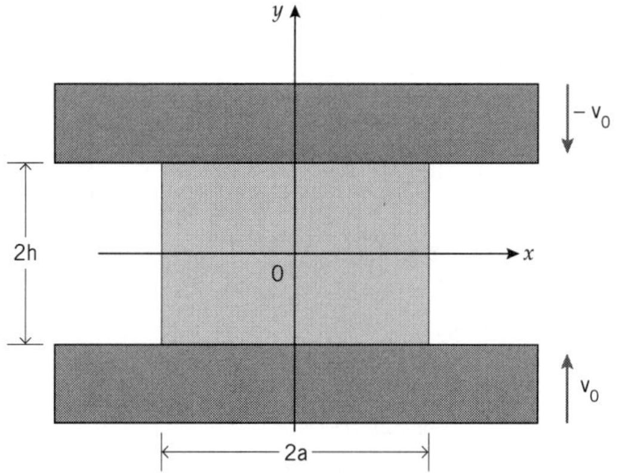

The velocity field at any moment is given by

$$v_x = +v_o \frac{x}{h}$$

$$v_y = -v_o \frac{y}{h}$$

where v_o is the (constant) upsetting velocity.

Express the kinematic energy K in the section, and the time derivative dK/dt. Calculate the local acceleration of any material point with coordinate (x, y), and the local density of the time derivative of kinematic energy. Verify that the integral of this density gives the previous value for dK/dt.

6. With the help of the general variational theory outlined in Section 5.5, verify that the functional Π defined by

$$\Pi(\mathbf{u}) = \frac{1}{2}\int_{\Omega}\left(\lambda\left(\sum_{i}\varepsilon_{ii}\right)^{2} + 2\,\mu\sum_{ij}\varepsilon_{ij}^{2}\right)dV - \int_{\partial\Omega^{s}}\left(\sum_{i}T_{i}\,u_{i}\right)dS$$

corresponds to the linear elastic problem for a material obeying Hooke's law (see Chapter 6), with a prescribed stress vector **T** on the boundary $\delta\Omega^s$.

7. A cylindrical sample is considered with length L and radius R, subjected to a prescribed tension force F at its right end, while the left one remains fixed (see following figure).

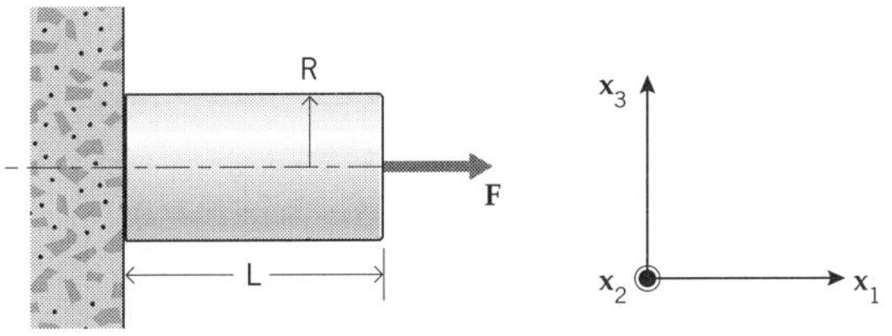

A linear displacement **u** is introduced with the form

$$\mathbf{u} \leftrightarrow \begin{bmatrix} u_1 = U\,x_1 \\ u_2 = V\,x_2 \\ u_3 = W\,x_3 \end{bmatrix}$$

Compute the energy functional Π (Eq. 5.104) for an elastic material with Lamé coefficients λ and μ. The minimization of Π allows explicit determination of the unknown parameters U, V, W. Show that the usual simple formulas are obtained when Young's modulus E and Poisson's ratio ν coefficient are used. Recall that the following equalities hold:

$$E = \frac{\mu\,(3\lambda + 2\mu)}{\lambda + \mu} \qquad \nu = \frac{\lambda}{2(\lambda + \mu)}$$

8. Find the viscoplastic potential for a material obeying the following constitutive equation:

$$\mathbf{s} = 2K(\sqrt{3})^{m-1}\left(\dot{\bar{\varepsilon}}_0^2 + \dot{\bar{\varepsilon}}^2\right)^{\frac{m-1}{2}} \dot{\boldsymbol{\varepsilon}}$$

where $\dot{\bar{\varepsilon}}_0$ is a (small) positive constant. Show that when $\dot{\bar{\varepsilon}} << \dot{\bar{\varepsilon}}_0$ the material constitutive equation is equivalent to a Newtonian behavior with the general form

$$\mathbf{s} = 2\eta\, \dot{\boldsymbol{\varepsilon}}$$

and that if $\dot{\bar{\varepsilon}} >> \dot{\bar{\varepsilon}}_0$, then it is equivalent to a Norton-Hoff behavior (Eq. 5.110).

B. Depth problems

9. A potential $\varphi(\dot{\boldsymbol{\varepsilon}})$ is convex if, for any $\dot{\boldsymbol{\varepsilon}}_1$ and $\dot{\boldsymbol{\varepsilon}}_2$, we have

$$\varphi(\lambda\, \dot{\boldsymbol{\varepsilon}}_1 + (1-\lambda)\dot{\boldsymbol{\varepsilon}}_2) \le \lambda\, \varphi(\dot{\boldsymbol{\varepsilon}}_1) + (1-\lambda)\, \varphi(\dot{\boldsymbol{\varepsilon}}_2) \text{ for } 0 \le \lambda \le 1$$

Apply the above inequality to $\dot{\boldsymbol{\varepsilon}}_1 = 0$ and $\dot{\boldsymbol{\varepsilon}}_2 = \dot{\boldsymbol{\varepsilon}}$, and express the derivative of each side at $\lambda = 0$ to show that

$$\frac{\partial \varphi}{\partial \dot{\boldsymbol{\varepsilon}}}(\dot{\boldsymbol{\varepsilon}}) : \dot{\boldsymbol{\varepsilon}} \ge \varphi(\dot{\boldsymbol{\varepsilon}}) \ge 0$$

10. Give the proof for the generalized expression of the variational formulation to mechanical problems in Section 5.5.

11. From the variational formulation given by Eq. 5.111, derive the constitutive equation and the equilibrium equation for a Norton-Hoff viscoplastic material (Eq. 5.112). To obtain the appropriate form, utilize the following mathematical property of an incompressible vector field **v**:

$$\text{if } \operatorname{div}(\mathbf{v}) = 0, \text{ then a vector potential } \boldsymbol{\xi} \text{ exists so that: } \mathbf{v} = \mathbf{curl}(\boldsymbol{\xi})$$

First, verify that

$$\operatorname{curl}\left[\operatorname{div}(\mathbf{s})\right] = 0$$

where **s** is the deviatoric stress tensor. With the help of the previous mathematical result, it is possible to show that div(**s**) is the gradient of a scalar field, which is identified with the pressure field p.

CHAPTER 6

Elasticity

This chapter brings in our first look at a real material response, albeit a very simple one. The physical picture of **linear elasticity** we present has changed little since the time of Hooke, who originally noted that load and extension are proportionally related in many materials, especially when the extension is not too large. We will follow this basic concept and frame it a bit more mathematically, but will avoid the complexities involved in large-strain pictures, and the difficulties of ensuring objectivity of the various constitutive formulations.

6.1 ONE-DIMENSIONAL ELASTICITY

The basic idea of **elasticity** is simple: the deformation of a body depends only on its current loading, and is independent of history, rate, time, path, and other variables. Extensions of the basic idea can be made to include thermal strains (**thermoelasticity**) or time-dependent effects (**viscoelasticity, anelasticity**). That is, an unloaded elastic body always returns to its original shape, no matter what happens to it between the two unloaded times. In 1-D, we can write the law simply:

$$\sigma = f(\varepsilon) \quad \text{or} \quad \varepsilon = f(\sigma) \tag{6.1}$$

where f is a single-valued function passing through the point ($\varepsilon = 0, \sigma = 0$) which is easily inverted. While we have shown the proper variables as Cauchy stress and small strain, the distinction among the various measures is lost when the difference in geometry between the initial or reference state and the deformed state is vanishingly small. For the elastic response of metals, this assumption is a very good one, since the elastic strains are usually of the order of 0.001. When the elastic response is superimposed onto a large deformation, or when the small elastic strains produce large deflections (for example, during the elastic bending of a long, slender rod), more care must be taken to account for the difference between original and current geometry.

Recognizing that the strains and displacements are usually small, Eq. 6.1 can be easily rewritten in terms of external variables, such as load (P) and extension (Δl):

$$P = f(\Delta l) \quad \text{or} \quad \Delta l = f(P) \tag{6.2}$$

The 1-D force law shown in Eq. 6.2 is a clear example of a **conservative force** (in physics terminology), which does work depending only on the starting and finishing position, independent of the path in between. It follows directly that the work is zero over a closed path:

$$dw = Fdl = V\sigma\, d\varepsilon \tag{6.3a}$$

$$w = \int_a^b Fdl = V\int_a^b \sigma d\varepsilon \tag{6.3b}$$

$$0 = \int_a^a Fdl = V\int_a^a \sigma d\varepsilon \tag{6.3c}$$

where V = volume. Equation. 6.3a is obtained by noting that $P = \sigma A$ and $dl = l d\varepsilon$ such that $Pdl = Al\sigma d\varepsilon = V\sigma d\varepsilon$, where V = material volume.

This energy-conserving nature of elastic constitutive laws is the distinguishing feature, expressed either in the form of Eq. 6.1, where F is understood to be a function only of ε, or Eqs. 6.3, where the elastic work is path-independent. (The path independence is often tested by determining that dε is an exact differential.) The path independence of elastic work must be exhibited by any properly constructed theory of elasticity.

For metals, again because the elastic strains are very small before plastic deformation intervenes, it is sufficient to consider the function f (Eq. 6.1) as a simple proportionality, which is **Hooke's law**:

$$\sigma = c\varepsilon \quad \text{or} \quad \varepsilon = s\sigma \tag{6.4}$$

where c and s are constants called the **elastic stiffness** and the **elastic compliance**, respectively. Unfortunately, the conventional naming has the letters reversed:

c = elastic stiffness, or elastic constant, or modulus

s = elastic compliance

While it is clear mathematically from the 1-D view of deformation (Eq. 6.4) that $c = 1/s$, c and s are not reciprocal in real 1-D tests of materials because c is determined when there is only strain in one direction (corresponding to several stress components) while s is determined when there is only one stress component (as in a tensile test). That is, straining in one dimension is not equivalent to stressing in one dimension, and it is impossible to accomplish both simultaneously in real tests. The reciprocal of s is known as Young's modulus (see Chapter 1).

6.2 ELASTICITY VS. HYPERELASTICITY VS. HYPOELASTICITY

While not significant for the vanishingly small displacements we assume throughout this chapter, there can be differences between Eqs. 6.1 and 6.3 in the large-strain case. Equation 6.1 represents the basic form of **elasticity**, where stress is related functionally to strain. Equation 6.3 represents a form of **hyperelasticity** where w is a potential function of strain that automatically ensures that Eqs. 6.3 are followed and that $d\varepsilon$ is exact. Equation 6.1 can also be written in differential form (i.e., $d\varepsilon = f(d\sigma)$), which is distinct in the case where strains can be noninfinitesimal. This last form is known as **hypoelasticity,** and care must be taken in formulating the theory to ensure that $d\varepsilon$ is exact. In this book, we will always assume that a potential function (Eq. 6.3) exists, and that the other forms are derived from it. This approach ensures that the constitutive equations meet the necessary test of path independence and energy conservation.

6.3 HOOKE'S LAW IN 3-D

The extension of linear elasticity to three dimensions is straightforward: we want stress and strain to be proportional, and recognize that each measure represents a second-ranked tensor:

$$\boldsymbol{\sigma} = \mathbf{c} \cdot \boldsymbol{\varepsilon} \tag{6.5a}$$

or

$$\sigma_{ij} = c_{ijkl}\, \varepsilon_{kl} \tag{6.5b}$$

Equations 6.5a and 6.5b represent either the tensorial (frame-invariant) form of Hooke's law or the component (frame-dependent) form. **c** is clearly a fourth-ranked tensor with components c_{ijkl} in a specified coordinate system.[1] Equation 6.5a emphasizes that **c** is independent of any coordinate system, whereas the numbers representing **c** (i.e., c_{ijkl}) will change as the coordinate system is altered. Since **c** is a material property (not an applied tensor like σ or ε), it is related only to the fixed orientation of the material, not to the applied stresses or strains.

> ***Note:*** *As is almost universally done in applied problems for metals, we will consider only* ***linear elasticity****, where* **c** *is a constant tensor. However, there is no conceptual difficulty in defining* **c** *(ε), where* **c** *is a material property that depends on the state of the material.*

Equation 6.5b shows that there are 81 elastic constants (i.e., 3 × 3 × 3 × 3 corresponding to the indices i, j, k, l) that must be specified for a material in 3-D. Conceptually, we can think of these arranged in a four-dimensional array where the first dimension relates to the stressed plane, the second dimension to the stress direction (these two corresponding to the indices i and j in σ_{ij}), and the third dimension to the relative displacement of a material vector defined by the fourth dimension (these corresponding to the indices k and l in ε_{kl}).

While the 3 × 3 × 3 × 3 scheme for elastic constants is simplest and most elegant, it is difficult to represent on a piece of 2-D paper. Thus manual manipulations (and teaching!) are usually done by writing Hooke's law on 2-D paper and assigning single indices corresponding to pairs 11, 22, and so forth. This change of index is conventionally carried out as follows:

	3×3×3×3 scheme		**9×9 scheme**
	11	→	1
	22	→	2
	33	→	3
	23	→	4
	13	→	5
	12	→	6
	32	→	7
	31	→	8
	21	→	9
i.e.,	σ_{ij}	→	σ_k
	ε_{ij}	→	ε_k
	c_{ijkl}	→	c_{mn}

(based on mapping shown) (6.6)

[1] The tensorial nature of **c** can be conceptually demonstrated with either of two nonrigorous arguments. In the first, a physical law such as Eq. 6.5a that relates two second-ranked tensor quantities must involve a fourth-ranked tensor. Secondly, since **c** is a material property, it cannot not depend on some external, arbitrary coordinate system.

With this substitution, the proper equations for Hooke's law can be easily (but lengthily) written on 2-D paper:

$$\begin{bmatrix} \sigma_1 \\ \sigma_2 \\ \sigma_3 \\ \sigma_4 \\ \sigma_5 \\ \sigma_6 \\ \sigma_7 \\ \sigma_8 \\ \sigma_9 \end{bmatrix} = \begin{bmatrix} c_{11} & c_{12} & c_{13} & c_{14} & c_{15} & c_{16} & c_{17} & c_{18} & c_{19} \\ c_{21} & c_{22} & c_{23} & c_{24} & c_{25} & c_{26} & c_{27} & c_{28} & c_{29} \\ c_{31} & c_{32} & c_{33} & c_{34} & c_{35} & c_{36} & c_{37} & c_{38} & c_{39} \\ c_{41} & c_{42} & c_{43} & c_{44} & c_{45} & c_{46} & c_{47} & c_{48} & c_{49} \\ c_{51} & c_{52} & c_{53} & c_{54} & c_{55} & c_{56} & c_{57} & c_{58} & c_{59} \\ c_{61} & c_{62} & c_{63} & c_{64} & c_{65} & c_{66} & c_{67} & c_{68} & c_{69} \\ c_{71} & c_{72} & c_{73} & c_{74} & c_{75} & c_{76} & c_{77} & c_{78} & c_{79} \\ c_{81} & c_{82} & c_{83} & c_{84} & c_{85} & c_{86} & c_{87} & c_{88} & c_{89} \\ c_{91} & c_{92} & c_{93} & c_{94} & c_{95} & c_{96} & c_{97} & c_{98} & c_{99} \end{bmatrix} \begin{bmatrix} \varepsilon_1 \\ \varepsilon_2 \\ \varepsilon_3 \\ \varepsilon_4 \\ \varepsilon_5 \\ \varepsilon_6 \\ \varepsilon_7 \\ \varepsilon_8 \\ \varepsilon_9 \end{bmatrix} \quad (6.7)$$

Using this conventional change of index, Hooke's law may be written

$$\sigma_i = c_{ij}\,\varepsilon_j \quad (6.8)$$

where i and j take the values 1 to 9, and whose meaning is defined by Eq. 6.6. Note, however, that the components in this conventional scheme are not tensor components as usually defined, because **c** is *fourth-ranked* in 3-D space instead of being second-ranked in 9-D space, as Eq. 6.8 implies. Thus **c** has components that transform according to the usual tensor transformation:

$$c'_{ijkl} = R_{im}\,R_{jn}\,R_{ko}\,R_{lp}\,c_{mnop} \quad (6.9)$$

6.4 REDUCTION OF ELASTIC CONSTANTS FOR A GENERAL MATERIAL

Equation 6.7 shows that there are 81 elastic constants in Hooke's law. They relate each component of strain with each component of stress. In fact, any column of elastic constants represents directly the stress that results when a unit strain is applied in the corresponding strain component while all other strains are constrained to be zero. For example, we can define c_{11}, c_{21}, c_{31},...c_{91} conceptually as the nine stress components (s_1, s_2, s_3,...s_9) that result from the strain state ($e_1 = 1$, other $e_i = 0$).

However, it should be immediately evident that the nine equations in Eq. 6.7 are not independent. In fact, since we know that the stress tensor is symmetric (Chapter 3), the equations represented by the seventh, eighth, and ninth rows are identical to the ones represented by the fourth, fifth, and sixth rows, respectively. (If any of these constants were different, σ_{ij} would not be equal to σ_{ji}.) Therefore, we need not keep the last three rows of the system of equations and will instead simply remember that $\sigma_4 = \sigma_7$, $\sigma_5 = \sigma_8$, and $\sigma_6 = \sigma_9$.

Secondly, $\varepsilon_{ij} = \varepsilon_{ji}$ (Chapter 4), so we don't need to retain the independent multiplications of the final three strain components: ε_7, ε_8, ε_9. Instead, we can keep all of the same information in Eq. 6.7 while ignoring these repeated strain components. We could do this in two ways: add the elastic constants that multiply the last three strains to the ones that multiply the 4, 5, and 6 components. That is, $c_{14}^{new} = c_{14}^{old} + c_{17} = 2\,c_{14}^{old}$. This procedure is usually not followed; instead, the elastic constants are kept the same and the shear strains are multiplied by two to keep the proper equations. (Note that c_{14} and c_{17}, for example, have no independent meaning because one relates σ_1 to ε_4, and the other relates σ_1 to ε_7; however, ε_4 and ε_7 are identical, so there is no way to measure c_{14} independently of c_{17}. Thus $c_{14} = c_{17}$ and there is no loss of material information in deleting the last three columns of [c].)

The doubled shear strains, $2\varepsilon_4$, $2\varepsilon_5$, $2\varepsilon_6$, are usually denoted by the engineering quantities γ_4, γ_5, γ_6 (corresponding in the $3 \times 3 \times 3 \times 3$ scheme to γ_{23}, γ_{13}, γ_{12}). Thus, Eq. 6.7 can be rewritten in the conventional 6×6 scheme to take advantage of the stress and strain symmetry.[2]

9 × 9 scheme

$$\begin{bmatrix} \sigma_1 \\ \sigma_2 \\ \sigma_3 \\ \sigma_4 \\ \sigma_5 \\ \sigma_6 \\ \sigma_7 \\ \sigma_8 \\ \sigma_9 \end{bmatrix} = \begin{bmatrix} c_{11} & c_{12} & c_{13} & c_{14} & c_{15} & c_{16} & c_{17} & c_{18} & c_{19} \\ c_{21} & c_{22} & c_{23} & c_{24} & c_{25} & c_{26} & c_{27} & c_{28} & c_{29} \\ c_{31} & c_{32} & c_{33} & c_{34} & c_{35} & c_{36} & c_{37} & c_{38} & c_{39} \\ c_{41} & c_{42} & c_{43} & c_{44} & c_{45} & c_{46} & c_{47} & c_{48} & c_{49} \\ c_{51} & c_{52} & c_{53} & c_{54} & c_{55} & c_{56} & c_{57} & c_{58} & c_{59} \\ c_{61} & c_{62} & c_{63} & c_{64} & c_{65} & c_{66} & c_{67} & c_{68} & c_{69} \\ c_{71} & c_{72} & c_{73} & c_{74} & c_{75} & c_{76} & c_{77} & c_{78} & c_{79} \\ c_{81} & c_{82} & c_{83} & c_{84} & c_{85} & c_{86} & c_{87} & c_{88} & c_{89} \\ c_{91} & c_{92} & c_{93} & c_{94} & c_{95} & c_{96} & c_{97} & c_{98} & c_{99} \end{bmatrix} \begin{bmatrix} \varepsilon_1 \\ \varepsilon_2 \\ \varepsilon_3 \\ \varepsilon_4 \\ \varepsilon_5 \\ \varepsilon_6 \\ \varepsilon_7 \\ \varepsilon_8 \\ \varepsilon_9 \end{bmatrix} \tag{6.10}$$

6 × 6 scheme

$$\begin{bmatrix} \sigma_1 \\ \sigma_2 \\ \sigma_3 \\ \sigma_4 \\ \sigma_5 \\ \sigma_6 \end{bmatrix} = \begin{bmatrix} c_{11} & c_{12} & c_{13} & c_{14} & c_{15} & c_{16} \\ c_{21} & c_{22} & c_{23} & c_{24} & c_{25} & c_{26} \\ c_{31} & c_{32} & c_{33} & c_{34} & c_{35} & c_{36} \\ c_{41} & c_{42} & c_{43} & c_{44} & c_{45} & c_{46} \\ c_{51} & c_{52} & c_{53} & c_{54} & c_{55} & c_{56} \\ c_{61} & c_{62} & c_{63} & c_{64} & c_{65} & c_{66} \end{bmatrix} \begin{bmatrix} \varepsilon_1 \\ \varepsilon_2 \\ \varepsilon_3 \\ \gamma_4 \\ \gamma_5 \\ \gamma_6 \end{bmatrix} \tag{6.11}$$

where $\gamma_{ij} = 2\,\varepsilon_{ij}$.

---

[2] It is equally possible to retain the tensor strain components in the 6×6 scheme by multiplying c_{i4}, c_{i5}, and c_{i6} by 2, but this leads to nonsymmetric and nontensor components of matrix [c], which we wish to avoid. Thus we will retain tensor components of **c** and multiply strains, losing the tensor character of strain components, when using the 6×6 scheme.

Finally, we consider the path independence of elasticity to show that some of the 36 components remaining in Eq. 6.11 are not independent. If $W_{elastic}$ is the path-independent work done from a stress-free state ($\sigma_{ij} = \varepsilon_{ij} = 0$) to a stressed state ($\sigma_{ij}$, ε_{ij}), we can write

$$W_{elastic} = \int_{path} dw = \int_0^{\varepsilon_i} \sigma_i d\varepsilon_i \tag{6.12}$$

where we have used the simple index notation of the 6 ¥ 6 scheme for simplicity. Then, by taking derivatives, we find that

$$\frac{\partial W_{elastic}}{\partial \varepsilon_i} = \sigma_i \tag{6.13}$$

$$\frac{\partial^2 W_{elastic}}{\partial \varepsilon_i \partial \varepsilon_j} = \frac{\partial \sigma_i}{\partial \varepsilon_j} = c_{ij} \tag{6.14}$$

The last equality is obtained by the definition of the elastic constant: $\sigma_i = c_{ij}\varepsilon_j$. However, since $W_{elastic}$ is a path-independent function of strain,

$$\frac{\partial^2 W}{\partial \varepsilon_i \partial \varepsilon_j} = \frac{\partial^2 W}{\partial \varepsilon_j \partial \varepsilon_i} \tag{6.15}$$

and, therefore

$$c_{ij} = c_{ji} \tag{6.16}$$

Thus the 36 elastic constants in the 6 × 6 scheme are reduced to 21 independent numbers that consist of the diagonal constants (c_{11}, c_{22}, etc.) and either the upper or lower triangle, with the other side following.

> ***Note:*** *The symmetry of the stress and strain tensors and the path independence of elastic work are frame-independent, so the symmetry of the 9 × 9 (and 6 × 6) matrix follows in every coordinate system.*

6.5 REDUCTION OF ELASTIC CONSTANTS BY MATERIAL SYMMETRY

The reduction of 81 independent elastic constants to 21 reflects the symmetry of the stress and strain tensors and the path independence of elastic work that occur because of our definitions of these quantities and our concept of ideal elasticity. It is also possible (and necessary) to reflect the properties of the specific material in the form of **c**.

In order to further illustrate the concepts of anisotropy and tensor symmetry, we will develop the rules governing c_{ij} in several special cases. We will assume that the material is in the form of a **single crystal**, which allows us to use the well-known crystal symmetry operations which leave **c** unchanged. Most real materials are **polycrystals** (assemblies of single crystals or "grains"), so the macroscopic properties represent averages of the individual grain properties. In the polycrystal case, the elastic constants may exhibit symmetry, but this will depend on the "texture"; that is, the distribution of preferred orientations of the individual crystals.

> ***Note**: For readers without a background in materials, it is essential to know only the following:* ***Crystals*** *are periodic arrays of atoms that exhibit certain kinds of symmetry. That is, certain directions in the crystal are indistinguishable from certain other directions. Another way to think about this is to imagine rotating the crystal from one orientation to another one. If the rotation is a* ***symmetry operation***, *the crystal in the two orientations will look identical, as will all of its properties. Therefore,* **c**, *which is a crystal property, will be identical after such a rotation. Imagine a square (2-D) array of atoms:*
>
> 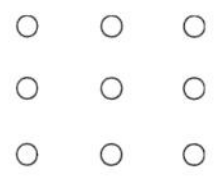
>
> *If it is rotated 90° or 180°, it is the same crystal. Any other rotation appears different:*
>
> 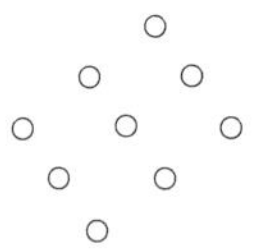
>
> *Thus the square is said to have* ***"fourfold" symmetry*** *(because any rotation by $\pi/2$, $3\pi/2$ or 2π is a symmetry operation).*

While there are several kinds of **crystal symmetry** in nature, we will illustrate the procedure using easily-computed ones. (We will choose convenient coordinate systems[3] and convenient crystal types to reduce complexity.).

[3] We will use the short-hand notation "xl" to indicate "crystal" when referring to standard coordinate axes aligned along certain crystallographic directions.

By doing this, we will avoid carrying out general, formal rotations of **c**, which are very time-consuming. However, we will show a formal rotation when reducing to the isotropic case.

First consider an orthogonal (**orthorhombic**) crystal, one which has a structural unit cell as shown in Figure 6.1.

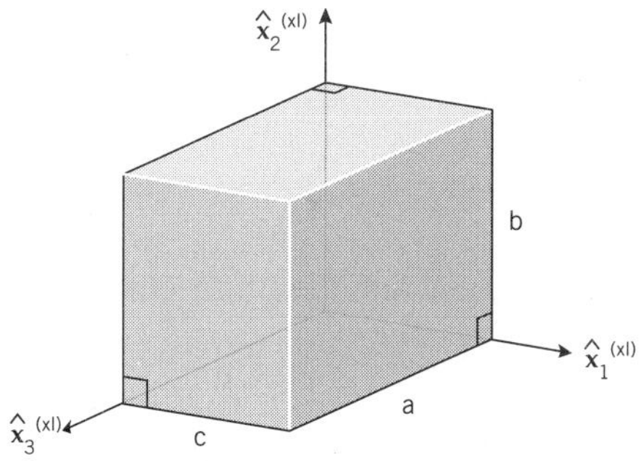

Figure 6.1 Orthogonal symmetry.

All three angles are right angles ($\theta, = \phi = \gamma = 90°$) and all three sides have different lengths ($a \neq b \neq c$).[4] We can capture the full symmetry of the orthogonal crystal by noting that $\hat{X}_1$, $\hat{X}_2$, and $\hat{X}_3$, are twofold axes. That is, the crystal is unchanged if we rotate it by 180° around any one or any combination of these axes. (Try it using Figure 6.1.) Of course, when we rotate the crystal, the elastic constant tensor must follow, since it is linked with the crystal itself. (Note: This application is a real rotation of the tensor itself, not a change of coordinate axes.)

Orthotropic Symmetry Reduction

Let's start with an example for the orthotropic case and use a semi-intuitive approach for the rotation in order to save time. Consider a 180° right-hand rotation of the crystal[5] about $\widehat{\mathbf{x}}_2$, so that the crystal edges (specified by the original orientation) exchange as follows:

xl edge directions		
original		rotated
$\hat{x}_1$	$\rightarrow$	$-\hat{x}_1$
$\hat{x}_2$	$\rightarrow$	$\hat{x}_2$
$\hat{x}_3$	$\rightarrow$	$-\hat{x}_3$

[4] This kind of symmetry is called "orthotropic" in general and it is found in polycrystalline sheets and plates that have been rolled extensively. The three symmetry directions in that case are the rolling direction (RD), long transverse direction (TD), and the thickness direction (STD = short transverse direction). See Chapter 7.

[5] Because **c** is a tensor property of a material, rotation of the material must rotate the tensor itself. This is quite different from finding two sets of components for the same tensor, reversed to different coordinate systems, although the algebra is identical.

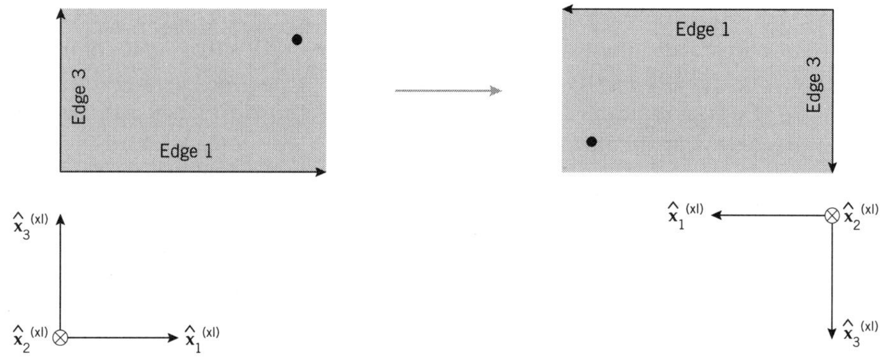

Figure 6.2 Crystal rotation about twofold symmetry axis x_2

(The dot is shown in Figure 6.2 so that the crystal rotation can be visualized. Otherwise, the crystal before and after is indistinguishable, as required by symmetry.) The transformation table $(\hat{x}_1 \rightarrow -\hat{x}_1,\ \hat{x}_2 \rightarrow \hat{x}_2,\ \hat{x}_3 \rightarrow -\hat{x}_3)$ can be used to find the new elastic constant tensor, **c′**, expressed in the old coordinate system $(\hat{x}_1, \hat{x}_2, \hat{x}_3)$. To visualize this better, remember that in the component representation, c_{ijkl}, ij refers to σ_{ij} and kl refers to ε_{kl}. So we can rotate these parts to see what happens to **c**. The transformation of components can be summarized as follows:

Rotation of 180° about $\hat{x}_2$

stresses and strains

vectors	*9×9*	*6×6*	***Elastic constants, all***	
$x_1 \rightarrow -x_1$	$\sigma_{11} \rightarrow \sigma_{11}$	$\sigma_1 \rightarrow \sigma_1$	all	$c_{ij} \rightarrow -c_{ij}$
$x_2 \rightarrow x_2$	$\sigma_{22} \rightarrow \sigma_{22}$	$\sigma_2 \rightarrow \sigma_2$	except	$c_{14} \rightarrow -c_{14}$
$x_3 \rightarrow -x_3$	$\sigma_{33} \rightarrow \sigma_{33}$	$\sigma_3 \rightarrow \sigma_3$	except	$c_{24} \rightarrow -c_{24}$
	$\sigma_{23} \rightarrow -\sigma_{23}$	$\sigma_4 \rightarrow -\sigma_4$	except	$c_{34} \rightarrow -c_{34}$
	$\sigma_{13} \rightarrow \sigma_{13}$	$\sigma_5 \rightarrow \sigma_5$	except	$c_{54} \rightarrow -c_{54}$
	$\sigma_{12} \rightarrow -\sigma_{12}$	$\sigma_6 \rightarrow -\sigma_6$	except	$c_{16} \rightarrow -c_{16}$
			except	$c_{26} \rightarrow -c_{26}$
			except	$c_{36} \rightarrow -c_{36}$
			except	$c_{56} \rightarrow -c_{56}$

The last column shows how the original elastic constant component is moved in c_{ij} when **c** and the crystal itself are rotated. Note that some simple, intuitive rules make this transformation easy to remember. When rotating about $\hat{x}_2$, only σ_{23} and σ_{12} (and ε_{23}, ε_{12}) change signs. This is because the normal components (σ_{ii}, ε_{ii}) always have both axes transforming to + or – of the same axis and thus the sign of the stress or strain component does not change. Refer to Chapters 2 and 3 to recall how the components are defined. Clearly $\sigma(\hat{x}_1,\hat{x}_1) = \sigma(-\hat{x}_1,-\hat{x}_1)$ because the force vectors on the opposite faces of the cube are opposite, but so is the outward area normal, so the stress component is equal, by definition. So the only stress and strain components that are altered by the 180° rotation are those that contain the 9×9 subscripts 1 and 3 *once and only once*. These components are σ_4, σ_6, ε_4, and ε_6. Similarly, for c_{ij}, the only components that change are those that have indices 4 and 6 *once and only once*. These components of **c** are c_{14}, c_{24}, c_{34}, c_{54}, c_{16}, c_{26}, c_{36}, and c_{56}.

Now, consider what we have done physically. The crystal was rotated and the new elastic constants for this rotated crystal (in the same coordinate system) have been found. If the rotation was a **symmetry operation**, then the crystals before and after rotation must be identical. Thus the elastic constants must not have changed, that is, **c** = **c**′ if the prime refers to the symmetry-rotated crystal. Thus we have the following situation:

By tensor rotation	**By symmetry definition**	**Conclusion**
$c_{14} = -c_{14}'$	$c_{14} = c_{14}'$	$c_{14} = 0$
$c_{24} = -c_{24}'$	$c_{24} = c_{24}'$	$c_{24} = 0$
$c_{34} = -c_{34}'$	$c_{34} = c_{34}'$	$c_{34} = 0$
$c_{54} = -c_{54}'$	$c_{54} = c_{54}'$	$c_{54} = 0$
$c_{16} = \ \ c_{16}'$	$c_{16} = c_{16}'$	$c_{16} = 0$
$c_{26} = \ \ c_{26}'$	$c_{26} = c_{26}'$	$c_{26} = 0$
$c_{36} = \ \ c_{36}'$	$c_{36} = c_{36}'$	$c_{36} = 0$
$c_{56} = \ \ c_{56}'$	$c_{56} = c_{56}'$	$c_{56} = 0$
other $c_{ij} = c_{ij}'$	other $c_{ij} = c_{ij}$	no conclusion

So, by a single symmetry rotation of an orthogonal crystal about the $\hat{x}_2$ symmetry axis, we have found that 8 of the 21 independent constants must be identically zero for this crystal.

A second symmetry rotation—say, about the $\hat{x}_3$ axis—can be carried out in precisely the same way with similar results:

Vectors	2nd-ranked components		4th-ranked components		Conclusion
$\hat{x}_1 \to -\hat{x}_1$	$\sigma_{11} \to \sigma_{11}$	$\sigma_1 \to \sigma_1$	all	$c_{ij} \to c_{ij}$	$c_{14} = 0$ *
$\hat{x}_2 \to -\hat{x}_2$	$\sigma_{22} \to \sigma_{22}$	$\sigma_2 \to \sigma_2$	except	$c_{14} \to -c_{14}$ *	$c_{24} = 0$ *
$\hat{x}_3 \to \hat{x}_3$	$\sigma_{33} \to \sigma_{33}$	$\sigma_3 \to \sigma_3$	except	$c_{24} \to -c_{24}$ *	$c_{34} = 0$ *
	$\sigma_{23} \to -\sigma_{23}$	$\sigma_4 \to -\sigma_4$	except	$c_{34} \to -c_{34}$ *	$c_{64} = 0$
	$\sigma_{13} \to -\sigma_{13}$	$\sigma_5 \to -\sigma_5$	except	$c_{64} \to -c_{64}$	$c_{15} = 0$
	$\sigma_{12} \to \sigma_{12}$	$\sigma_6 \to \sigma_6$	except	$c_{15} \to -c_{15}$	$c_{25} = 0$
			except	$c_{25} \to -c_{25}$	$c_{35} = 0$
			except	$c_{35} \to -c_{35}$	$c_{65} = 0$ *
			except	$c_{65} \to -c_{65}$ *	

*These components were already found to be zero by the $\hat{x}_2$ crystal rotation.

The final symmetry operation (about $\hat{x}_1$) is redundant, simply reproducing the information obtained by the first two operations. This result can easily be anticipated by performing the three rotations on the unit cell in Figure 6.1 and observing that the crystal is returned to its original position (the dot on one corner can be used to distinguish the arbitrary orientation).

We are left with the elastic constants for an **orthogonal crystal**, as expressed in a conventional coordinate system fixed along the symmetry axes:

$$\begin{bmatrix} c_{11} & c_{12} & c_{13} & 0 & 0 & 0 \\ c_{12} & c_{22} & c_{23} & 0 & 0 & 0 \\ c_{13} & c_{23} & c_{33} & 0 & 0 & 0 \\ 0 & 0 & 0 & c_{44} & 0 & 0 \\ 0 & 0 & 0 & 0 & c_{55} & 0 \\ 0 & 0 & 0 & 0 & 0 & c_{66} \end{bmatrix} \quad (6.17)$$

(orthotropic elastic constants referred to symmetry axes)

Cubic Symmetry Reduction

Most metals are cubic, so it is most instructive to find a form of c_{ij} for this case. A **cubic crystal** is clearly a special case of the orthotropic crystal, because the angles remain 90° but the cell edge lengths have the relationship a = b = c. Thus all of the previous symmetry is retained and additional kinds will be found. Therefore, to save time, let's start from our result for the orthotropic crystal and find the additional rules.

We note that the additional symmetry is represented by fourfold axes along the cell edges. Since each face is a square, a 90° rotation preserves its appearance. Let's follow our previous procedure for a 90° rotation of the crystal about $\hat{x}_2$:

xl edge directions		
orig.		rotated
$\hat{\mathbf{x}}_1$	$\rightarrow$	$\hat{\mathbf{x}}_3$
$\hat{\mathbf{x}}_2$	$\rightarrow$	$\hat{\mathbf{x}}_2$
$\hat{\mathbf{x}}_3$	$\rightarrow$	$-\hat{\mathbf{x}}_1$

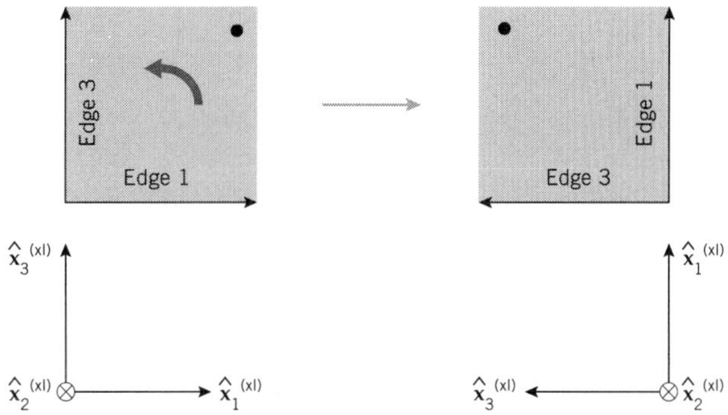

Figure 6.3 Rotation of a cubic crystal by 90° about $\hat{\mathbf{x}}_2$.

The following transformation is obtained:

Vectors	2nd–ranked components 9×9	2nd–ranked components 6×6	4th–ranked components
	$\sigma_{11} \rightarrow \sigma_{33}$	$\sigma_1 \rightarrow \sigma_3$	$c_{11} \rightarrow c_{33}$
$\hat{\mathbf{x}}_1 \rightarrow \hat{\mathbf{x}}_3$	$\sigma_{22} \rightarrow \sigma_{22}$	$\sigma_2 \rightarrow \sigma_2$	$c_{33} \rightarrow c_{11}$
$\hat{\mathbf{x}}_2 \rightarrow \hat{\mathbf{x}}_2$	$\sigma_{33} \rightarrow \sigma_{11}$	$\sigma_3 \rightarrow \sigma_1$	$c_{12} \rightarrow c_{23}$
$\hat{\mathbf{x}}_3 \rightarrow -\hat{\mathbf{x}}_1$	$\sigma_{23} \rightarrow -\sigma_{12}$	$\sigma_4 \rightarrow -\sigma_6$	$c_{23} \rightarrow c_{12}$
	$\sigma_{13} \rightarrow -\sigma_{13}$	$\sigma_5 \rightarrow -\sigma_5$	$c_{44} \rightarrow c_{66}$
	$\sigma_{12} \rightarrow \sigma_{23}$	$\sigma_6 \rightarrow \sigma_4$	$c_{66} \rightarrow c_{44}$

As before, the crystal rotation leaves the physical situation unchanged, so the following conditions are found:

By tensor rotation	By symmetry definition	Conclusion
$c_{11} = c_{33}'$	$c_{33} = c_{33}'$	$c_{11} = c_{33}'$
$c_{33} = c_{11}'$	$c_{11} = c_{11}'$	$c_{33} = c_{11}'$
$c_{12} = c_{23}'$	$c_{23} = c_{23}'$	$c_{12} = c_{23}'$
$c_{23} = c_{12}'$	$c_{12} = c_{12}'$	$c_{23} = c_{12}'$
$c_{44} = c_{66}'$	$c_{66} = c_{66}'$	$c_{44} = c_{66}'$
$c_{66} = c_{44}'$	$c_{44} = c_{44}'$	$c_{66} = c_{44}'$

By carrying out a second 90° rotation about either of the remaining symmetry axes ($\hat{\mathbf{x}}_1$ or $\hat{\mathbf{x}}_3$), the following additional conditions are found: $c_{11} = c_{22} = c_{33}$, $c_{44} = c_{66} = c_{55}$, and $c_{12} = c_{23} = c_{13}$. (As before, the third symmetry operation reproduces the first two.) Thus, the elastic constant matrix, as expressed in a coordinate system fixed on the symmetry (cubic edge) axes, is as follows:

$$\begin{bmatrix} c_{11} & c_{12} & c_{12} & 0 & 0 & 0 \\ c_{12} & c_{11} & c_{12} & 0 & 0 & 0 \\ c_{12} & c_{12} & c_{11} & 0 & 0 & 0 \\ 0 & 0 & 0 & c_{44} & 0 & 0 \\ 0 & 0 & 0 & 0 & c_{44} & 0 \\ 0 & 0 & 0 & 0 & 0 & c_{44} \end{bmatrix} \tag{6.18}$$

(cubic elastic constants referred to symmetry axes)

Typical elastic constants for a few metal crystals are reproduced in Table 6.1.[6]

TABLE 6.1 Typical Metal Crystal Elastic Constants and Anisotropy Ratios (A)

	c_{11} (GPa)	c_{12} (GPa)	c_{44} (GPa)	A
Fe	242	147	112	2.36
Cu	168	121	75	3.21
Al	108	61	29	1.21
Na	6	5	6	8.15
Nb	25	13	29	0.51

Isotropic Symmetry Reduction

There is one final reduction that can be made: When *all* rotations reproduce the same material properties, then the material is said to be **isotropic**. As previously, we start from the cubic case and then add the new symmetry operations needed to reveal the remaining simplifications. Since we have already enforced all of the 180° and 90° rotations, we will now have to consider other, not-as-simple rotations. We will perform this reduction more formally in order to illustrate it. All of the previous rotations could have been carried out this way (if we had enough patience), and this method is needed to investigate general symmetry (three-fold axes, five-fold axes, etc.). However, we will still make use of some physical reasoning to avoid carrying out

[6] Values in Table 6.1 are reproduced from J. P. Hirth and J. Lothe, *Theory of Dislocations*, (New York: McGraw-Hill, 1968), 762.

the arbitrary rotation of **c**, which would involve 26,244 multiplications[7] (and a lot of page space in this book!).

To limit the calculation, let's imagine that a cubic crystal is strained in one direction (say $\hat{\mathbf{X}}_1$) only, such that $\sigma_{11} = \varepsilon$ and other $\varepsilon_{ij} = 0$. Let's look at the response before and after a 45° rotation of the material or crystal, Fig. 6.4.

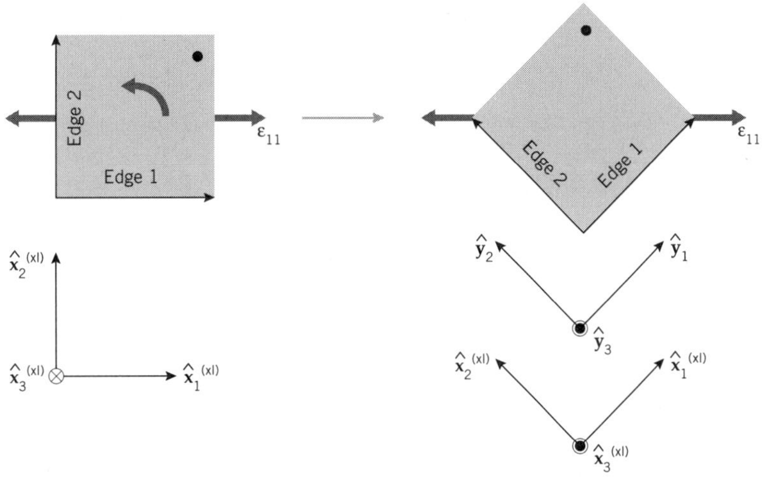

Figure 6.4 Rotation of an isotropic material by 45° about $\hat{\mathbf{x}}_3$.

Before the rotation, Hooke's law (for an assumed cubic crystal) gives us the stresses:

$$\text{before rotation: } \varepsilon_{11} = \varepsilon,\ \text{other } \varepsilon_{ij} = 0,\ \sigma_1 = c_{11}\varepsilon,\ \sigma_2 = c_{12}\varepsilon,\ \sigma_3 = c_{12}\varepsilon. \qquad (6.19)$$

After we rotate the crystal, the strain occurs along the diagonal. In order to find the stresses after the rotation, let's refer to a new coordinate system, one in which we *know* the components of **c** (i.e., we consider a coordinate system that rotated with the crystal and tensor **c** itself, so that the components of the tensor don't change). In the $\hat{\mathbf{y}}_1$ coordinate system, the components of the applied strain are found by rotating ε_{11}:

$$R = \begin{bmatrix} \frac{1}{\sqrt{2}} & \frac{1}{\sqrt{2}} \\ -\frac{1}{\sqrt{2}} & \frac{1}{\sqrt{2}} \end{bmatrix} \qquad \begin{aligned} \varepsilon_1 &= \varepsilon_{11}\ (\text{in } \hat{\mathbf{y}}_i) = R_{11}R_{11}\varepsilon_{11} = \frac{1}{2}\varepsilon \\ \varepsilon_2 &= \varepsilon_{22}\ (\text{in } \hat{\mathbf{y}}_i) = R_{21}R_{21}\varepsilon_{11} = \frac{1}{2}\varepsilon \\ \varepsilon_6 &= \varepsilon_{12}\ (\text{in } \hat{\mathbf{y}}_i) = R_{11}R_{21}\varepsilon_{11} = -\frac{1}{2}\varepsilon \end{aligned} \qquad (6.20)$$

[7] Each new term of c_{ijkl} will require 4 × 81 multiplications, the 4 representing R_{im}, R_{jn}, R_{ko}, and R_{lp}, and the 81 representing the 81 elastic constants in the 9 × 9 scheme (including zeros).

The corresponding stresses, again in the $\hat{\mathbf{y}}_i$ coordinate system, are found from Hooke's law:

$$\sigma_1(\text{in } \hat{\mathbf{y}}_i) = c_{11}\,\varepsilon_1 + c_{12}\,\varepsilon_2 = (c_{11} + c_{12})\,\frac{\varepsilon}{2} \tag{6.21a}$$

$$\sigma_2(\text{in } \hat{\mathbf{y}}_i) = c_{12}\,\varepsilon_1 + c_{22}\,\varepsilon_2 = (c_{12} + c_{22})\,\frac{\varepsilon}{2} \tag{6.21b}$$

$$\sigma_6(\text{in } \hat{\mathbf{y}}_i) = 2c_{44}(-\frac{1}{2})\varepsilon = -c_{44}\varepsilon \tag{6.21c}$$

To complete the comparison, let's rotate the stress components back to the original coordinate system:[8]

$$\sigma_{11}'(\text{in } \hat{\mathbf{x}}_i) = \frac{1}{2}\sigma_1(\text{in } \hat{\mathbf{y}}_i) + \frac{1}{2}\sigma_2(\text{in } \hat{\mathbf{y}}_i) - \sigma_6(\text{in } \hat{\mathbf{y}}_i) \tag{6.22a}$$

$$\sigma_1'(\text{in } \hat{\mathbf{x}}_i) = \left[\frac{1}{2}(c_{11} + c_{12}) + \frac{1}{2}(c_{12} + c_{11}) + 2c_{44}\right]\frac{\varepsilon}{2} = \left(\frac{1}{2}c_{11} + \frac{1}{2}c_{12} + c_{44}\right)\varepsilon \tag{6.22b}$$

Now we can complete the development by noting that σ_1 must be equal to σ_1', where σ_1 is derived from Hooke's law in the original coordinate system ($= c_{11}\varepsilon$), and σ_1' is derived from Hooke's law for the rotated material, in the same coordinate system, for the same applied strain:

$$\sigma_1\ (\text{in } \hat{\mathbf{x}}_i) = \sigma_1'\ (\text{in } \hat{\mathbf{x}}_i) \tag{6.23a}$$

$$c_{11}\,\varepsilon = \frac{1}{2}(c_{11} + c_{12} + c_{44})\,\varepsilon \tag{6.23b}$$

$$\text{or,}\quad c_{44} = c_{11} - c_{12} \tag{6.23c}$$

[8] Note: Care must be taken to carry out the tensor rotations properly, using the 9 × 9 scheme.

With this additional result, the isotropic elastic constant may be written as follows, in *any* coordinate system:

$$\begin{bmatrix} c_{11} & c_{12} & c_{12} & 0 & 0 & 0 \\ c_{12} & c_{11} & c_{12} & 0 & 0 & 0 \\ c_{12} & c_{12} & c_{11} & 0 & 0 & 0 \\ 0 & 0 & 0 & c_{11}-c_{12} & 0 & 0 \\ 0 & 0 & 0 & 0 & c_{11}-c_{12} & 0 \\ 0 & 0 & 0 & 0 & 0 & c_{11}-c_{12} \end{bmatrix} = \begin{bmatrix} \lambda+2\mu & \lambda & \lambda & 0 & 0 & 0 \\ \lambda & \lambda+2\mu & \lambda & 0 & 0 & 0 \\ \lambda & \lambda & \lambda+2\mu & 0 & 0 & 0 \\ 0 & 0 & 0 & 2\mu & 0 & 0 \\ 0 & 0 & 0 & 0 & 2\mu & 0 \\ 0 & 0 & 0 & 0 & 0 & 2\mu \end{bmatrix} \tag{6.24}$$

where the traditional isotropic elastic Lame's constants λ and μ have been substituted:

$$\mu = c_{44} = \frac{1}{2}(c_{11} - c_{12}) \tag{6.25}$$

$$\lambda = c_{12} \tag{6.26}$$

$$\lambda + 2\mu = c_{11} \tag{6.27}$$

The **anisotropy ratio** for anisotropic crystals is defined as

$$A = \frac{2\, c_{44}}{c_{11} - c_{12}} \tag{6.28}$$

which departs from unity as the crystal becomes more anisotropic.

Hooke's law for isotropic materials may be written compactly in indicial notation using Eq. 6.24:

$$\sigma_{ij} = \left[\lambda\, \delta_{ij}\, \delta_{kl} + \mu\left(\delta_{ik}\, \delta_{jl} + \delta_{il}\, \delta_{jk}\right)\right] \varepsilon_{kl} \tag{6.29}$$

where

$$c_{ijkl} = \lambda\, \delta_{ij}\, \delta_{kl} + \mu\left(\delta_{ik}\, \delta_{jl} + \delta_{il}\, \delta_{jk}\right) \tag{6.30}$$

or it may be written in conventional nonmatrix form:

$$\sigma_{11} = (\lambda + 2\mu)\,\varepsilon_{11} + \lambda\,\varepsilon_{22} + \lambda\,\varepsilon_{33} \tag{6.31a}$$

$$\sigma_{22} = \lambda\,\varepsilon_{11} + (\lambda + 2\mu)\,\varepsilon_{22} + \lambda\,\varepsilon_{33} \tag{6.31b}$$

$$\sigma_{33} = \lambda\,\varepsilon_{11} + \lambda\,\varepsilon_{22} + (\lambda + 2\mu)\,\varepsilon_{33} \tag{6.31c}$$

$$\sigma_{23} = 2\mu\,\varepsilon_{23} \tag{6.31d}$$

$$\sigma_{31} = 2\mu\,\varepsilon_{31} \tag{6.31e}$$

$$\sigma_{12} = 2\mu\,\varepsilon_{12} \tag{6.31f}$$

The isotropic elastic constant matrix may also be inverted readily in order to find the elastic compliances, s_{ij}, such that:

$$\varepsilon_i = s_{ij}\,\sigma_j \quad \text{or} \quad \varepsilon_{ij} = s_{ijkl}\,\sigma_{kl} \tag{6.32}$$

The work is left to the following exercise.

Exercise 6.1 **Invert the isotropic elastic constant matrix in Eq. 6.24 to obtain the compliance matrix.**

We note that [c] is diagonal, in either the 6×6 or 9×9 scheme, except for the first three rows and columns. The inverses of the diagonal elements are simply reciprocals, except for the 3×3 filled matrix. For the first three rows and columns, we invert using the signed cofactor method presented in Chapter 2. First, the determinant is found:

$$\Delta = (c_{12} - c_{11})^2\ (2c_{12} + c_{11}) \tag{6.1-1}$$

Then we find the cofactors and new elements:

$$s_{11} = \frac{\text{cofactor } c_{11}}{\Delta} = \frac{c_{11} + c_{12}}{(c_{11} - c_{12})\,(2c_{12} + c_{11})} \tag{6.1-2}$$

$$s_{12} = \frac{\text{cofactor } c_{12}}{\Delta} = \frac{c_{12}}{(c_{12} - c_{11})(2c_{12} + c_{11})} \tag{6.1-3}$$

and the matrix separately is preserved, so Hooke's law for an isotropic material may be written in inverse form:

$$\begin{bmatrix} \varepsilon_{11} \\ \varepsilon_{22} \\ \varepsilon_{33} \\ \varepsilon_{23} \\ \varepsilon_{13} \\ \varepsilon_{12} \end{bmatrix} = \begin{bmatrix} s_{11} & s_{12} & s_{12} & 0 & 0 & 0 \\ s_{12} & s_{11} & s_{12} & 0 & 0 & 0 \\ s_{12} & s_{12} & s_{11} & 0 & 0 & 0 \\ 0 & 0 & 0 & 2(s_{11} - s_{12}) & 0 & 0 \\ 0 & 0 & 0 & 0 & 2(s_{11} - s_{12}) & 0 \\ 0 & 0 & 0 & 0 & 0 & 2(s_{11} - s_{12}) \end{bmatrix} \begin{bmatrix} \varepsilon_{11} \\ \varepsilon_{22} \\ \varepsilon_{33} \\ \varepsilon_{23} \\ \varepsilon_{13} \\ \varepsilon_{12} \end{bmatrix} \tag{6.1-4}$$

The elastic compliances for an isotropic material are usually written in terms of conventional constants:

$$s_{ij} = \begin{bmatrix} 1/E & -\nu/E & -\nu/E & 0 & 0 & 0 \\ -\nu/E & 1/E & -\nu/E & 0 & 0 & 0 \\ -\nu/E & -\nu/E & 1/E & 0 & 0 & 0 \\ 0 & 0 & 0 & 1/\mu & 0 & 0 \\ 0 & 0 & 0 & 0 & 1/\mu & 0 \\ 0 & 0 & 0 & 0 & 0 & 1/\mu \end{bmatrix} \tag{6.1-5}$$

6.6 TRADITIONAL ISOTROPIC ELASTIC CONSTANTS

A simple way to introduce traditional isotropic elastic constants is to consider the behavior of a simple 1-D sample subjected to a tension, a test that must give identical results whatever direction it has in the material. The initial length of the sample is l_0, its initial cross-section area is A_0, and the traction force is F, as shown in Figure 6.5.

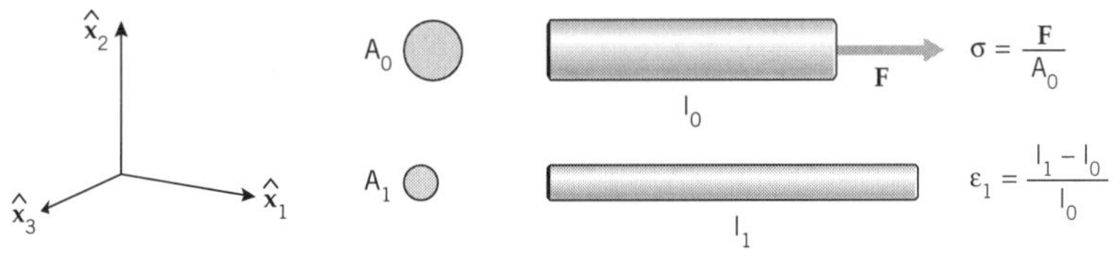

Figure 6.5 Elastic deformation of a 1-D sample.

In this case we need only to describe the elongation of the bar and the variation of its cross section. The first condition is expressed in terms of 1-D stress and strain by the linear relation

$$\sigma_1 = E\,\varepsilon_1 \quad \text{or} \quad \varepsilon_1 = \frac{1}{E}\,\sigma_1 \tag{6.33}$$

where the parameter E is called **Young's modulus**. Taking advantage of the symmetry around the $\hat{\mathbf{x}}_1$ axis for an isotropic material, the second condition will be written

$$\varepsilon_2 = \varepsilon_3 = -\nu\,\varepsilon_1 \tag{6.34}$$

so that the new cross- sectional area is

$$A_1 = A_0\,(1 - 2\,\nu\,\varepsilon_1) \tag{6.35}$$

where 6rv is the **Poisson coefficient,** which ranges between 0 and 0.5, the value 0.5 corresponding to incompressibility, as is easily seen by substitution:

$$\varepsilon_1 + \varepsilon_2 + \varepsilon_3 = (1 - 2\,\nu)\,\varepsilon_1 = 0 \tag{6.36}$$

Comparison of Eqs. 6.27 and 6.1-4 reveals that E is equal to

$$\frac{1}{s_{11}} = \frac{(c_{11} - c_{12})\,(2c_{12} + c_{11})}{c_{11} + c_{12}} \tag{6.37}$$

The **shear modulus**, G, is defined in simple shear such that $J_{ij} = G\gamma_{ij} = 2G\varepsilon_{ij}(i \neq j)$. Therefore it is clear that $G = \mu$. With these few traditional isotropic definitions, it is possible to rearrange all isotropic elastic expressions in terms of any two conventional constants (unless they are dependent, such as μ and G). A few simple expressions may prove helpful:

$$\mu = G = \frac{E}{2(1+\nu)} = c_{44} \tag{6.38}$$

$$\lambda = \frac{\nu E}{(1+\nu)\,(1-2\nu)} = \frac{2\nu\mu}{1-2\nu} \tag{6.39}$$

$$B = \text{bulk modulus}^9 = \lambda + \frac{2}{3}G = \frac{E}{3\,(1-2\nu)} \tag{6.40}$$

[9] The bulk modulus is the ratio of the pressure to the volume contraction: $B = \dfrac{\frac{1}{3}(\sigma_1 + \sigma_2 + \sigma_3)}{\varepsilon_{11} + \varepsilon_{22} + \varepsilon_{33}} = \dfrac{1}{K}$ where K = **compressibility**.

$$\nu = \frac{3B - 2\mu}{2\,(3B + \mu)} = \frac{\lambda}{2(\mu + \lambda)} = \frac{E - 2\mu}{2\mu} = \frac{c_{12}}{c_{12} + c_{11}} \tag{6.41}$$

$$E = \frac{\mu\,(3\lambda + 2\mu)}{\mu + \lambda} = \frac{9\,\mu\,B}{3B + \mu} = 2\mu\,(1 + \nu) \tag{6.42}$$

Typical traditional elastic constants for polycrystalline metals[10] are presented in Table 6.2.

TABLE 6.2 Typical Polycrystal Elastic Constants for Bulk Metals

	E (GPa)	G (GPa)	ν
Steel	200	76	0.3
Copper	110	41	0.35
Aluminum	69	26	0.33

6.7 ELASTIC PROBLEMS: AIRY'S STRESS FUNCTION

Solving elasticity problems has little interest for metal forming, except for some specialized applications combined with large-strain plasticity, such as **springback,**[11] or deformation of tooling. However, since we will proceed with a numerical approach throughout the remainder of the book, it is worthwhile introducing the differential equilibrium and compatibility equations, and their combination in a form suitable for solving analytically.

In order to simplify the presentation, we will consider only 2-D problems, that is, ones in which the field variables (ε, σ) depend only on coordinates x_1 and x_2. For example:

$$\varepsilon_{ij} = \varepsilon_{ij}(x_1\ x_2)$$
$$\sigma_{ij} = \sigma_{ij}(x_1\ x_2)$$

Furthermore, we need not consider ε_3 or σ_3 explicitly, because we will assume that one of these two is given by the boundary conditions (e.g., $\varepsilon_3 = 0$ for **plane strain,** or $\sigma_3 = 0$ for **plane stress**). The one that is not specified can be found after solution by substitution into Hooke's law.

---

[10] Values for Table 6.2 are taken from L. E. Malvern, *Introduction to the Mechanics of a Continuous Medium* (New York: Prentice-Hall, 1969)p. 293.2

[11] Springback is the elastic deflection that occurs after forming when the loads are removed. It occurs because of residual stresses introduced during the operation.

We can start with the 2-D equilibrium equations and combine them to make a single equation (using the notation x, y in place of x_1, y_2):

$$x: \quad \frac{\partial \sigma_x}{\partial x} + \frac{\partial \sigma_{xy}}{\partial y} = 0 \quad \rightarrow \quad \frac{\partial^2 \sigma_x}{\partial x^2} + \frac{\partial^2 \sigma_{xy}}{\partial x \partial y} = 0 \tag{6.43}$$

$$y: \quad \frac{\partial \sigma_y}{\partial y} + \frac{\partial \sigma_{xy}}{\partial x} = 0 \quad \rightarrow \quad \frac{\partial^2 \sigma_y}{\partial y^2} + \frac{\partial^2 \sigma_{xy}}{\partial x \partial y} = 0 \tag{6.44}$$

$$\frac{\partial^2 \sigma_{xy}}{\partial x \partial y} = -\frac{1}{2}\left(\frac{\partial^2 \sigma_x}{\partial x^2} + \frac{\partial^2 \sigma_y}{\partial y^2}\right) \tag{6.45}$$

Equation 6.45 is the equilibrium condition for 2-D problems. Note that it represents a relationship among the derivatives of three functions: $\sigma_x(x,y)$, $\sigma_y(x,y)$, $\sigma_{xy}(x,y)$.

Equations 6.43 and 6.44 can be automatically satisfied by noting that a certain substitution ("trick") leads to an identity:

$$\text{Let:} \quad \frac{\partial^2 \psi}{\partial x^2} = \sigma_y, \qquad \frac{\partial^2 \psi}{\partial y^2} = \sigma_x, \qquad \frac{\partial^2 \psi}{\partial x \partial y} = -\sigma_{xy} \tag{6.46}$$

where Ψ is an arbitrary, differentiable function of x and y. Rewriting Equations 6.43 to 6.45 using these substitutions obtains:

$$\frac{\partial \sigma_x}{\partial x} + \frac{\partial \sigma_{xy}}{\partial y} = \frac{\partial^3 \Psi}{\partial x \partial y^2} - \frac{\partial^3 \psi}{\partial x \partial y^2} = 0 \tag{6.47a}$$

$$\frac{\partial \sigma_{xy}}{\partial x} + \frac{\partial \sigma_y}{\partial y} = \frac{-\partial^3 \Psi}{\partial x^2 \partial y} + \frac{\partial^3 \psi}{\partial x^2 \partial y} = 0 \tag{6.47b}$$

$$\frac{\partial^4 \psi}{\partial x^2 \partial y^2} = \frac{1}{2}\left(\frac{\partial^4 \psi}{\partial x^2 \partial y^2} + \frac{\partial^4 \psi}{\partial y^2 \partial x^2}\right) \tag{6.47c}$$

which is identically true if Ψ is differentiable:

$$\frac{\partial^2 \psi}{\partial x \partial y} = \frac{\partial^2 \psi}{\partial y \partial x} \tag{6.48}$$

Therefore, any Ψ continuous and differentiable will satisfy Eq. 6.37, and we need not concern ourselves with equilibrium. Ψ is called an **Airy stress function** after Sir George Airy (1801–1892), a famous English astronomer and mathematician.

6.8 COMPATIBILITY CONDITION

Our original problem statement requires that all motion descriptions be derived from two displacement functions, $u_1(x, y)$ and $u_2(x, y)$. It so happens that our material behavior is expressed in terms of strains:

$$\varepsilon_x = \frac{\partial u_x}{\partial x} \quad \varepsilon_y = \frac{\partial u_y}{\partial y} \quad \varepsilon_{xy} = \frac{1}{2}\left(\frac{\partial u_x}{\partial y} + \frac{\partial u_y}{\partial x}\right) \tag{6.49}$$

We must therefore insure that our strains can be derived from just two displacements. This places an additional restriction on the allowable variation of strain components in space:

$$\frac{\partial^2 \varepsilon_{xx}}{\partial y^2} = \frac{\partial^3 u_x}{\partial x \partial y^2}, \quad \frac{\partial^2 \varepsilon_{yy}}{\partial x^2} = \frac{\partial^3 u_y}{\partial y \partial x^2}, \quad \frac{\partial^2 \varepsilon_{xy}}{\partial x \partial y} = \frac{1}{2}\left(\frac{\partial^3 u_x}{\partial x \partial y^2} + \frac{\partial^3 u_y}{\partial y \partial x^2}\right) \tag{6.50}$$

But $u_i = u_i(x,y)$ must be continuous and differentiable:

$$\frac{\partial^3 u_x}{\partial x \partial y^2} = \frac{\partial^3 u_x}{\partial y^2 \partial x} \tag{6.51}$$

That is, the order of differentiation is not important, such that

$$\frac{\partial^2 \varepsilon_{xx}}{\partial y^2} + \frac{\partial^2 \varepsilon_{yy}}{\partial x^2} = 2\frac{\partial^2 \varepsilon_{xy}}{\partial x \partial y} \tag{6.52}$$

Equation 6.52, the **compatibility condition**, shows the required relationship among derivatives of strain components, but we frequently need to make the connection to stresses to get a single, required relationship. We consider the two usual choices:

plane stress

$$\varepsilon_{xx} = \frac{1}{E}(\sigma_{xx} - \nu\sigma_{xx})$$
$$\varepsilon_{yy} = \frac{1}{E}(\sigma_{yy} - \nu\sigma_{xx})$$
$$\varepsilon_{xy} = \frac{1+\nu}{E}\sigma_{xy}$$

plain strain[12]

$$\varepsilon_{xx} = \frac{1}{E}\left[\left(1-\nu^2\right)\sigma_{xx} - \nu(1+\nu)\sigma_{yy}\right]$$
$$\varepsilon_{yy} = \frac{1}{E}\left[\left(1-\nu^2\right)\sigma_{yy} - \nu(1+\nu)\sigma_{xx}\right]$$
$$\varepsilon_{xy} = \frac{1+\nu}{E}\sigma_{xy}$$

(6.53)

Let's assume plane-stress conditions since they are shorter to write. Rewriting terms in Eq. 6.52 obtains

$$\frac{\partial^2 \varepsilon_{xx}}{\partial y^2} = \frac{1}{E}\frac{\partial^2}{\partial y^2}\left(\sigma_{xx} - \nu\sigma_{yy}\right) \tag{6.54a}$$

$$\frac{\partial^2 \varepsilon_{yy}}{\partial x^2} = \frac{1}{E}\frac{\partial^2}{\partial x^2}\left(\sigma_{yy} - \nu\sigma_{xx}\right) \tag{6.54b}$$

$$\frac{\partial^2 \varepsilon_{xy}}{\partial x \partial y} = \frac{1+\nu}{E}\frac{\partial^2}{\partial x \partial y}\sigma_{xy} \tag{6.54c}$$

Now we can rewrite the **compatibility condition in stress-equivalent form**:

$$\frac{\partial^2}{\partial y^2}\left(\sigma_{xx} - \nu\sigma_{yy}\right) + \frac{\partial^2}{\partial x^2}\left(\sigma_{yy} - \nu\sigma_{xx}\right) = 2\,(1+\nu)\frac{\partial^2}{\partial x \partial y}\sigma_{xy} \tag{6.55}$$

and we introduce the equilibrium condition:

$$\frac{\partial^2}{\partial y^2}\left(\sigma_{xx} - \nu\sigma_{yy}\right) + \frac{\partial^2}{\partial x^2}\left(\sigma_{yy} - \nu\sigma_{xx}\right) = -\,(1+\nu)\left(\frac{\partial^2 \sigma_{xx}}{\partial x^2} + \frac{\partial^2 \sigma_{yy}}{\partial y^2}\right) \tag{6.56}$$

---

[12] While the plane-stress relations are obvious from Eqs. 6.31 by setting $\sigma_{33} = 0$, it is necessary in the plane-strain case to substitute $\varepsilon_{33} = 0$ into Eq. 6.31c, find σ_{33}, and then solve for ε_{ij} in terms of σ_{11} and σ_{22}.

$$\text{or: } \frac{\partial^2}{\partial y^2}\left(\sigma_{xx} - \nu\sigma_{yy} + (1+\nu)\sigma_{yy}\right) + \frac{\partial^2}{\partial x^2}\left[\sigma_{yy} - \nu\sigma_{xx} + (1+\nu)\,\sigma_{xx}\right] = 0 \tag{6.57}$$

$$\text{or: } \left(\frac{\partial^2}{\partial x^2} + \frac{\partial^2}{\partial y^2}\right)\left(\sigma_{xx} + \sigma_{yy}\right) = 0 \tag{6.58}$$

Equation 6.58 made use of both equilibrium and compatibility but did not ensure that each was satisfied. Let's go back to the Ψ form, which ensures equilibrium automatically.

Substituting Eqs. 6.46 into Eq. 6.58 obtains

$$\left(\frac{\partial^2}{\partial x^2} + \frac{\partial^2}{\partial y^2}\right)\left(\frac{\partial^2 \psi}{\partial y^2} + \frac{\partial^2 \psi}{\partial x^2}\right) = 0 \tag{6.59}$$

$$\frac{\partial^4 \psi}{\partial x^2 \partial y^2} + \frac{\partial^4 \psi}{\partial x^4} + \frac{\partial^4 \psi}{\partial y^4} + \frac{\partial^4 \psi}{\partial x^2 \partial y^2} = 0 \tag{6.60}$$

$$\left(\frac{\partial^2}{\partial x^2} + \frac{\partial^2}{\partial y^2}\right)^2 \psi = 0 \tag{6.61}$$

$$\text{or: } \nabla^4 \psi = 0 \tag{6.62}$$

Thus any continuous and differentiable function Ψ that satisfies Eq. 6.62 will also satisfy both equilibrium and compatibility, and Hooke's law for material behavior. The trick, as always, is locating a differentiable Ψ that satisfies the specified boundary values or conditions.

CHAPTER 6 PROBLEMS

A. Proficiency Problems

1. Consider a cubic crystal with $c_{11} = 2$, $c_{12} = 1$, and $c_{44} = 1.5$.

   a. Find c_{12}', referred to the new coordinate system, if the $\hat{x}_i'$ axes are rotated 30° counterclockwise about $\hat{x}_3$. The $\hat{x}_i$ axes are aligned along the crystal axes.

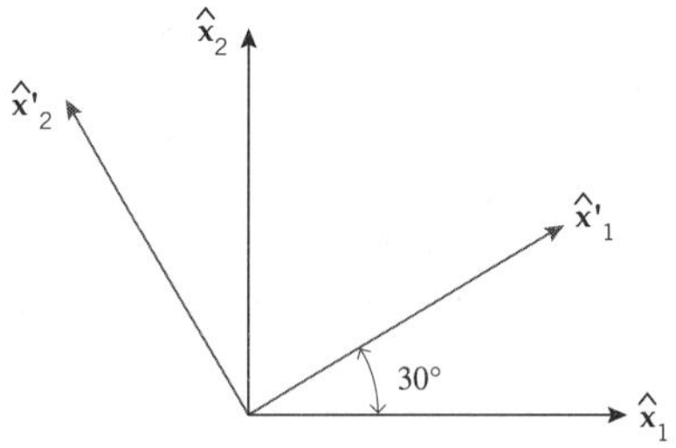

b. What is the anisotropy parameter for this crystal?

2. Repeat Problem 1b for $c_{11} = 1$, $c_{12} = 2$, $c_{44} = 0.5$.

3. Use symmetry to reduce the general 6 × 6 elastic constant matrix to the proper form for a tetragonal cell as shown. (Note: All angles are 90°, and a = b, so there is 90° rotational symmetry about $\hat{x}_2$, and 180° rotational symmetry about $\hat{x}_3$.)

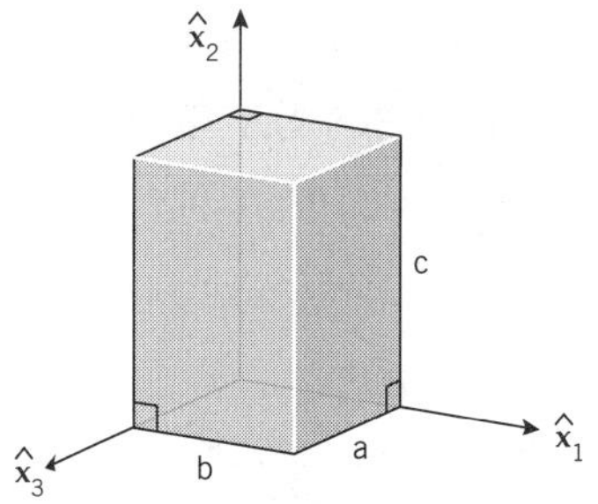

4. a. Is $\psi(x, y) = x^4 + y^4 - 12x^2y^2$ a valid Airy stress function in terms of both equilibrium and compatibility requirements? Show why or why not.

   b. Assuming that the $\psi(x, y)$ in part a is correct, find the stress components at the point (1, 2).

   c. Find the strain components for the point (1, 2) assuming plane-stress conditions ($\sigma_{3i} = 0$). Leave your answers in terms of general Young's modulus (E) and Poisson's ratio (ν).

5. Given: $\psi = 3xy^5 + 3x^5y - 10x^3y^3$

   a. Verify that ψ meets the condition for a proper Airy stress function. Show your work.

   b. Find the stresses as functions of x and y that correspond to ψ.

   c. Check to see if the stresses in part b satisfy equilibrium. Show your work.

   d. With a little manipulation using the stresses in part b and Hooke's law for plane stress, obtain the strain functions shown below. Verify that these strain functions satisfy the compatibility conditions for strains. Show your work.

$$\varepsilon_{xx} = 4\,b\left(xy^3 - x^3y\right)$$
$$\varepsilon_{yy} = 4\,b\left(x^3y - xy^3\right)$$
$$\varepsilon_{xy} = b\left(-x^4 - y^4 + 6x^2y^2\right)$$
$$\text{where}: \quad b \equiv 15\,\frac{(1+\nu)}{E}$$

6. Verify that Eqs. 6.38–6.42 follow from the traditional definitions of elastic constants.

7. Solve Eqs. 6.31 for strains and verify that the result is identical to Eqs. 6.1–5.

B. Depth Problems

8. Discuss any material restrictions that apply to the compatibility conditions Eqs. 6.52 and 6.55. Consider anisotropy, elasticity vs. plasticity or other constitutive equations, possible presence of body forces, and the possibility of voids and cracks developing during deformation.

9. a. Show that the elastic work done during small straining is given by

$$w = \frac{1}{2} c_{ijkl}\ \varepsilon_{ij}\ \varepsilon_{kl}$$

   b. Write the elastic work in terms of stresses alone.

   c. Write results for parts a and b for the isotropic case.

10. a. Show that Hooke's law for an isotropic material may be written in the following form:

$$\sigma_{ij} = 2\mu\left[\varepsilon_{ij} + \frac{\nu}{1-2\nu}\,\varepsilon_{ij}\ \delta_{ij}\right]$$

   b. Find the equilibrium equation in indicial form in terms of strains alone.

11. Consider a plane-strain isotropic linear elastic problem defined in a given domain. Assume the displacement vector field takes the form

$$\mathbf{u} \leftrightarrow \begin{bmatrix} u_1 = a\,x_1^2 + 2b\,x_1x_2 + c\,x_2^2 \\ u_2 = a'\,x_1^2 + 2b'\,x_1x_2 + c'\,x_2^2 \end{bmatrix}$$

   a. Write the strain tensor as a function of x_1 and x_2.

b. Compute the stress tensor as a function of x_1 and x_2 with the isotropic Hooke's law.

c. Write the stress equilibrium equation assuming no body forces, and determine the relation between the coefficients a, b, c, a′, b′, and c′ that we must impose.

d. Calculate the stress vector on each side of a square (see following figure).

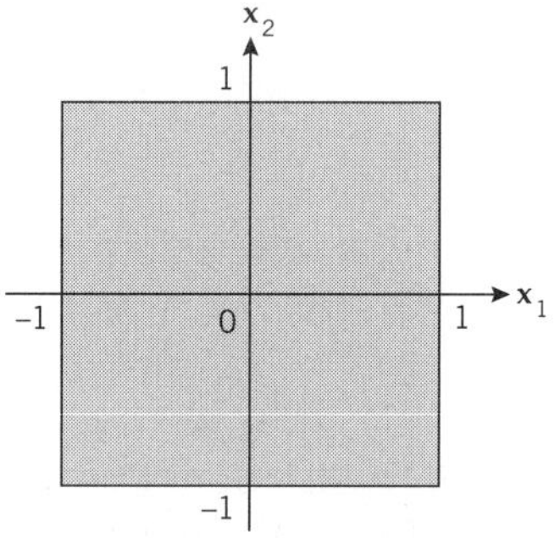

Verify that the external forces on the square are in equilibrium when the condition defined in part c is fulfilled.

12. Consider three elastic bars of equal length that are pinned at the ends as shown in the following illustration, and to which a vertical force **F** is applied.

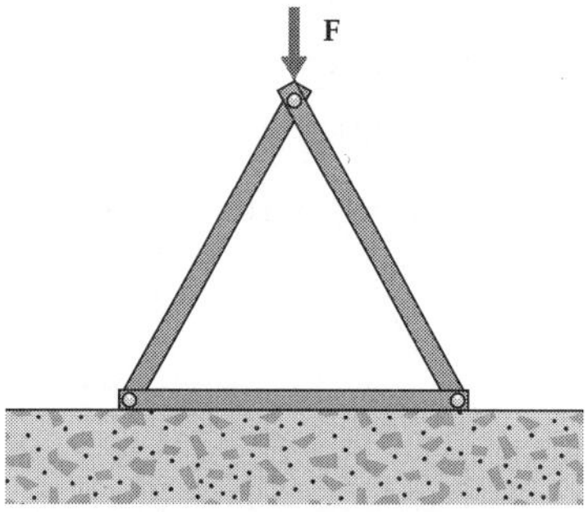

Each bar has a length l, a section area s, and a Young's modulus E. We assume the links between the bars do not permit any torque, the bars remain straight, and the contact with the horizontal plane is frictionless. Compute the variation of length of each bar, with the hypothesis of small displacements. Consider the case where the variation of length is no longer negligible compared to the initial geometry.

13. Consider a plane-strain compression test of a sample with height h, imposing a displacement u at the top as shown in the figure, and assuming no friction on the plates.

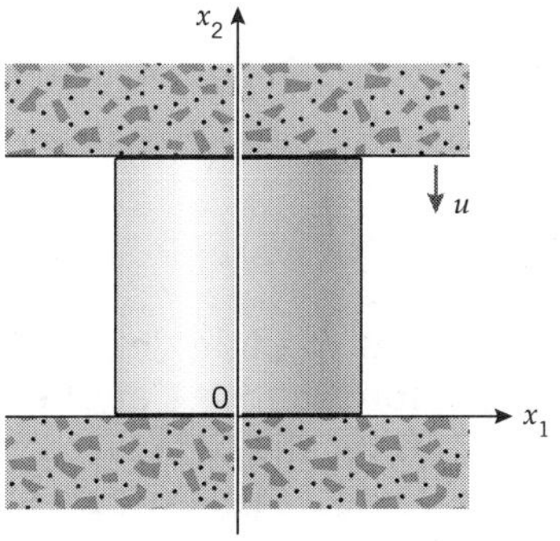

Write the displacement field using a linear expression and taking into account the boundary conditions and the symmetry of the problem. An unknown parameter α should be introduced in the x_1 component. Calculate the strain tensor, the stress tensor with the isotropic Hooke's law, and the elastic energy. Show that the solution of the problem can be obtained by minimizing the elastic energy with respect to the unknown parameter α. Verify that the stress on the vertical sides is equal to zero.

14. Repeat Problem 13 for a cubic crystal where $\mathbf{x}_1$ and $\mathbf{x}_2$ are oriented along fourfold symmetry axes.

15. Repeat Problem 13 with a 3-D sample and an orthotropic material.

16. Show that the *only* incompressible isotropic elastic medium is a liquid by computing the shear modulus, μ.

17. The Airy stress function for a screw dislocation in an isotropic crystal is given by[13]

$$\psi = \frac{\mu b y}{4\pi(1-\nu)} \ln\left(x^2 + y^2\right)$$

where b = Burger's vector, a small constant.

   a. Find the stresses and strains for the screw dislocation.
   b. Is compatibility satisfied everywhere? Why or why not?
   c. Do plane-strain or plane-stress conditions hold?

18. The elastic fields of a screw dislocation are most simply derived by considering the displacements in polar coordinates:

$$u_z(r,\theta) = \frac{b\theta}{2\pi} = \frac{b}{2\pi}\tan^{-1}\frac{y}{x}$$

where b = Burger's vector, a small constant.

---

[13] J. P. Hirth and J. Lothe, *Theory of Dislocations* (New York: McGraw-Hill, 1968)P. 74.

a. Find the stresses and strains for the screw dislocation.
b. Is compatibility satisfied everywhere? Why or why not?
c. Do plane-strain or plane-stress conditions hold?

C. Numerical Problems

19. Consider the material triangle defined in Chapter 4, Problem 16. Find a matrix expression for the forces applied to each corner of the triangle in order to deform it elastically such that small displacements $\Delta\mathbf{a}, \Delta\mathbf{b}, \Delta\mathbf{c}$ occur at each corner. Assume a unit thickness of material. You may wish to use an equivalent work principle or make other approximations and assumptions about how the stress felt by the triangle relates to the forces applied at the corner.

CHAPTER 7

Plasticity

In this chapter, we introduce the most important elements of classical plasticity, the main material behavior used in analyzing large-strain forming operations. Because of its central importance to the field, we develop the ideas in some detail, starting with the concept of a yield surface, the normality condition, the flow curve, and strain hardening. We finish this chapter by examining some classical yield functions, and their advantages and disadvantages, and leave more complex presentations to the next chapter.

We consider in this chapter only the **flow theory of plasticity** (also known as the classical theory); that is, a formulation in which the current strain rates[1] (or infinitesimal increments) depend on the stress. A distinct theory, the so-called **deformation theory** (also known as Hencky theory), relates total strain to stress, and as such represents a version of nonlinear elasticity. Only along proportional strain and stress paths do the two approaches coincide. When unloading occurs, the results of the deformation theory are in clear error when compared with experience (i.e., only elastic materials return to their original shape when the load is removed.) For this reason, deformation theory is used only when the convenient mathematical form is required to construct closed-form solutions to fairly simple problems. This was largely the motivation in using the theory originally, long before high-speed digital computation allowed the use of flow theory for realistic problems of interest.

The theory of plasticity is in poor condition relative to linear elasticity. All existing formulations are approximate at best, and few have any connection to the fundamental (micromechanical) material mechanisms responsible. For this

[1] The actual strain rate (i.e., $\dot{\boldsymbol{\varepsilon}} = \mathbf{D}$) is not important in time-independent plasticity, but we must distinguish between large strains and the current infinitesimal increment of strain, $d\varepsilon$. For reasons of convenience, this infinitesimal increment is often called the strain rate, with the understanding that the time scale is not important, only that we watch the current change of strain over a small increment of displacement, time, loading, or other path-defining parameter.

reason, all plasticity "theories" must be considered empirical. Even worse, detailed testing of a plasticity theory requires extensive, careful experimentation using special machinery and precise interpretation over many paths and histories. This requirement has slowed development of the field greatly.

The **history dependence** of plasticity also means that virtually no analytical solutions exist for real forming problems using realistic materials. Thus analysis of forming usually proceeds numerically, except for certain steady-state or 1-D problems with idealized material response. This is in sharp contrast to elasticity and elastic problems, where most developments have been made analytically, with numerical solutions reserved for complex structures.

We now proceed in an intuitive manner to develop the basic ideas of plasticity, by considering the simplest model that has a reasonable connection to the physical reality. We begin from the most basic concepts of flow plasticity: (1) that stress induces strain rates (analogous to pressure and velocity in liquids), and (2) that a two-stage behavior (elastic and plastic) is implied. We develop these ideas by means of thought experiments.

7.1 THE YIELD SURFACE

Let's start by extending our idea of a tensile to two dimensions. Imagine that we can load and unload a square sheet of material in two directions, in any proportion that we choose (Figure 7.1).

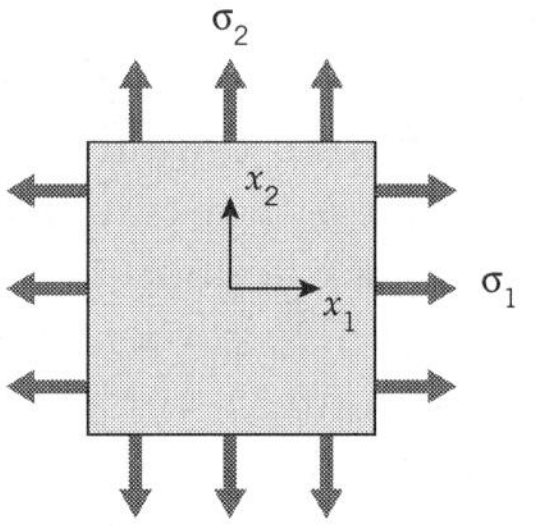

Figure 7.1 Biaxially loaded sheet.

Let's choose a specific **stress path**: a fixed $\sigma_2 = \sigma_2'$ that does not cause plastic yielding, combined with various s_1's that are applied, as shown in Figure 7.2. We load to σ_2', $\sigma_1(1, 2, 3,...)$, as shown by the e's and p's in Figure 7.2*a*, and unload each time, looking for **permanent deformation** remaining in the unloaded state. Figure 7.2*b* shows the measured stress and strain along x_1. We can see that plastic deformation began at a stress (or strain) between $\sigma_1(7)$ and $\sigma_1(8)$. Before this, the deformation was fully elastic. Thus, within the uncertainty defined by $\sigma_1(8) - \sigma_1(7)$, we have measured one yield stress under combined σ_1-σ_2 loading. We can, of course, improve on the accuracy of the measurement by taking smaller $\Delta\sigma_1$, as long as we can measure the small $\Delta\varepsilon_{plastic}$ that results.

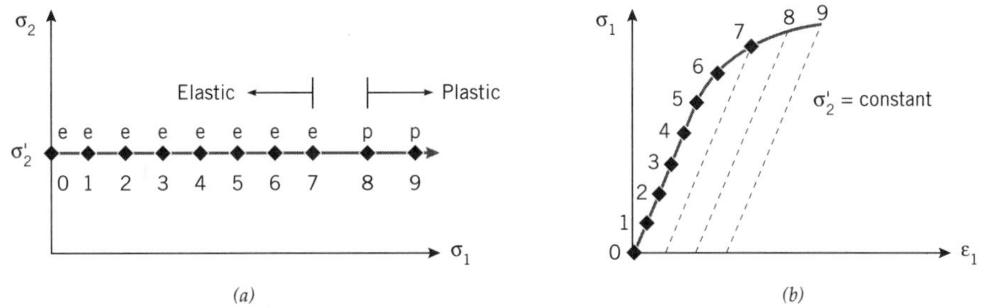

Figure 7.2 Elastic-plastic transition with constant σ_2.

We can continue this procedure for a different value of $\sigma_2 = \sigma_2''$ and identify another **elastic-plastic transition** within the accuracy of $\Delta\sigma_1$. If we repeat the process for many values, we can obtain a 2-D mapping of the discrete stress states that first cause plastic deformation (Figure 7.3*a*).

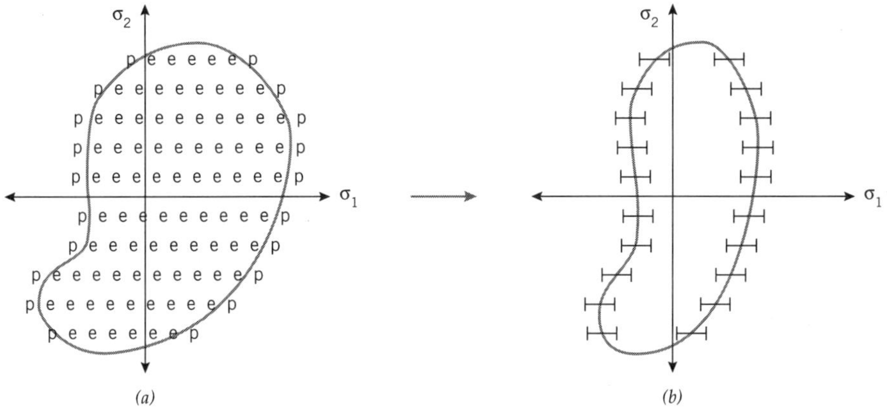

Figure 7.3 Stress states for first observation of plastic deformation.

When we interpolate between the last stress inducing purely elastic deformation and the first one inducing plastic deformation, we obtain a plot like the one shown in Figure 7.3*b*. Finally, by connecting the measured points with a smooth curve, we can arrive at a surface that represents, within the experimental limitations, the loci of all stress combinations that first cause plastic deformation. This curve or surface is called the **yield surface**.

There are several questions and criticisms that arise immediately from the procedure followed. There is limited accuracy unless a great many experiments are done to reduce $\Delta\sigma_1$ and $\Delta\sigma_2$. Given the distinct points that are found, it is impossible to say confidently that the entire locus of such points is smooth. It could be that the pattern of yield stresses between points is extremely rough, as shown in Figure 7.4.

If such behavior were discovered, a great many additional experiments would be needed. As it is, the presence of strain hardening requires a new, virgin specimen for each path locating one yield-stress combination. (Otherwise, the small plastic deformation induced will cause hardening and will change the form of the yield surface being sought.)

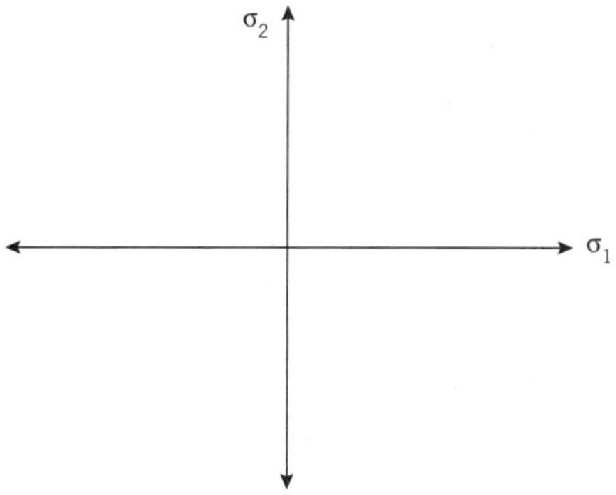

Figure 7.4 Possible shape of yield surface determined from discrete measurements.

The last conceptual drawback to the measurement envisioned is that the yield surface shape may depend on the direction of the stress or **strain path** used to probe.[2] In this case, we are on firmer ground because any path *inside* the yield surface is elastic, and so the response should be **path-independent** until the yield surface is reached. In fact, most measurements of yield surfaces are made with **radial paths** (also called **proportional paths**)—Figure 7.5*a*—but other paths can be followed, as illustrated in Figure 7.5*b*.

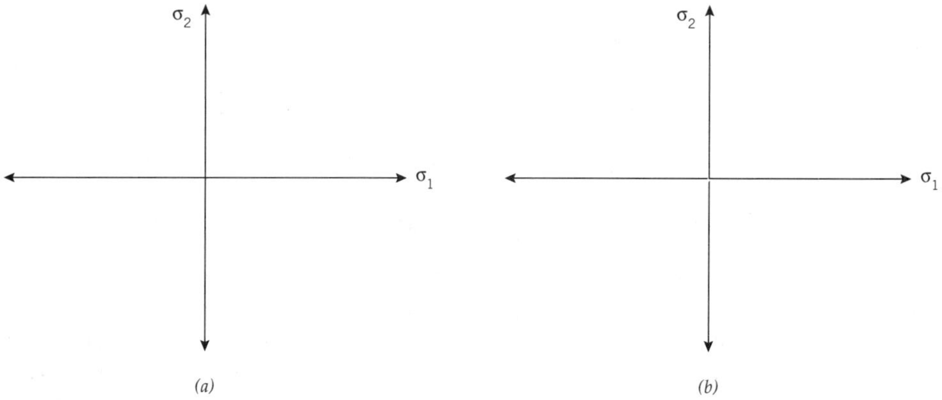

Figure 7.5 Radial and other stress paths for probing plastic yielding.

7.2 A SIMPLE YIELD FUNCTION

Many forms of the yield function (which defines the yield surface) have been proposed. Many of the older ones are only of historical interest, because they did not meet the test of reflecting the basic ideas of a simple plastic response.

---

[2] Such a possibility is contrary to our ideal view of elastic material response, but real materials may behave differently.

In order to avoid this pitfall, we construct the simplest yield function that meets basic criteria; again, we resort to intuition.

We start by assuming that the yield surface is a closed, smooth surface of nonpathological shape. At an instant of time (i.e., at one material *state*), the yield surface is defined in the usual way:

$$f(\sigma_{ij}) = f(\sigma_{11}, \sigma_{22}, \sigma_{33}, \sigma_{23}, \sigma_{13}, \sigma_{12}) = k \tag{7.1}$$

The surface is thus represented in 6-D space, where each dimension refers to one of the stress components.

Let's restrict our attention to an **isotropic material**; that is, one that has the same properties in all directions. In this case, we can write everything in terms of principal stresses[3] (σ_1, σ_2, σ_3) and reduce the surface to a three-dimensional one. Unfortunately, as shown in previous chapters, it is necessary to solve a cubic equation to relate the stress components to the principal components, so it is simpler for purposes of discussion to consider the stress invariants rather than the principal stresses. (Recall that J_1, J_2, J_3 or λ_1, λ_2, λ_3, directly represent the stress state via the cubic equations.)

$$\lambda^3 - J_1\lambda^2 - J_2\lambda - J_3 = 0 \quad \text{or} \quad (\sigma-\lambda_1)(\sigma-\lambda_2)(\sigma-\lambda_3) = 0 \tag{7.2}$$

The stress invariants are simply related to the principal stresses:

$$J_1 = \sigma_1 + \sigma_2 + \sigma_3 \tag{7.3a}$$

$$J_2 = -(\sigma_1\sigma_2 + \sigma_2\sigma_3 + \sigma_1\sigma_3) \tag{7.3b}$$

$$J_3 = \sigma_1\sigma_2\sigma_3 \tag{7.3c}$$

So, to incorporate this material simplification into our yield function under construction, we write it in terms of either triplet:

$$\text{isotropic: } k = f(J_1, J_2, J_3) \quad \text{or} \quad k = f(\sigma_1, \sigma_2, \sigma_3) \tag{7.4}$$

---

[3] This is possible because all stress tensors have only three components in the properly chosen, principal coordinate system. Because in an isotropic material we cannot distinguish one direction from another, we may freely choose this convenient coordinate system that captures the essence of the stress state while losing the orientation information, which is irrelevant.

As a second reduction in complexity, we note that, to a very high accuracy, plastic deformation is **pressure-independent.**[4] We can test this by noting that the plastic volume change is nearly zero, and that solids under hydrostatic pressure do not deform plastically. Micromechanically, dislocations are moved in response to shear stresses, which are not present in a hydrostatic state.

To ensure that our yield function under construction is pressure-independent, we can write it in terms of reduced stress variables, which have the pressure removed. That is, we must map the principal stress components to three other components that differ only by a pressure term:

$$\sigma_1 \rightarrow \sigma_1' = \sigma_1 + p/3 \tag{7.5a}$$

$$\sigma_2 \rightarrow \sigma_2' = \sigma_2 + p/3 \tag{7.5b}$$

$$\sigma_3 \rightarrow \sigma_3' = \sigma_3 + p/3 \tag{7.5c}$$

Because the hydrostatic pressure is $-\frac{\sigma_1 + \sigma_2 + \sigma_3}{3}$ (or $-\frac{\sigma_{11} + \sigma_{22} + \sigma_{33}}{3}$), the mapping is as follows:

$$\sigma_1' = \sigma_1 - \frac{\sigma_1 + \sigma_2 + \sigma_3}{3} = \frac{2}{3}\sigma_1 - \frac{1}{3}\sigma_2 - \frac{1}{3}\sigma_3 \tag{7.6a}$$

$$\sigma_2' = \sigma_2 - \frac{\sigma_1 + \sigma_2 + \sigma_3}{3} = \frac{2}{3}\sigma_2 - \frac{1}{3}\sigma_1 - \frac{1}{3}\sigma_3 \tag{7.6b}$$

$$\sigma_3' = \sigma_3 - \frac{\sigma_1 + \sigma_2 + \sigma_3}{3} = \frac{2}{3}\sigma_3 - \frac{1}{3}\sigma_1 - \frac{1}{3}\sigma_2 \tag{7.6c}$$

The deviatoric stress invariants follow directly from these results. Clearly, σ_i' are the **deviatoric stress components,** as defined in Chapter 3. Thus we can write our yield function in terms of deviatoric stress to avoid any dependence on pressure:

$$\text{Isotropic, pressure-independent:}\ f(J_2', J_3') \tag{7.7}$$

J_1' is identically zero, because $\sigma_1' + \sigma_2' + \sigma_3' = 0$. Therefore, we have reduced our yield surface to a function of two variables: J_2' and J_3'.

In addition to isotropy and pressure-independence, the plastic response of metals is often observed to be nearly the same in tension and compression. That

---

[4] However, at very high pressures, there is some small effect on the yield surfaces of metals, presumably related to the effect of pressure on dislocation cores and subsequent mobility.

is, there is typically no **Bauschinger** effect, or a very small effect, particularly after a small transitional strain following the path change.[5] In the case of no Bauschinger effect, the sign of the stress is unimportant; that is, σ_{ij} should be equivalent to $-\sigma_{ij}$. By examining the form of the remaining stress variables, J_2' and J_3' (Eqs. 7.3b and 7.3c), we can make additional reductions:

$$J_2'(\sigma'_{ij}) = J_2'(-\sigma'_{ij}) \tag{7.8a}$$

$$J_3'(\sigma'_{ij}) = -J_3'(-\sigma'_{ij}). \tag{7.8b}$$

Thus we can see that lack of a Bauschinger effect provides no additional restriction on the use of J_2', but we must insure that f is an even function of J_3'; otherwise, there will always be a built-in Bauschinger effect. We could simply insist that J_3' enter any yield surface in squared form $(J_3')^2$, but this would involve powers of 6 of the stress components, which would lead away from the simplicity we seek. Instead, we can eliminate Bauschinger effects simply and effectively by ignoring J_3' altogether. Then the form of our simple, physically realistic yield function becomes

$$c = f(J_2') \tag{7.9}$$

(isotropic, pressure-independent, no Bauschinger effect)

Referring still to principal stress axes, we can back-substitute the definition of the second stress invariant (Eq. 7.3b) and deviatoric stress components (Eqs. 7.6) to obtain the yield function represented in Eq. 7.9 in terms of stress components:

$$J_2' = -(\sigma_1'\sigma_2' + \sigma_2'\sigma_3' + \sigma_1'\sigma_3') \tag{7.10}$$

$$= -\frac{1}{9}\left[(2\sigma_1 - \sigma_2 - \sigma_3)(2\sigma_2 - \sigma_1 - \sigma_3) + (2\sigma_2 - \sigma_1 - \sigma_3)(2\sigma_3 - \sigma_1 - \sigma_2) + (2\sigma_1 - \sigma_2 - \sigma_3)(2\sigma_3 - \sigma_1 - \sigma_2)\right] \tag{7.11}$$

$$= \frac{1}{3}\left[\sigma_1^2 + \sigma_2^2 + \sigma_3^2 - \sigma_1\sigma_2 - \sigma_1\sigma_3 - \sigma_2\sigma_3\right] \tag{7.12}$$

$$J_2' = \frac{1}{6}\left[(\sigma_1 - \sigma_2)^2 + (\sigma_1 - \sigma_3)^2 + (\sigma_2 - \sigma_3)^2\right] \tag{7.13}$$

[5] The Bauschinger effect loses significance when strain paths are nearly proportional, as they are for many forming operations. Only when the strains remain near the elastic-plastic limit, and when the sign of the stress reverses sign, are Bauschinger effects dominant. For most forming analysis, we will assume that there is no Bauschinger effect.

Thus Eq. 7.9 becomes

$$c = (\sigma_1 - \sigma_2)^2 + (\sigma_1 - \sigma_3)^2 + (\sigma_2 - \sigma_3)^2 = f(\sigma_i) \tag{7.14a}$$

or, in a general (not principal) coordinate system:

$$c = (\sigma_{11} - \sigma_{22})^2 + (\sigma_{11} - \sigma_{33})^2 + (\sigma_{22} - \sigma_{33})^2 + 6\sigma_{12}^2 + 6\sigma_{13}^2 + 6\sigma_{23}^2 \tag{7.14b}$$

where the factor of $\frac{1}{6}$ has been incorporated into the arbitrary constant, c. Equations 7.14 represent the well-known **von Mises yield function**. By inspection, we can see that it is isotropic (written in principal stresses, *and* each principal stress enters f equivalently); pressure-independent (adding a constant to the principal stresses does not change the function); and has no Bauschinger effect (the sign of the stresses does not alter the function).

For plane-stress plasticity problems, we assume that $\sigma_3 = 0$, and the von Mises yield condition becomes

$$c = (\sigma_1 - \sigma_2)^2 + \sigma_1{}^2 + \sigma_2{}^2 \tag{7.15a}$$

or

$$\frac{c}{2} = \sigma_1{}^2 + \sigma_2{}^2 - \sigma_1\sigma_2 \tag{7.15b}$$

> **Note**: *Although we used the ideas of deviatoric stresses and stress invariants throughout the thought derivation of this yield function, the result can be expressed in real stress components, without reference to these special, reduced representations. For nearly all applications, it is more logical to deal with stresses directly, ignoring the special forms.*

The yield surface represented by Eqs. 7.14 and 7.15 is sketched in Figure 7.6. Note that in 2-D ($\sigma_3 = 0$, or σ_3 = another constant) the yield surface is a closed curve (an ellipse with major axis along $\sigma_1 = \sigma_2$), and in 3-D is an infinite cylinder with generator along the $x_1 = x_2 = x_3$ axis. The different appearance is related to the pressure-independence. In 2-D, the specification of $\sigma_3 = 0$ means that each choice of σ_1 and σ_2 has a unique pressure. When all three principal stress components may vary independently, there are many equivalent stress states, all of which differ by a fixed addition, that is, $\sigma_1, \sigma_2, \sigma_3$ is equivalent to σ_1 - p, σ_2 - p, σ_3 - p, for all p. Thus the $\sigma_1 = \sigma_2 = \sigma_3$ axis must be a symmetry one for translations along it. As a result, all the hydrostatic stress states (up to any pressure) are elastic, and must remain inside the yield surface.

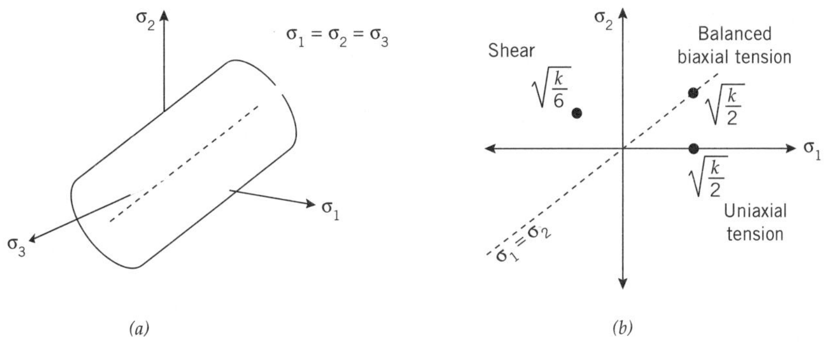

Figure 7.6 Schematic of von Mises yield surface in terms of three principal stresses.

The values of the yield surface points shown Figure 7.6*b* are easily obtained by substituting stress ratios in Eqs. 7.15:

$$\text{uniaxial tension: } \sigma_1 = \sigma,\ \sigma_2 = \sigma_3 = 0 \quad \sigma = \sqrt{\frac{C}{2}} \tag{7.16}$$

$$\text{balanced biaxial tension: } \sigma_1 = \sigma_2 = \sigma,\ \sigma_3 = 0 \quad \sigma = \sqrt{\frac{C}{2}} \tag{7.17}$$

$$\text{pure shear:}^{6}\ \sigma_{12} \rightarrow \sigma_2 = -\sigma_1 = \sigma,\ \sigma_3 = 0 \quad \sigma = \sqrt{\frac{C}{6}} = k \tag{7.18}$$

Where we have conventionally used k to represent the sheer yield stress, reference to Fig. 7.6*b* shows that the ellipse has a ratio of major to minor axes of $\sqrt{2/3}$, which results from passing a plane through the 3-D cylinder, 45° to the generator axis.

7.3 EQUIVALENT OR EFFECTIVE STRESS

For application of plasticity to real problems, it is necessary to find the value of the arbitrary constant k, which represents the *size* of the yield surface, as

[6] Because shear stresses are not shown on our plot, we must first rotate the coordinate system by 45° to find the principal stresses corresponding to pure shear (or, we could simply solve the eigenvalue problem). The result shows that $\sigma_{12} = \sigma$ is equivalent to $\sigma_1 = \sigma$, $\sigma_2 = -\sigma$. This transformation is legitimate because the material is isotropic, so the material response will be the same in every orientation.

opposed to the *shape* (which is fixed by the form of the equation). Physically, k represents the hardness[7] of the material; its elastic-plastic transition stress. Because this quantity is unknown and depends on the state of the material (microstructure, heat treatment, prior processing, chemistry, grain size, etc.), we must rely on simple experiments to find its value at a given state of the material.

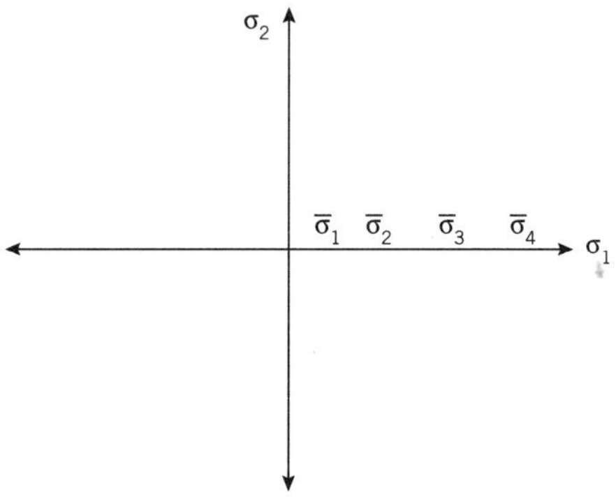

Figure 7.7 Conceptual view of subsequent yield surfaces defined by effective stress.

The tensile test is usually employed for this purpose, although other tests have advantages in providing data at higher strains. The yield stress in tension is defined to be $\bar{\sigma}$, the **equivalent stress**, or **effective stress**. The words "equivalent" or "effective" refer to the yield surface as an iso-state surface, representing all of the combinations of stress that represent the elastic-plastic transition. A single number, $\bar{\sigma}$, in conjunction with the form of the yield surface, provides all such combinations for one state, whereas all values of $\bar{\sigma}$ represent sets of allowable yield functions at various states, as in Figure 7.7.

To evaluate k from a given test, we simply substitute the measured yield stresses into the yield function (Eq. 7.14) and solve. For a tensile test ($\sigma_1 = \bar{\sigma}$, $\sigma_2 = \sigma_3 = 0$), we find

$$k = \bar{\sigma}^2 + \bar{\sigma}^2, \text{ so } k = 2\bar{\sigma}^2 \tag{7.19}$$

in terms of the material effective stress (hardness), the von Mises yield function becomes

$$\bar{\sigma}^2 = \frac{1}{2}\left[(\sigma_1 - \sigma_2)^2 + (\sigma_2 - \sigma_3)^2 + (\sigma_2 - \sigma_3)^2\right] \tag{7.20a}$$

or

[7] We use the term "hardness" in a general sense without reference to specific test procedure. We mean that a harder material requires larger stresses to undergo first plastic deformation.

$$\sqrt{2}\,\bar{\sigma} = \left[(\sigma_1-\sigma_2)^2+(\sigma_1-\sigma_3)^2+(\sigma_2-\sigma_3)^2\right]^{1/2} \quad (7.20b)$$

In this book, we will always assume a tensile state of stress to define $\bar{\sigma}$, but alternate tests provide similar results (following Eqs. 7.17 and 7.18, e.g.):

balanced biaxial tension: same as uniaxial tension (7.21a)

$$\text{pure shear: } \bar{\sigma}'^2 = k/6$$

so

$$\sqrt{6}\,\bar{\sigma}' = \left[(\sigma_1-\sigma_2)^2+(\sigma_1-\sigma_3)^2+(\sigma_2-\sigma_3)^2\right]^{1/2} \quad (7.21b)$$

7.4 NORMALITY AND CONVEXITY, MATERIAL STABILITY

The yield surface tells us only the combinations of stresses that initiate plastic flow, analogous in 1-D to the yield stress. In order to build a constitutive equation for a material, we must connect the yielding with plastic-strain increments (or "rates"). The situation is analogous to fluid flow. We must find the velocities (strain rates) knowing the pressures (stresses). Until the 1950s, such relationships were proposed without theoretical basis. Starting then, however, Drucker showed that stable plastic materials require certain relationships between $\dot{\boldsymbol{\varepsilon}}$ and $\mathbf{s}$, which restrict the form of the yield function. We follow Drucker's approach in the following discussion.

Stability

The idea of **first-order mechanical stability** can be understood intuitively. If we picture a sharp pencil on a table, the three cases of stability can be readily seen, as illustrated in Figure 7.8.

To probe for stability, we imagine perturbing the position of the pencil by δx and examining the behavior. In fact, we can look at the work done, δw, by the force (f) required to move the pencil by δx:

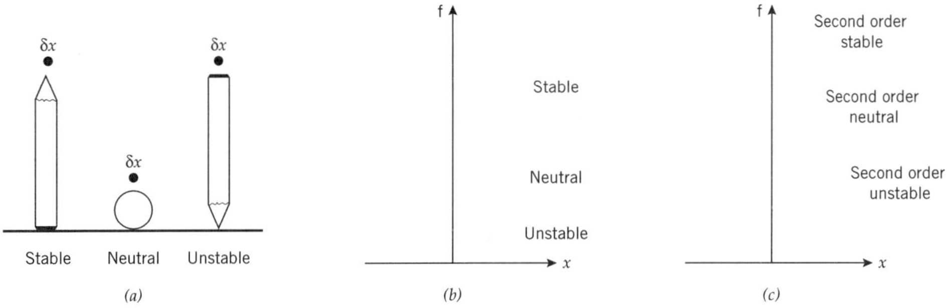

Figure 7.8 Conceptual views of first-order and second-order mechanical stability.

$$\text{stable: f parallel to } \delta x,\ f\delta x > 0 \tag{7.22a}$$

$$\text{neutral: } f = 0,\ f\delta x = 0 \tag{7.22b}$$

$$\text{unstable: f opposite to } \delta x,\ f\delta x < 0 \tag{7.22c}$$

In the stable case, the pencil tends to return to its original position. In the neutral case, there is no response, so the pencil will continue to move if moving. In the unstable case, the pencil will tend to move in the direction of δx. Equations 7.22 define the first order stability of a mechanical system in 1-D. This picture corresponds to a material's elastic response, for example, where the forces (stresses) depend on displacements (strains).

Now, to consider **second order stability**, imagine that a work absorber (friction, for example, or any kind of dissipative absorber will do) is added to the pencil, such that $f\delta x$ is automatically positive for any of the three cases in Figure 7.8 The result is automatically stable because the pencil comes to rest at δx. The question is, can we now distinguish among the three cases? The answer is yes, but we must ask whether the force applied to move the pencil by δx increases or decreases as the distance δx is moved. That is, does the pencil become *more stable* or *less stable* when the δx is applied? Clearly, we can check this by asking whether the force (work gradient) increases or decreases:

$$\delta f\ \delta x > 0 \quad \text{second order stable} \tag{7.23a}$$

$$\delta f\ \delta x = 0 \quad \text{second order neutral} \tag{7.23b}$$

$$\delta f\ \delta x < 0 \quad \text{second order unstable} \tag{7.23c}$$

The second-order situation, where friction automatically absorbs the work done, is shown in Figure 7.8*c*.

We can easily examine material laws with stability in mind by considering 1-D illustrations.

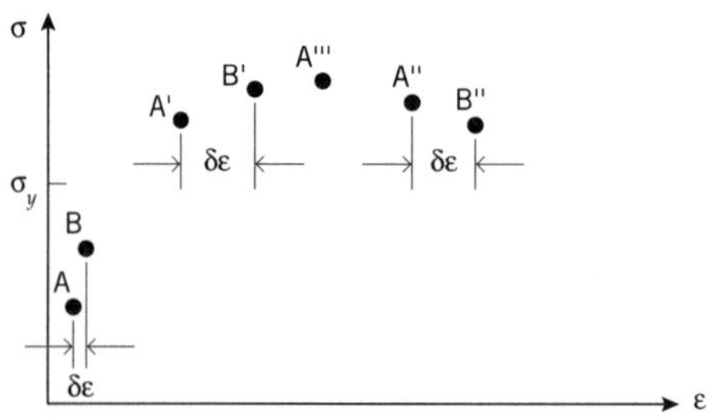

Figure 7.9 Conceptual elastic-plastic response under uniaxial tension.

For elasticity (Figure 7.9, $\sigma < \sigma_y$), we know the fundamental principle is that no work is done around a closed path; that is, that there exists a potential function. The work done by a small displacement ($\delta\varepsilon$) is $\sigma\delta\varepsilon$, where $\sigma = c\delta\varepsilon$, so $w = 1/2\, c\varepsilon^2$. We require this work to be positive, thus c must be positive. The second-order stability is irrelevant in this case because no work is dissipated.

For a plastic response, all of the work done is dissipated (or stored, but not returned as work), so first order stability is automatically satisfied and we must use the idea of second order stability to distinguish the two plastic cases shown in Figure 7.9. Clearly, the stress increases in applying $\delta\varepsilon$ from A´ to B´, while the stress decreases from A´´ to B´´. (The As and Bs were chosen such that $\delta\varepsilon_{A\to B}$ is positive, so that we can examine the sign of δf.) Two cases emerge:

$$A' \to B', \quad \delta f\,\delta\varepsilon > 0, \text{ second order stable} \longrightarrow \text{strain hardening} \tag{7.24a}$$

$$A'' \to B'', \quad \delta f\,\delta\varepsilon < 0, \text{ second order unstable} \longrightarrow \textbf{strain softening} \tag{7.24b}$$

$$A''', \quad \delta f\,\delta\varepsilon = 0, \text{ second order neutral} \longrightarrow \text{no strain hardening} \tag{7.24c}$$

We can see that strain hardening is equivalent to second-order material stability. Without this material property, we cannot solve plasticity problems that have applied forces for strains, because, for a given loading, the material continues to deform without stopping. Thus the strains increase to infinity. (**Creep** is a phenomenon with this kind of behavior, but rate effects limit the total strains at a given time.)

> ***Note:*** *We will restrict our attention to strain-hardening materials only, so that stable solutions for strains may be obtained when forces are applied. This restriction is not severe, as most metals are found to behave in this manner, except for local exceptions such as* ***Luder's bands.***

To complete the 1-D picture a bit more mathematically, consider a closed deformation path originating at any elastic state (σ^A), as in Figure 7.10. We can easily write the work done along the path (it is simply the area under the σ-ε curve for unit material volume):

$$w = (\sigma^B - \sigma^A)\, d\varepsilon^p + \frac{1}{2}\, d\sigma\, d\varepsilon^p = w_1 + w_2 \tag{7.25}$$

For a purely elastic response, $d\varepsilon^p = 0$, so no net work is done. Since we started at an elastic state, $(\sigma^B - \sigma^A)\, d\varepsilon^p$ is strictly positive for the plastic deformation, so first-order stability is satisfied. If $(\sigma^B - \sigma^A)\, d\varepsilon^p$ was less than zero, we could extract plastic work from the material. The second term, $1/2\, d\sigma\, d\varepsilon^p$, represents the test for second order stability, and we see that it is positive for a strain-hardening material.

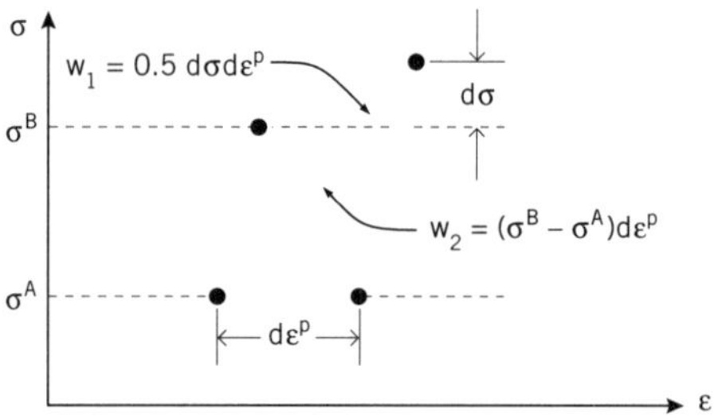

Figure 7.10 Work done during an elastic-plastic cycle in one dimension.

Note that the only requirement we imposed on this test was an initial elastic state and a closed stress path back to this stress. Thus we can consider the two work terms independently. If we let σ^A approach σ^B, the first-order work is zero and only w_2 survives. In general, of course, $d\sigma d\varepsilon << \sigma d\varepsilon$, so we can examine only the first-order work, w_1. For the material to follow our ideas of a stable plastic response, we require that $w_1 > 0$ and $w_2 > 0$ for a closed plastic path. The extension of these ideas to three dimensions is straightforward. The work done around a general, infinitesimal cycle starting from an elastic state requires consideration of all the stress terms and all of the resulting strains:

$$w = (\sigma_{ij}{}^B - \sigma_{ij}{}^A)\, d\varepsilon_{ij}{}^P + \frac{1}{2}\, d\sigma_{ij}\, d\varepsilon_{ij}{}^P = w_1 + w_2 \tag{7.26}$$

As in 1-D, we require that both w_1 and w_2 be greater than zero for an arbitrary path starting from an elastic state in order to ensure that the material response is stable. If we start from the second-order work term (i.e., choose $\sigma^A = \sigma^B$ for the path), we see that $d\sigma_{ij} d\varepsilon_{ij}{}^P > 0$ is required. Geometrically, it is easiest to think of $d\sigma$ and $d\varepsilon$ as vectors (in 6-D or 9-D space). In such a case, the condition that

$$d\sigma \cdot d\varepsilon > 0 \tag{7.27}$$

requires that the vectors $d\sigma$ and $d\varepsilon$ have a positive projection on one another (i.e., that the angle between the two be less than 90°).

Furthermore, from the fundamental definition of flow theory plasticity, the strain increment direction can depend only on the stress σ, not on the stress increment $d\sigma$.[8] Finally, we require $d\sigma$ to have a component outward from the elastic region such that plastic deformation is induced.[9] A 2-D representation of the situation is shown in Figure 7.11.

---

[8] For example, consider general fluid flow, which depends on total pressures, not on dp. The direction of water flowing in a pipe (analogous to dislocations flowing in a metal) depends on p, not small perturbations of p.

[9] A stress increment $d\sigma$ oriented inward (i.e., toward the elastic region) represents an elastic increment. Similarly, a $d\sigma$ tangent to the yield surface induces no hardening and is also elastic.

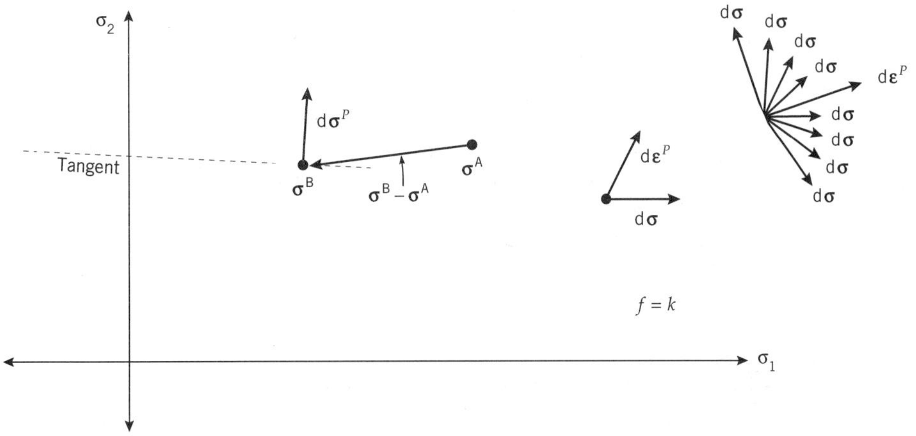

Figure 7.11 Geometry of yield surface and material response vectors for elastic-plastic cycle.

Our idea of stability (based on strain hardening) requires that θ be less than 90°, for any choice of **dσ** that produces plastic deformation (see inset, Fig. 7.11). However, because the *direction* of **dε** is independent of dσ (but does depend on σ, and thus on f), the only choice for **dε** satisfying these two conditions is a vector normal to the yield surface f. This choice of dε is called the **normality rule** or **normality condition**. It follows directly from the strain-hardening assumption and the independence of direction of **dε** and **dσ**. Mathematically, the normality condition is usually written

$$d\varepsilon_{ij} = d\lambda \frac{\partial f}{\partial \sigma_{ij}} \tag{7.28}$$

where dλ is an arbitrary constant indicating that the size of dε is not determined by the normality condition. It has been assumed throughout that f is differentiable in the vicinity of σ, thus ignoring the special case of yield surface **virtices**)

Now that we have found the required direction of the plastic-strain increment, we reexamine the first-order stability condition (in vector form): $(\boldsymbol{\sigma}^B - \boldsymbol{\sigma}^A)\cdot d\boldsymbol{\varepsilon}^P > 0$ Remembering that σ^A is *any choice of* elastic stress state, we can see one choice of σ^A and σ^B in Figure 7.11. Clearly, the vector $(\sigma^B - \sigma^A)$ has a negative projection onto dε. Thus work can be extracted from this system ($w_1 < 0$). Since this is forbidden by stability, no elastic states can be available outside of the tangent line (shown dashed at σ^B). The only way to forbid this happening is to insure that the surface f = k be convex outward at *every* point. A local convexity is not sufficient because the yield surface could pass through the tangent plane far away from the σ^B of interest if convexity were ever violated.

In summary, it is important to note the basis for the foregoing arguments and to recognize under what conditions the two results might be violated. The normality condition requires a strain-hardening material and a constitutive law with $d\sigma$ and $d\varepsilon$ independent. Few materials violate the first assumption locally,[10] but the independence of $d\sigma$ and $d\varepsilon$ is a relatively untested assumption arising from the analogy with fluid flow, and our understanding of single dislocation motion. It is easy to imagine a plasticity law where **de** is related both to σ and to $d\sigma$. Of course, if a unique normal cannot be defined (at a vertex, e.g.) then the strain increment may adopt a range of directions. If normality is violated, then the basis of convexity is destroyed. Instead, the stability of first-order work would require a kind of combined convexity relative to $d\varepsilon$ and to the yield surface.

Exercise 7.1 **Find the direction of the strain increment for any principal stress state (σ_1, σ_2, σ_3) for a von Mises material. Then, find real strain and stress ratios for three typical states (with $\sigma_3 = 0$): uniaxial tension, balanced biaxial tension, and plane-strain tension. Finally, write the strain increment direction for any arbitrary coordinate system.**

Starting from the von Mises yield function, in principal stress components, Eq 7.14a:

$$\bar{\sigma}^2 = \frac{1}{2}\left[(\sigma_1-\sigma_2)^2+(\sigma_1-\sigma_3)^2+(\sigma_2-\sigma_3)^2\right] \tag{7.1-1}$$

and applying the normality condition (Eq 7.28) yields:

$$d\varepsilon_1 = d\lambda\frac{\partial\bar{\sigma}^2}{\partial\sigma_1} = \frac{d\lambda}{2}\left[2(\sigma_1-\sigma_2)+2(\sigma_1-\sigma_3)\right] = (2\sigma_1-\sigma_2-\sigma_3)d\lambda \tag{7.1-2}$$

$$d\varepsilon_2 = d\lambda\frac{\partial\bar{\sigma}^2}{\partial\sigma_2} = \frac{d\lambda}{2}\left[-2(\sigma_1-\sigma_2)+2(\sigma_2-\sigma_3)\right] = (2\sigma_2-\sigma_1-\sigma_3)d\lambda \tag{7.1-3}$$

$$d\varepsilon_3 = d\lambda\frac{\partial\bar{\sigma}^2}{\partial\sigma_3} = \frac{d\lambda}{2}\left[-2(\sigma_1-\sigma_3)-2(\sigma_2-\sigma_3)\right] = (2\sigma_3-\sigma_1-\sigma_2)d\lambda \tag{7.1-4}$$

The result clearly shows that the material volume is unchanged during plastic deformation $(d\varepsilon_1+d\varepsilon_2+d\varepsilon_3=0)$. For uniaxial tension in the x_1 direction: $\sigma_1=\sigma, \sigma_2=\sigma_3=0$:

$$d\varepsilon_1 = 2\,\sigma d\lambda$$

[10] Macroscopic violations usually involve geometric or thermal effects, such as Luder's bands, adiabatic shear bands, and so on.

$$d\varepsilon_2 = -\sigma d\lambda \qquad \text{or} \qquad \varepsilon_1 / \varepsilon_2 / \varepsilon_3 = 2 / -1 / -1$$

$$d\varepsilon_3 = -\sigma d\lambda \tag{7.1-5}$$

For balanced biaxial tension, $\sigma_1 = \sigma_2 = \sigma$, $\sigma_3 = 0$:

$$d\varepsilon_1 = \sigma d\lambda$$

$$d\varepsilon_2 = \sigma d\lambda \qquad \text{or} \qquad \varepsilon_1 / \varepsilon_2 / \varepsilon_3 = 1 / 1 / -2$$

$$d\varepsilon_3 = -2\,\sigma d\lambda \tag{7.1-6}$$

Note that except for a sign change and a change of axis label, uniaxial tension represents the same state as balanced biaxial tension.

For plane-strain tension, $d\varepsilon_1 = d\varepsilon$, $d\varepsilon_2 = 0$, $d\varepsilon_3 = -d\varepsilon$:

$$d\varepsilon_2 = 0 = (2\sigma_2 - \sigma_1)\, d\lambda, \text{ so } \sigma_2 = \frac{\sigma_1}{2} \tag{7.1-7}$$

and the plane-strain stress ratios are found: $1 : \frac{1}{2} : 0$.

The direction of the strain increment expressed in a general (i.e. not principal) coordinate system is found by applying the normality condition (Eq. 7.28) to the general non Mises yield surface (Eq. 7.14b):

$$d\varepsilon_{11} = d\lambda\,(2\sigma_{11} - \sigma_{22} - \sigma_{33})$$

$$d\varepsilon_{22} = d\lambda\,(2\sigma_{22} - \sigma_{11} - \sigma_{33})$$

$$d\varepsilon_{33} = d\lambda\,(2\sigma_{33} - \sigma_{11} - \sigma_{22})$$

$$d\varepsilon_{12} = d\lambda\,(6\,\sigma_{12})$$

$$d\varepsilon_{13} = d\lambda\,(6\,\sigma_{13})$$

$$d\varepsilon_{23} = d\lambda\,(6\,\sigma_{23})$$

7.5 EFFECTIVE OR EQUIVALENT STRAIN

Now that we have found the direction of plastic-strain increment, we must consider its magnitude, $|d\varepsilon^p|$. None of the stability ideas illuminate this problem, which is illustrated in Figure 7.12 for the 1-D case.

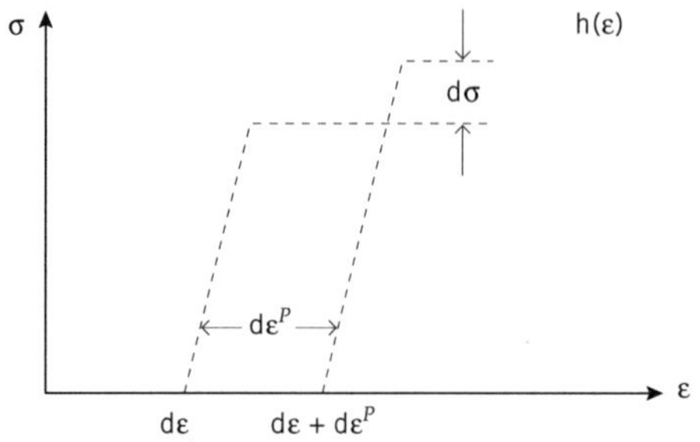

Figure 7.12 Relationship between plastic-strain increment and stress increment in one dimension.

We can see that while the direction of strains is determined by the yield function (i.e., as shown in Exercise 7.1), the size of the strain increment is related to $|d\sigma|$ and $h(\varepsilon)$, the **flow curve**. (The flow curve specifies how the **flow stress**, or current yield stress, varies with total plastic strain in 1-D tension.) In fact, $d\varepsilon^p = d\sigma/h'$.

Similar to the development for effective stress, we need to relate multiaxial strain increments to an **equivalent** increment in a uniaxial tensile test. (The biaxial situation is illustrated in Fig. 7.13.) To do this, we invoke the principle of **equivalent plastic work**, which means that we set the plastic work done in 1-D equal to the plastic work done in a general state in order to define the equivalent or effective strain increment, $d\bar{\varepsilon}$:

$$\bar{\sigma}d\bar{\varepsilon} = \sigma_{ij}\, d\varepsilon_{ij} = \sigma_i\, d\varepsilon_i = \sigma_1\, d\varepsilon_1 + \sigma_2\, d\varepsilon_2 + \sigma_3\, d\varepsilon_3 = \boldsymbol{\sigma} \cdot\, d\boldsymbol{\varepsilon} \tag{7.29}$$

where the last term refers to principal axes. This leads to an immediate definition of $d\bar{\boldsymbol{\varepsilon}}$ in terms of the current effective stress, the stress increment, and the strain increments,

$$d\bar{\varepsilon} = \frac{\sigma_{ij}}{\bar{\sigma}} d\varepsilon_{ij} \tag{7.30}$$

However, this form is not useful until we can relate $d\bar{\varepsilon}$ to a change in σ_{ij} or $d\varepsilon_{ij}$ alone. For this purpose, we use the normality condition again. Combining Eqs. 7.28 and 7.30 provides an equation for finding $d\bar{\varepsilon}$ in terms of σ_{ij} and f alone:

$$d\bar{\varepsilon} = \frac{\sigma_{ij} \dfrac{\partial f}{\partial \sigma_{ij}} d\lambda}{\bar{\sigma}} \tag{7.31}$$

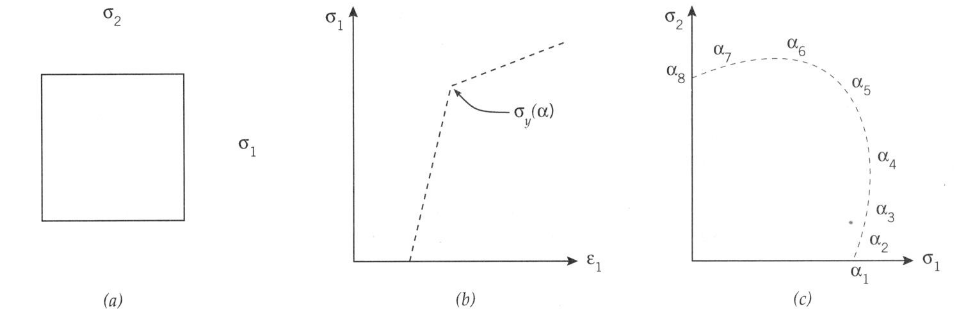

Figure 7.13 Two-dimensional yielding and plastic work.

Most of the time, $d\bar{\varepsilon}$ is needed in terms of $d\varepsilon_{ij}$, so that the total equivalent strain can be evaluated directly from the geometric strain path, without reference to the stresses involved. Finding this form, $d\bar{\varepsilon}(d\varepsilon_{ij})$, depends on inverting the normality conditions to obtain $\sigma_{ij} = f(d\varepsilon_{ij})$ and substituting into Eq. 7.30. In general, the inverse normality condition cannot be written explicitly, thus numerical evaluation is required to evaluate $d\bar{\varepsilon}$ in terms of $d\varepsilon_{ij}$. This difficulty has favored the use of certain yield functions that involve easily inverted normality conditions.[11] An illustration of the solving procedure to find $d\bar{\varepsilon}(d\varepsilon_{ij})$ is presented in Exercise 7.2.

Exercise 7.2 **Find an expression relating $d\bar{\varepsilon}$ to $d\varepsilon_{ij}$ for the von Mises yield function using principal stresses and strains. Also write this quantitiy for arbitrary (not principle) coordinate systems, and find an expression relating $d\varepsilon$ to σ and $d\bar{\sigma}$ alone, and σ to ε and $d\bar{\varepsilon}$ alone.**

The basic equations are as follows:

$$d\bar{\varepsilon} = \frac{1}{\bar{\sigma}}(\sigma_1 d\varepsilon_2 + \sigma_2 d\varepsilon_2 + \sigma_3 d\varepsilon_3) \qquad \text{Equivalent plastic work} \qquad (7.2\text{-}1)$$

$$\begin{aligned} d\varepsilon_1 &= (2\sigma_1 - \sigma_2 - \sigma_3)\, d\lambda \\ d\varepsilon_2 &= (2\sigma_2 - \sigma_1 - \sigma_3)\, d\lambda \\ d\varepsilon_3 &= (2\sigma_3 - \sigma_1 - \sigma_2)\, d\lambda \end{aligned} \qquad \text{Normality} \qquad (7.2\text{-}2)$$

We note that the three normality equations are not independent, because $d\varepsilon_1 + d\varepsilon_2 + d\varepsilon_3$ is identically zero.[12] So we subtract pairs of equations to arrive at two independent ones involving all three strains:

---

[11] In fact, the normality conditions need not be fully inverted, but a form suitable for substitution into Eq. 6.30 must be found to have a closed form $d\bar{\varepsilon}(d\varepsilon_{ij})$.

[12] Another way of looking at the problem is to notice that σ_i can never be found uniquely from $d\varepsilon_i$ because many states of stress (i.e., those that differ only by a pressure term,) are identical in terms of any pressure-independent yield function. Hence only differences of stresses, $\sigma_i - \sigma_j$, can be determined uniquely.

$$(d\varepsilon_2 - d\varepsilon_3) = (3\sigma_2 - 3\sigma_3)d\lambda \tag{7.2-3}$$

$$(d\varepsilon_1 - d\varepsilon_3) = (3\sigma_1 - 3\sigma_3)d\lambda$$

and rewrite the definition of $d\bar{\varepsilon}$ by substituting for one of the strains in terms of the others; say, $d\varepsilon_3 = -(d\varepsilon_1 + d\varepsilon_2)$:

$$d\bar{\varepsilon} = \frac{1}{\bar{\sigma}}\left[\sigma_1\, d\varepsilon_1 + \sigma_2\, d\varepsilon_2 - \sigma_3(d\varepsilon_1 + d\varepsilon_2)\right]$$

$$d\bar{\varepsilon} = \frac{1}{\bar{\sigma}}\left[(\sigma_1 - \sigma_3)d\varepsilon_1 + (\sigma_2 - \sigma_3)d\varepsilon_2\right] \tag{7.2-4}$$

Then we can substitute the inverse normality conditions, Eqs. 7.2-3, into the definition of $d\bar{\varepsilon}$ (Eq. 7.2-1) to obtain

$$d\bar{\varepsilon} = \frac{1}{3\bar{\sigma}d\lambda}\left[(d\varepsilon_1 - d\varepsilon_2)d\varepsilon_1 + (d\varepsilon_2 - d\varepsilon_3)d\varepsilon_2\right]$$

$$d\bar{\varepsilon} = \frac{1}{3\bar{\sigma}d\lambda}\left[d\varepsilon_1{}^2 + d\varepsilon_2{}^2 + d\varepsilon_3{}^2\right] \tag{7.2-5}$$

We have found the required form except for the term $\bar{\sigma}d\lambda$, which we require in terms of $d\bar{\varepsilon}$ or $d\varepsilon_{ij}$ to complete the formulation in terms of strain alone. To do this, we substitute the forward normality conditions into the definition of $d\bar{\varepsilon}$ to find a relationship among $d\bar{\varepsilon}$, $d\lambda$, and $\bar{\sigma}$:

$$d\bar{\varepsilon} = \frac{1}{\bar{\sigma}}\left[\sigma_1\,(2\sigma_1 - \sigma_2 - \sigma_3)d\lambda + \sigma_2\,(2\sigma_2 - \sigma_1 - \sigma_3)d\lambda + \sigma_3\,(2\sigma_3 - \sigma_1 - \sigma_2)d\lambda\right]$$

$$\frac{\bar{\sigma}\,d\bar{\varepsilon}}{d\lambda} = (\sigma_1 - \sigma_2)^2 + (\sigma_1 - \sigma_3)^2 + (\sigma_2 - \sigma_3)^2 = 2\bar{\sigma}^2$$

so

$$d\lambda = \frac{d\bar{\varepsilon}}{2\bar{\sigma}} \tag{7.2-6}$$

Finally, substituting the expression for $d\lambda$ (Eq. 7.2-6) into Eq. 7.2-5 obtains the desired result:

$$d\bar{\varepsilon} = \frac{2\bar{\sigma}}{3\bar{\sigma}\,d\bar{\varepsilon}}\left[d\varepsilon_1{}^2 + d\varepsilon_2{}^2 + d\varepsilon_3{}^2\right]$$

or

$$d\bar{\varepsilon} = \left\{\frac{2}{3}\left[d\varepsilon_1{}^2 + d\varepsilon_2{}^2 + d\varepsilon_3{}^2\right]\right\}^{\frac{1}{2}} \tag{7.2-7}$$

So a closed-form solution of $d\bar{\varepsilon}$ in terms of $d\varepsilon_i$ does exist for von Mises materials.

For the case of stress components expressed in a general coordinate system, an expression equivalent to Eq. 7.2-7 may be derived:

$$d\bar{\varepsilon} = \left[\frac{2}{3}(d\varepsilon_{11}^2 + d\varepsilon_{22}^2 + d\varepsilon_{33}^2 + 2d\varepsilon_{12}^2 + 2d\varepsilon_{13}^2 + 2d\varepsilon_{23}^2)\right]^{\frac{1}{2}}$$

or, some authors prefer to use an alternate expression in terms of strain rates (although we have assumed that the actual plasticity law is independent of the time rate of straining):

$$\dot{\bar{\varepsilon}} = \left[\frac{2}{3}(\dot{\varepsilon}_{11}^2 + \dot{\varepsilon}_{22}^2 + \dot{\varepsilon}_{33}^2 + 2\dot{\varepsilon}_{12}^2 + 2\dot{\varepsilon}_{13}^2 + 2\dot{\varepsilon}_{23}^2)\right]$$

By simple substitution, it is possible to write $d\boldsymbol{\varepsilon}$ in terms of $\boldsymbol{\sigma}$ and $d\boldsymbol{\sigma}$ if the 1-D flow curve is known. Since

$$\bar{\sigma} = h(\bar{\varepsilon}) \quad \text{and} \quad \bar{\varepsilon} = h^{-1}(\bar{\sigma}) \quad \text{(flow curve)}, \tag{7.2-8}$$

$$\frac{d\bar{\sigma}}{d\bar{\varepsilon}} = h'(\bar{\varepsilon}) \qquad \text{or} \qquad h'(\bar{\sigma}) = h'[\bar{\varepsilon}(\bar{\sigma})] = h'\left[h^{-1}(\bar{\sigma})\right] \tag{7.2-9}$$

Then, a substitution for $d\lambda$ is available:

$$d\lambda = \frac{d\bar{\varepsilon}}{2\bar{\sigma}} = \frac{d\bar{\sigma}}{2\ \bar{\sigma}\ h'(\bar{\sigma})}, \tag{7.2-10}$$

and, from normality conditions:

$$[d\varepsilon] = [K][\sigma] \quad \text{or} \tag{7.2-11}$$

$$\begin{bmatrix} d\varepsilon_1 \\ d\varepsilon_2 \\ d\varepsilon_3 \end{bmatrix} = \left(\frac{d\bar{\sigma}}{2\bar{\sigma}\ h'(\bar{\sigma})}\right)\begin{bmatrix} 2 & -1 & -1 \\ -1 & 2 & -1 \\ -1 & -1 & 2 \end{bmatrix}\begin{bmatrix} \sigma_1 \\ \sigma_2 \\ \sigma_3 \end{bmatrix} \tag{7.2-12}$$

Note that the direction of $d\varepsilon$ depends only on the *direction* of σ, while the magnitude of $d\varepsilon$ depends only on the scalar quantities $d\bar{\sigma}$ and $\bar{\sigma}$.

However, most finite element programs use displacements or strains as the principal variables, such that the inverse of Eq. 7.2-12 is wanted. For all incompressible plastic laws (such as von Mises, Hill, and many others), $d\varepsilon_1 + d\varepsilon_2 + d\varepsilon_3 \equiv 0$, such that [K] is not invertible (check that $|K| = 0$). It is sometimes possible to take advantage of some natural problem condition to invert Eq. 7.2-12. In the case of thin-sheet forming, the stress normal to the sheet plane is nearly zero ($\sigma_3 = 0$). With this additional condition, the inverse of Eqs. 7.2-11 and 7.2-12 may be uniquely determined.

$$d\varepsilon_1 = (2\sigma_1 - \sigma_2)\, d\lambda \qquad (7.2\text{-}13a)$$

$$d\varepsilon_2 = (2\sigma_2 - \sigma_1)\, d\lambda \qquad (7.2\text{-}13b)$$

or,

$$\sigma_1 = \frac{1}{3d\lambda}(2d\varepsilon_1 + d\varepsilon_2) \qquad (7.2\text{-}14a)$$

$$\sigma_2 = \frac{1}{3d\lambda}(2d\varepsilon_2 + d\varepsilon_1) \qquad (7.2\text{-}14b)$$

which is obtained by solving Eqs. 7.2-13 for σ, and σ_2. $d\lambda$ may be rewritten in terms of $\bar{\varepsilon}$ and $d\bar{\varepsilon}$:

$$d\lambda = \frac{d\bar{\varepsilon}}{2\bar{\sigma}} = \frac{d\bar{\varepsilon}}{2h(\bar{\varepsilon})} \qquad (7.2\text{-}15)$$

to complete the desired inversion:

$$\begin{bmatrix} \sigma_1 \\ \sigma_2 \end{bmatrix} = \left(\frac{2h(\bar{\varepsilon})}{3d\bar{\varepsilon}}\right) \begin{bmatrix} 2 & 1 \\ 1 & 2 \end{bmatrix} \begin{bmatrix} d\varepsilon_1 \\ d\varepsilon_2 \end{bmatrix} \qquad (7.2\text{-}16)$$

$$[\sigma_3 = 0, \quad d\varepsilon_3 = -(d\varepsilon_1 + d\varepsilon_2)]$$

Eqs. 7.2-12 and 7.2-16 are the most convenient forms for interpretation in finite element programs. Similar forms may be formed for other yield functions.

Equation 7.2-7 shows that the determination of the total effective strain requires evaluation of a path or line integral:

$$\Delta\bar{\varepsilon} = \int_{\bar{\varepsilon}_{(1)}}^{\bar{\varepsilon}_{(2)}} d\bar{\varepsilon} = \left(\frac{2}{3}\right)^{\frac{1}{2}} \int_{\substack{\text{strain path,} \\ \text{from 1 to 2}}} (d\varepsilon_1^{\,2} + d\varepsilon_2^{\,2} + d\varepsilon_3^{\,2})^{\frac{1}{2}} \qquad (7.32)$$

The plastic of the equation, with strain increments squared, indicates that *all changes* of strain cause an increase in $\bar{\varepsilon}$. Thus, for example, two paths originating and terminating at the same geometries can cause different changes in $\bar{\varepsilon}$, and a strain path originating and terminating at the same geometry will not produce a zero change in $\bar{\varepsilon}$. The behavior of $\bar{\varepsilon}$ (or $\bar{\sigma}$, which is directly related to $\bar{\sigma}$ by $\bar{\sigma} = h(\bar{\varepsilon})$, as measured in a tensile test) follows observation: the hardness of material increases with all straining, even for a stretch and compression back to the original geometry.

Exercise 7.3 **Find the effective strain for a general proportional stress path where $\varepsilon_2 = \alpha\,\varepsilon_1$ and $\varepsilon_3 = \beta\,\varepsilon_1$.**

We substitute the known path into Eq. 7.32 to obtain the immediate result:

$$\Delta\bar{\varepsilon} = \left(\frac{2}{3}\right)^{\frac{1}{2}} \int_0^{\Delta\varepsilon_1} \left(d\varepsilon_1{}^2 + \alpha^2 d\varepsilon_1{}^2 + \beta^2 d\varepsilon_1{}^2\right)^{\frac{1}{2}} \tag{7.3-1}$$

$$= \left(\frac{2}{3}\right)^{\frac{1}{2}} \left(1 + \alpha^2 + \beta^2\right)^{\frac{1}{2}} \int_0^{\Delta\bar{\varepsilon}_1} \left|d\varepsilon_1\right| \tag{7.3-2}$$

$$\Delta\bar{\varepsilon} = \left[\frac{2}{3}\left(\Delta\varepsilon_1{}^2 + \Delta\varepsilon_2{}^2 + \Delta\varepsilon_3{}^2\right)\right]^{\frac{1}{2}} \quad \text{(Proportional path)} \tag{7.3-3}$$

7.6 STRAIN HARDENING, EVOLUTION OF YIELD SURFACE

Our discussion of plasticity, in particular the von Mises yield function, has not explicitly mentioned models of strain hardening, or how the yield surface evolves with straining. In fact, the foregoing discussion presumed that the yield function was of the same form except for its size, which increased correspondingly to the flow stress in a tension test. This idea, known as **isotropic hardening,**[13] was implicit in our definitions of effective stress and effective strain, where we assumed the basic form of the yield function was identical at each time except for the parameter $\bar{\sigma}$, which set its size. The conceptual view of isotropic hardening is presented in Figure 7.14*a*. Only a single parameter ($\bar{\sigma}$, for example) is needed to describe the yield surface completely.

The other standard and simple concept of hardening, called **kinematic hardening**, is illustrated in Figure 7.14*b*. In kinematic hardening, the yield

[13] The term *isotropic hardening* is a poor one because it suggests a connection with isotropy, which is not the case. Instead, isotropic hardening means that the yield surface remains the same shape and has the same origin during hardening, with only the size changing. In a general sense, this involves an *isotropic expansion* of the yield surface.

surface size and shape remain constant, but the location translates according to the current stress vector, **σ**.[14] Basic kinematic hardening also requires only a single parameter, in this case a tensor, that describes the current origin of the yield surface. Kinematic hardening is mainly of interest in cases where a

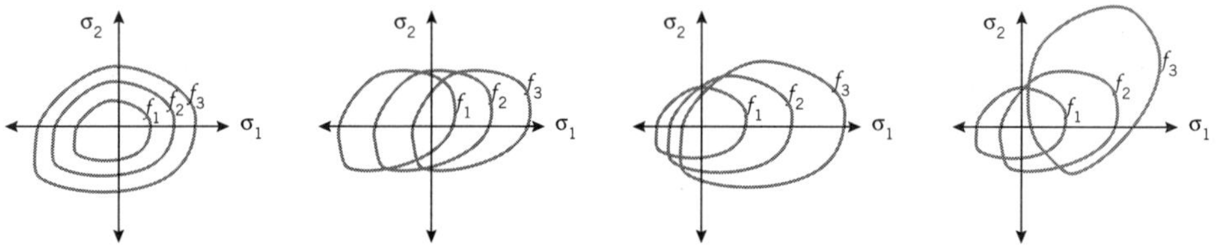

(a) Isotropic hardening *(b)* Kinematic hardening *(c)* Combined hardening *(d)* General hardening

Figure 7.14 Various conceptual models of strain hardening where f_1, f_2, and f_3 are sequential yield surfaces as straining progresses.

Bauschinger effect is important; for example, when abrupt reversals of strain path take place, and one wishes to simulate the small-strain behavior ($\varepsilon \approx 0.01$) immediately following such a change. Since this is not the usual situation in metal-forming analysis, where the strain paths are generally monotonic and we are concerned with large strains, kinematic hardening is seldom used for this application.

Figure 7.14*c* shows schematically a combination of isotropic and kinematic hardening that has been proposed by several authors. It is possible to make such hardening laws quite general by various kinds of scaling between the two components of hardening and the relationship to applied stress and strain.

Figure 7.14*d* illustrates the most general case of hardening, where the yield surface *evolves* in a general way. In this general case, the current yield surface could depend on the entire strain and stress history starting from an initial yield function. Experimental measurement of such general hardening is clearly a task of astronomical proportions given the range of strain and stress paths that would require probing (each with a new, identical specimen!). However, crystal-based plasticity calculations lead to general kinds of evolutions that can then be tested against a limited number of experiments for calibration and verification.

> ***Note:*** *Unless otherwise specified explicitly, we shall always assume that materials harden isotropically, and thus that a single hardness parameter* ($\bar{\sigma}$) *is sufficient to describe the evolution (size) of the yield surface.*

---

[14] However, more complicated versions of the general kinematic hardening approach have been proposed. In these cases, the yield surface may translate according to a complicated function of stress and strain, the only requirement being *consistency*; that is, that a hardening increment of stress lies at some location on the new yield surface.

7.7 PLASTIC ANISOTROPY

A discussion of the most general forms of plastic anisotropy is beyond the scope of this book because these general forms have seldom found practical application or verification under forming conditions. (See, however, Chapter 8, which provides an introduction to general forms of anisotropy based on crystal plasticity.) A few particularly simple yield surfaces have been used widely, mainly in sheet-forming analysis, where a natural set of material axes exist in which to express anisotropy. Figure 7.15 shows the conventional sheet material axes, related to the rolling process used in making the sheets. The rolling direction (RD) is often taken as $\mathbf{x}_1$, the transverse direction (TD) is taken as $\mathbf{x}_2$, and, conventionally, the through-thickness direction of the sheet is $\mathbf{x}_3$.

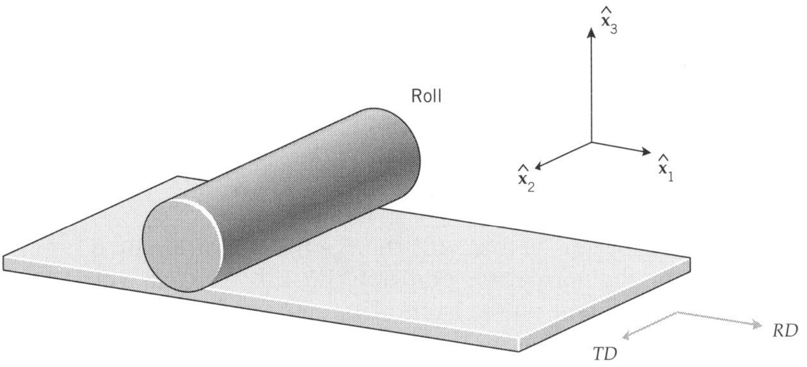

Figure 7.15 Sheet rolling and conventional material directions.

Hill's Quadratic Yield Criterion

The most widely used anisotropic yield surface is known as **Hill's quadratic yield function**, or **Hill's 1948 yield criterion**,[15] which may be expressed as follows:

$$f(\sigma) = F(\sigma_2 - \sigma_3)^2 + G(\sigma_1 - \sigma_3)^2 + H(\sigma_1 - \sigma_2)^2 \tag{7.33}$$

where F, G, and H are constants and the subscripts 1, 2, and 3 serve two purposes: σ_i are the three principal stresses, whose vectors are supposed to be parallel to $\mathbf{x}_i$, where $\mathbf{x}_1$ = RD, $\mathbf{x}_2$ = TD, and $\mathbf{x}_3$ = thickness direction (as shown in Figure 7.15).

[15] R. Hill, *Mathematical Theory of Plasticity*, (London, Oxford University Press, 1950) Chapter 12.

This means that Eq. 7.33 can be used only for problems where the principal stress axes are always parallel to the material anisotropy axes.[16] Equation 7.33 was constructed primarily to account for the fact that sheet tensile tests are known to exhibit anisotropy such that the lateral strains are not equal. We will show how the constants F, G, and H are fit to such data. Note that Hill's quadratic yield function is pressure-independent and does not allow for a Bauschinger effect.

We proceed to derive the necessary equations for Hill's quadratic yield function following the procedures introduced for the von Mises yield criterion. Let's first define the **effective stress** in terms of a tensile test along the RD ($\mathbf{x}_1$) direction, such that $\sigma_1 = \bar{\sigma}, \sigma_2 = \sigma_3 = 0$:

$$f = F(0-0)^2 + G(\bar{\sigma}-0)^2 + H(\bar{\sigma}-0)^2 = (G+H)\bar{\sigma}^2 \tag{7.34}$$

such that the yield condition becomes

$$\bar{\sigma}^2 = \frac{F}{G+H}(\sigma_2-\sigma_3)^2 + \frac{G}{G+H}(\sigma_1-\sigma_3)^2 + \frac{H}{G+H}(\sigma_1-\sigma_2)^2 \tag{7.35}$$

The choice of tensile axis will change the definition of $\bar{\sigma}$.

Using the normality condition (Eq. 7.28), we can solve for the strain ratios in two equivalent ways. In the first approach, proceeding as we did for von Mises yield, we consider the yield function in arbitrary form (Eq. 7.33) and find the **normality condition** with arbitrary $d\lambda$:

$$d\varepsilon_1 = d\lambda\left[G(\sigma_1-\sigma_3) + H(\sigma_1-\sigma_2)\right] \tag{7.36a}$$

$$d\varepsilon_2 = d\lambda\left[F(\sigma_2-\sigma_3) - H(\sigma_1-\sigma_2)\right] \tag{7.36b}$$

$$d\varepsilon_3 = d\lambda\left[-F(\sigma_2-\sigma_3) - G(\sigma_1-\sigma_3)\right] \tag{7.36c}$$

In this approach, we calibrate for $d\lambda$ by consideration of the equivalent work definition of **effective strain increment**, $d\bar{\varepsilon}$, using Eq. 7.29:

---

[16] It is possible to use Eq. 7.33 in approximate form by ignoring the dual nature of the subscripts, instead only originally orienting these directions approximately parallel and then following the principal stress directions wherever they may lie. Alternatively, it is possible to construct generalizations of Eq. 7.33 in terms of arbitrary σ_{ij} expressed in $\mathbf{x}_i$ (anisotropy axes) such that when the principal axes coincide, the original form is recovered. Hill's original form includes six constants when the principal directions are not coincident with symmetry axes x, y, and z:

$$f(\boldsymbol{\sigma}) = F(\sigma_y-\sigma_z)^2 + G(\sigma_z-\sigma_x)^2 + H(\sigma_x-\sigma_y)^2 + 2L\tau_{yz}{}^2 + 2M\tau_{zx}{}^2 + 2N\tau_{xy}{}^2$$

$$\bar{\sigma} d\bar{\varepsilon} = \sigma_1 \, d\varepsilon_1 + \sigma_2 \, d\varepsilon_2 + \sigma_3 \, d\varepsilon_3$$

and substitute Eqs. 7.36 for the strain increments:

$$\frac{\bar{\sigma} d\bar{\varepsilon}}{d\lambda} = \sigma_1[G(\sigma_1 - \sigma_3) + H(\sigma_1 - \sigma_2)] + \sigma_2[F(\sigma_2 - \sigma_3) - H(\sigma_1 - \sigma_2)] +$$

$$\sigma_3[-F(\sigma_2 - \sigma_3) - G(\sigma_1 - \sigma_3)] \qquad (7.37)$$

The right-hand side of Eq. 7.37 can be rearranged algebraically to obtain

$$\frac{\bar{\sigma} d\bar{\varepsilon}}{d\lambda} = F(\sigma_2 - \sigma_3)^2 + G(\sigma_1 - \sigma_3)^2 + H(\sigma_1 - \sigma_2)^2 \qquad (7.38)$$

which is simplified using the definition of equivalent stress, Eq. 7.35, to find $d\lambda$:

$$\frac{\bar{\sigma} d\bar{\varepsilon}}{d\lambda} = (G + H)\bar{\sigma}^2 \qquad (7.39)$$

$$d\lambda = \frac{1}{(G+H)} \frac{d\bar{\varepsilon}}{\bar{\sigma}} \quad \left(\text{for } d\varepsilon_{ij} = \frac{\partial f}{\partial \sigma_{ij}} d\lambda, \text{ Eqs. 7.37}\right) \qquad (7.40)$$

The specific normality conditions can then be rewritten without the undetermined constant $d\lambda$:

$$d\bar{\varepsilon}_1 = \frac{1}{(G+H)} \frac{d\bar{\varepsilon}}{\bar{\sigma}} [G(\sigma_1 - \sigma_3) + H(\sigma_1 - \sigma_2)] \qquad (7.41a)$$

$$d\bar{\varepsilon}_2 = \frac{1}{(G+H)} \frac{d\bar{\varepsilon}}{\bar{\sigma}} [F(\sigma_2 - \sigma_3) - H(\sigma_1 - \sigma_2)] \qquad (7.41b)$$

$$d\bar{\varepsilon}_3 = \frac{1}{(G+H)} \frac{d\bar{\varepsilon}}{\bar{\sigma}} [-F(\sigma_2 - \sigma_3) - G(\sigma_1 - \sigma_3)] \qquad (7.41c)$$

Equations 7.41 can be obtained more elegantly by starting from the yield function written in terms of the effective stress such as Eq. 7.35. In general terms,

$$\bar{\sigma} = \bar{\sigma}(\sigma_{ij}) \tag{7.42}$$

Then the normality conditions are derived using equations of this form:

$$d\varepsilon_{ij} = \frac{\partial \bar{\sigma}}{\partial \sigma_{ij}} d\lambda \tag{7.43}$$

and the definite form of $d\lambda$ can be found by considering the stress state that defines $\bar{\sigma}$ uniaxial tension in our case. For $\mathbf{x}_1$ tension, $\sigma_1 = \bar{\sigma}$, other $\sigma_{ij} = 0$. Then the equivalent plastic work definition of $d\bar{\varepsilon}$ is used to find $d\lambda$:

$$\bar{\sigma} d\bar{\varepsilon} = \sigma_{ij}\, d\,\varepsilon_{ij} = \sigma_{ij} \frac{\partial \bar{\sigma}}{\partial \sigma_{ij}} d\lambda \qquad \text{(Eq. 7.29)}$$

$$\bar{\sigma} d\bar{\varepsilon} = \bar{\sigma} \frac{\partial \bar{\sigma}}{\partial \bar{\sigma}} d\lambda = \bar{\sigma} d\lambda \text{ (uniaxial tension)} \tag{7.44}$$

$$d\lambda = d\bar{\varepsilon} \quad \left(\text{for } d\varepsilon_{ij} = \frac{\partial \bar{\sigma}}{\partial \sigma_{ij}} d\lambda\right) \tag{7.45}$$

Note that in deriving Eq. 7.45, we did not consider any particular form of the yield function. In fact, it is true for any yield function written in terms of $\bar{\sigma} = \bar{\sigma}(\sigma_{ij})$. Equation 7.45 lets us express the normality condition without arbitrary constants:

$$d\varepsilon_{ij} = \frac{\partial \bar{\sigma}}{\partial \sigma_{ij}} d\bar{\varepsilon} \tag{7.46}$$

Equation 7.46 may be used directly to define the normality conditions without an arbitrary constant for Hill's quadratic yield function, as illustrated in the following exercise.

Exercise 7.4 Find the nonarbitrary form of normality conditions for Hill's quadratic yield function directly from the yield function written in terms of effective stress.

To derive the required result, we could rewrite Eq. 7.35 in order to find the desired form, $\bar{\sigma} = \bar{\sigma}(\sigma_{ij})$:

$$\bar{\sigma} = \left[\frac{F}{G+H}(\sigma_2 - \sigma_3)^2 + \frac{G}{G+H}(\sigma_1 - \sigma_3)^2 + \frac{H}{G+H}(\sigma_1 - \sigma_2)^2 \right]^{\frac{1}{2}} \tag{7.4-1}$$

and then differentiate according to Eq. 7.46. However, it is more elegant (and much shorter in writing) to do the first manipulation symbolically:

$$\bar{\sigma}^2 = f(\sigma_{ij}), \quad \text{where f is the RHS of Eq. 7.35} \tag{7.4-2}$$

$$2\bar{\sigma}\frac{\partial\bar{\sigma}}{\partial\sigma_{ij}} = \frac{\partial f}{\partial\sigma_{ij}} \tag{7.4-3}$$

$$\text{so,} \quad \frac{\partial\bar{\sigma}}{\partial\sigma_{ij}} = \frac{1}{2\bar{\sigma}}\frac{\partial f}{\partial\sigma_{ij}} \tag{7.4-4}$$

Now, we can obtain the desired form simply:

$$d\varepsilon_1 = \frac{1}{(G+H)}\frac{d\bar{\varepsilon}}{\bar{\sigma}}\left[G(\sigma_1 - \sigma_3) + H(\sigma_1 - \sigma_2)\right] \tag{7.4-5a}$$

$$d\varepsilon_2 = \frac{1}{(G+H)}\frac{d\bar{\varepsilon}}{\bar{\sigma}}\left[F(\sigma_2 - \sigma_3) - H(\sigma_1 - \sigma_2)\right] \tag{7.4-5b}$$

$$d\varepsilon_3 = \frac{1}{(G+H)}\frac{d\bar{\varepsilon}}{\bar{\sigma}}\left[-F(\sigma_2 - \sigma_3) - G(\sigma_1 - \sigma_3)\right] \tag{7.4-5c}$$

Inspection shows that this result is identical to Eqs. 7.41, which were derived from the general form of f and then calibrated to uniaxial tension.

To complete the derivation of the general expressions needed for Hill's quadratic yield function, we obtain the **associated flow law** form from the normality conditions:

$$\frac{d\varepsilon_1}{G(\sigma_1-\sigma_3)+H(\sigma_1-\sigma_2)} = \frac{d\varepsilon_2}{F(\sigma_2-\sigma_3)-H(\sigma_1-\sigma_2)} = \frac{d\varepsilon_3}{-F(\sigma_2-\sigma_3)-G(\sigma_1-\sigma_3)} \tag{7.47}$$

by equating $d\lambda$ among the three expressions in Eqs. 7.36, or equating $d\bar{\varepsilon}$ among the three expressions in Eqs. 7.41.

The final, and usually most difficult, step in applying a yield criterion lies in deriving an expression for $d\bar{\varepsilon}$ in terms of $d\varepsilon_{ij}$. We begin by rearranging the specific normality conditions (Eqs. 7.41) to find stress differences. To do this, we simply multiply each of Eqs. 7.41 by the constant that does not appear in the brackets and then subtract, obtaining

$$\sigma_2 - \sigma_3 = \frac{(G+H)\bar{\sigma}}{(FG+FH+GH)\,d\bar{\varepsilon}}(Gd\varepsilon_2 - Hd\varepsilon_3) \tag{7.48a}$$

$$\sigma_1 - \sigma_3 = \frac{(G+H)\bar{\sigma}}{(FG+FH+GH)\,d\bar{\varepsilon}}(Fd\varepsilon_1 - Hd\varepsilon_3) \tag{7.48b}$$

$$\sigma_1 - \sigma_2 = \frac{(G+H)\bar{\sigma}}{(FG+FH+GH)\,d\bar{\varepsilon}}(Fd\varepsilon_1 - Gd\varepsilon_2) \tag{7.48c}$$

We then substitute these into the yield condition in terms of $\bar{\sigma}$, Eq. 7.35:

$$\bar{\sigma}^2 = \left[\frac{(G+H)\bar{\sigma}}{(FG+FH+GH)\,d\bar{\varepsilon}}\right]^2$$

$$\left[\frac{F}{G+H}(Gd\varepsilon_2 - Hd\varepsilon_3)^2 + \frac{G}{G+H}(Fd\varepsilon_1 - Hd\varepsilon_3)^2 + \frac{H}{G+H}(Fd\varepsilon_1 - Gd\varepsilon_2)^2\right] \tag{7.49}$$

which reduces to the desired form:

$$d\bar{\varepsilon}^2 = \frac{(G+H)}{(FG+FH+GH)^2}\left[F(Gd\varepsilon_2 - Hd\varepsilon_3)^2 + G(Fd\varepsilon_1 - Hd\varepsilon_3)^2 + H(Fd\varepsilon_1 - Gd\varepsilon_2)^2\right] \tag{7.50a}$$

or

$$d\bar{\varepsilon}^2 = \frac{(G+H)}{(FG+FH+GH)^2}\left[F^2(G+H)d\varepsilon_1{}^2 + G^2(F+H)d\varepsilon_2{}^2 + H^2(F+G)d\varepsilon_3{}^2\right] \tag{7.50b}$$

Fitting Hill's Quadratic Yield Function to Experimental Data

In the foregoing, we derived the most important equations related to Hill's quadratic yield function, as expressed in principal stress axes lying on the material anisotropy axes, Eqs. 7.35, 7.41, and 7.50. However, it remains to find a connection between the undetermined constants F, G, and H, and measured material behavior. (Or, for the more general form where principal axes do not coincide with material symmetry axes, it is necessary to find six constants: F, G, H, L, M, and N.)

While this fitting could take place by measuring yield stresses in various directions and under various states of stress,[17] this procedure is seldom followed. Instead, F, G, and H are usually found indirectly, via the normality condition, by measuring strain ratios in tensile tests.

Consider first an RD (x_1) tensile test, as shown in Figure 7.16. We introduce the conventional **plastic anisotropy parameter**, *r*, for characterizing strain ratios in sheet tensile tests:

$$r = \frac{\text{true width strain}}{\text{true thickness strain}} = \frac{\varepsilon_{width}}{\varepsilon_{thickness}} \tag{7.51a}$$

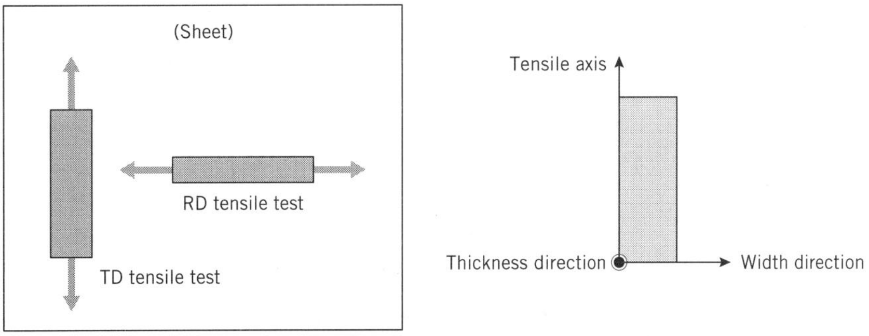

Figure 7.16 Orientation of conventional sheet tensile tests

Note that r is equal to 1 for an isotropic material (von Mises or otherwise) because the width and thickness directions are unloaded and are thus equivalent. The strain ratios for the two tensile tests shown in Figure 7.16 are distinguished by the direction of the test:

$$r_{RD} = r_1 = r_0 \tag{7.51b}$$

$$r_{TD} = r_2 = r_{90} \tag{7.51c}$$

where subscripts 1 and 2 refer to the conventional axes $\mathbf{x}_1$ and $\mathbf{x}_2$ (Fig. 7.15), and 0 and 90 refer to the angle (in degrees) that the tensile axis makes with the RD.

[17] See, for example, W. F. Hosford and R. M. Caddell; *Metal Forming Mechanics and Metallurgy*, (Englewood Cliffs, NJ: Prentice-Hall, 1983), pp. 266-269.

We can find the relationship between the plastic anisotropy parameters (r_{RD}, r_{TD}) and Hill's constants (F, G, and H) by considering the normality condition, Eqs. 7.41, of standard tensile (or compression) tests in RD and TD:

$$\text{RD}: \qquad \sigma_1 = \sigma, \quad \sigma_2 = \sigma_3 = 0$$

$$r_{RD} \equiv \frac{d\varepsilon_{width}}{d\varepsilon_{thickness}} = \frac{d\varepsilon_2}{d\varepsilon_3} = \frac{H}{G} \tag{7.52}$$

$$\text{TD}: \qquad \sigma_2 = \sigma, \quad \sigma_1 = \sigma_3 = 0$$

$$r_{TD} \equiv \frac{d\varepsilon_{width}}{d\varepsilon_{thickness}} = \frac{d\varepsilon_1}{d\varepsilon_3} = \frac{H}{F} \tag{7.53}$$

Substitution of these relationships (F = r_{RD}, G = r_{TD}, H = $r_{RD}r_{TD}$) into Eq. 7.35 obtains Hill's quadratic yield function in terms of the plastic anisotropy ratios:

$$\bar{\sigma}^2 = \frac{1}{r_{TD}(1+r_{RD})}\left[r_{RD}(\sigma_2-\sigma_3)^2 + r_{TD}(\sigma_1-\sigma_3)^2 + r_{RD}r_{TD}(\sigma_1-\sigma_2)^2\right] \tag{7.54}$$

This is the form most frequently used with this theory. The other important equations for Hill's quadratic yield function may be written directly by substitution:

Normality:

$$d\varepsilon_1 = \frac{d\bar{\varepsilon}}{\bar{\sigma}}\frac{1}{r_{TD}(1+r_{RD})}\left[r_{TD}(\sigma_1-\sigma_3) + r_{RD}r_{TD}(\sigma_1-\sigma_2)\right] \tag{7.55a}$$

$$d\varepsilon_2 = \frac{d\bar{\varepsilon}}{\bar{\sigma}}\frac{1}{r_{TD}(1+r_{RD})}\left[r_{RD}(\sigma_2-\sigma_3) - r_{RD}r_{TD}(\sigma_1-\sigma_2)\right] \tag{7.55b}$$

$$d\varepsilon_3 = \frac{d\bar{\varepsilon}}{\bar{\sigma}}\frac{1}{r_{TD}(1+r_{RD})}\left[-r_{RD}(\sigma_2-\sigma_3) - r_{TD}(\sigma_1-\sigma_3)\right] \tag{7.55c}$$

$$d\bar{\varepsilon}^2 = \left[\frac{r_{TD}(1+r_{RD})}{\left(r_{RD}r_{TD} + r_{RD}{}^2 r_{TD} + r_{RD}r_{TD}{}^2\right)^2}\right.$$

$$\left[r_{RD}\left(r_{TD}d\varepsilon_2 - r_{RD}r_{TD}d\varepsilon_3\right)^2 + r_{TD}\left(r_{RD}d\varepsilon_1 - r_{RD}r_{TD}d\varepsilon_3\right)^2 + r_{RD}r_{TD}\left(r_{RD}d\varepsilon_1 - r_{TD}d\varepsilon_2\right)^2 \right] \tag{7.56}$$

Normal Anisotropy and Plane Stress

It is frequently convenient to analyze sheet-forming operations under the assumption that all directions lying in the plane are equivalent, whereas the through-thickness direction has different properties. For example, in membrane- and shell-finite elements, the through-thickness direction is treated specially anyway, while the in-plane directions are indistinguishable except by material properties. This is a special case of orthotropic symmetry known as **normal anisotropy**, where "normal" refers to the direction perpendicular to, or normal to, the sheet plane.[18]

In terms of Hill's quadratic yield function, this assumption is equivalent to setting $r_{RD} = r_{TD} = r$, where r need not be equal to 1 because of the different properties through the thickness. In fact, nearly all sheet materials exhibit some anisotropy in the sheet plane, so some kind of averaging is needed to find a value of *r* typical of all sheet directions. Exact averaging would involve an infinite number of tensile tests, equally weighted by tensile direction:

$$\bar{r} = \int_{\theta=0}^{\theta=360^\circ} r_\theta \, d\theta \tag{7.57}$$

which is, of course, impractical. Usually three directions are probed within a 90° arc (if orthotropy is accurate, this range represents 360° by symmetry) to obtain r_{RD}, r_{TD}, and r_{45}. In order to weigh these results equally by the angle represented,[19] the average plastic anisotropy ratio is

$$\bar{r} = \frac{r_0 + 2r_{45} + r_{90}}{4} \tag{7.58}$$

A second parameter indicative of the degree of planar anisotropy is also frequently quoted:

$$\Delta r = \frac{r_0 + r_{90} - 2r_{45}}{2} \tag{7.59}$$

which is a measure of how different the 45° directions are from the symmetry axes.

---

[18] Anisotropy in the plane of the sheet is sometimes called **planar anisotropy**.

[19] The 45° tests represent an angular range of 45° whereas the RD and TD tests represent 22.5° each because these directions lie on the symmetry lines. Consider a full range of 360° to convince yourself that this is the case.

Hill's quadratic yield function is most often used in normal anisotropic form because of the convenience it introduces in the analytical procedures. The important relationships are easily derived by setting $r_{RD} = r_{TD} = r$:

Yield:

$$\bar{\sigma}^2 = \frac{1}{(1+r)}\left[(\sigma_2-\sigma_3)^2+(\sigma_1-\sigma_3)^2+r(\sigma_1-\sigma_2)^2\right] \tag{7.60}$$

Normality:

$$d\varepsilon_1 = \frac{d\bar{\varepsilon}}{\bar{\sigma}}\frac{1}{(1+r)}\left[(\sigma_1-\sigma_3)+r(\sigma_1-\sigma_2)\right] \tag{7.61a}$$

$$d\varepsilon_2 = \frac{d\bar{\varepsilon}}{\bar{\sigma}}\frac{1}{(1+r)}\left[(\sigma_2-\sigma_3)-r(\sigma_1-\sigma_2)\right] \tag{7.61b}$$

$$d\varepsilon_3 = \frac{d\bar{\varepsilon}}{\bar{\sigma}}\frac{1}{(1+r)}\left[-(\sigma_2-\sigma_3)-(\sigma_1-\sigma_3)\right] \tag{7.61c}$$

Effective strain increment:

$$d\bar{\varepsilon}^2 = \frac{(1+r)}{(1+2r)^2}\left[(d\varepsilon_2-rd\varepsilon_3)^2+(d\varepsilon_1-rd\varepsilon_3)^2+r(d\varepsilon_1-d\varepsilon_2)^2\right] \tag{7.62}$$

While some sheet-forming operations rely on compression through the sheet thickness (rolling and ironing are common examples), press forming of sheet metal parts is conducted by tension in the sheet plane and virtually no loading through the thickness. Therefore, Hill's quadratic yield function (and many others for sheet metal forming) are often written under the assumption that $\sigma_3 = 0$; that is., that a condition of **plane stress** exists throughout. Similarly, it is convenient to eliminate $d\varepsilon_3$ from the equations by the plastic volume constraint ($d\varepsilon_3 = -\,d\varepsilon_1 - d\varepsilon_2$) for press-forming applications because the strains in such operations are measured from grids printed onto the sheet surface. See Chapter 17, for example.

Under plane-stress conditions and normal anisotropy, the important equations for Hill's quadratic yield function are as follows:

Yield:

$$\bar{\sigma}^2 = \left(\sigma_1^2+\sigma_2^2-\frac{2r}{1+r}\sigma_1\sigma_2\right) \tag{7.63}$$

Normality:

$$d\varepsilon_1 = \frac{d\bar{\varepsilon}}{\bar{\sigma}}\left(\sigma_1 - \frac{r}{1+r}\sigma_2\right) \tag{7.63a}$$

$$d\varepsilon_2 = \frac{d\bar{\varepsilon}}{\bar{\sigma}}\left(\sigma_2 - \frac{r}{1+r}\sigma_1\right) \tag{7.63b}$$

$$d\varepsilon_3 = \frac{d\bar{\varepsilon}}{\bar{\sigma}(1+r)}(-\sigma_1 - \sigma_2) \tag{7.63c}$$

Effective strain increment:

$$d\bar{\varepsilon}^2 = \frac{(1+r)^2}{(1+2r)}\left[d\varepsilon_1{}^2 + d\varepsilon_2{}^2 + \frac{2r}{1+r}d\varepsilon_1\, d\varepsilon_2\right] \tag{7.64}$$

Nonquadratic Yield Functions

It is possible to generalize Hill's quadratic yield criterion to general exponents:

$$F\left|\sigma_2 - \sigma_3\right|^M + G\left|\sigma_1 - \sigma_3\right|^M + H\left|\sigma_1 - \sigma_2\right|^M \tag{7.65}$$

where the absolute values are necessary for definition of the quantities for noninteger M when $\sigma_2 < \sigma_3$, $\sigma_1 < \sigma_3$, $\sigma_1 < \sigma_2$. The motivation for making such a generalization arose from numerous experiments that showed that the quadratic form often did not fit yielding in plane strain or balanced biaxial tension. As shown schematically in Figure 7.17, fitting the yield stresses and the strain ratios from tensile tests often did not assure a fit away from these strain states. By adjusting the exponent in Eq. 7.65, it is possible to fit a wider range of experimental data.

At least three authors seem to have proposed yield functions of the general type of Eq. 7.65, either motivated by experimental results like those shown in Figure 7.17, or by crystallographic calculations that produced yield loci different from quadratic representations. We will mention all three briefly, each in the normal anisotropic form that has been most tested.

Hill introduced several plane-stress variants of nonquadratic yield, only one of which, case IV,[20] has received much attention:

---

[20] Hill's nonquadratic yield function, case IV, was first presented by Mellor and Parmer:

P. B. Mellor and A. Parmer, *Mechanics of Sheet Metal Forming*, (New York, Plenum Press, 1978): p. 53.

A. Parmer and P. B. Mellor, *Int. J. Mech. Sci.* 20 (1978): 385.

A. Parmer and P. B. Mellor, *Int. J. Mech. Sci.* 20 (1978): 707 and was later presented in more general form by Hill directly in R. Hill, *Math Proc. Camb. Phil. Soc.* 85 (1979): 179–191.

$$f = k\left|\sigma_1 - \sigma_2\right|^M + \left|\sigma_1 + \sigma_2\right|^M \tag{7.66}$$

Following the procedures outlined for quadratic yield functions, and restricting our attention to the first octant,[21] where $\sigma_1 \geq \sigma_2 \geq 0$ the following important equations can be derived (see also problems at the end of this chapter):

Effective stress:

$$\bar{\sigma}^M = \frac{2r+1}{2(1+r)}\left(\sigma_1 - \sigma_2\right)^M + \frac{1}{2(1+r)}\left(\sigma_1 + \sigma_2\right)^M \tag{7.67}$$

Normality:

$$d\varepsilon_1 = \frac{d\bar{\varepsilon}}{\bar{\sigma}^{M-1}}\left[\frac{2r+1}{2(1+r)}\left(\sigma_1 - \sigma_2\right)^{M-1} + \frac{1}{2(r+1)}\left(\sigma_1 + \sigma_2\right)^{M-1}\right] \tag{7.68a}$$

$$d\varepsilon_2 = \frac{d\bar{\varepsilon}}{\bar{\sigma}^{M-1}}\left[\frac{-(2r+1)}{2(1+r)}\left(\sigma_1 - \sigma_2\right)^{M-1} + \frac{1}{2(r+1)}\left(\sigma_1 + \sigma_2\right)^{M-1}\right] \tag{7.68b}$$

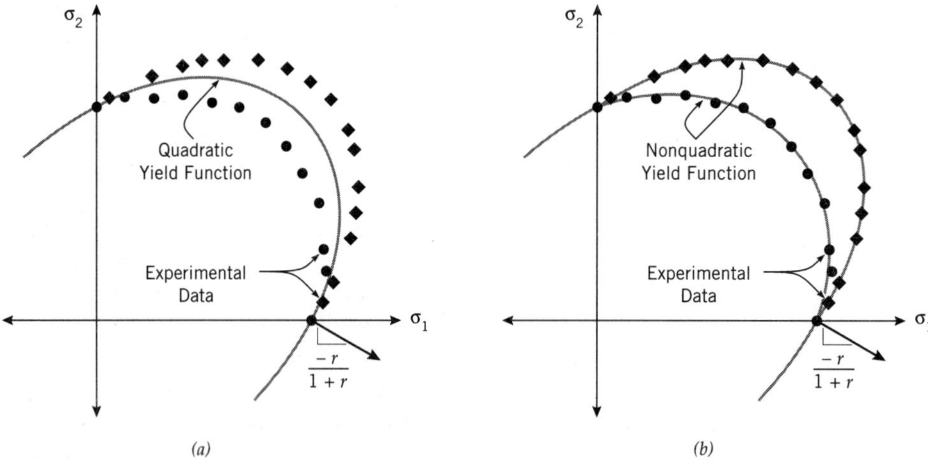

Figure 7.17 Schematic representation of quadratic and nonquadratic yield functions fit to experimental data.

---

[21] Other octants can be considered separately to construct the full range of application, thus avoiding difficulties in handling the absolute values of Eq. 7.66.

$$d\varepsilon_3 = \frac{d\bar{\varepsilon}}{\bar{\sigma}^{M-1}}\left[-\frac{1}{1+r}(\sigma_1+\sigma_2)^{M-1}\right] \tag{7.68c}$$

Effective strain increment:

$$d\bar{\varepsilon} = \frac{[2(1+r)]^{\frac{1}{M}}}{2}\left[\frac{1}{(1+r)^{\frac{1}{M-1}}}(\varepsilon_1-\varepsilon_2)^{\frac{M}{M-1}}+(\varepsilon_1+\varepsilon_2)^{\frac{M}{M-1}}\right]^{\frac{M-1}{M}} \tag{7.69}$$

Hosford[22] introduced a plane-stress yield function inspired by crystallographic calculations of polycrystalline yield functions, which showed variations similar to those depicted in Figure 7.17. The yield function has the form

$$\bar{\sigma}^M = \frac{1}{r+1}\left(|\sigma_1|^M+|\sigma_2|^M+r|\sigma_1-\sigma_2|^M\right) \tag{7.70}$$

where the recommendation is to use M = 6 for metals with body-centered cubic crystal structures, and M = 8 for metals with face-centered cubic crystal structures. Conveniently enough, use of these even, integer powers for M allows the elimination of the absolute value signs. As with Hill's nonquadratic yield criterion, it is convenient to consider only the first octant $\sigma_1 > \sigma_2 > 0$ and to extend the computations to other octants as necessary. The normality equations are

$$d\varepsilon_1 = \frac{d\bar{\varepsilon}}{(r+1)\,\bar{\sigma}^{M-1}}\left[\sigma_1^{M-1}+r(\sigma_1-\sigma_2)^{M-1}\right] \tag{7.71a}$$

$$d\varepsilon_2 = \frac{d\bar{\varepsilon}}{(r+1)\,\bar{\sigma}^{M-1}}\left[\sigma_2^{M-1}-r(\sigma_1-\sigma_2)^{M-1}\right] \tag{7.71b}$$

$$d\varepsilon_3 = \frac{d\bar{\varepsilon}}{(r+1)\,\bar{\sigma}^{M-1}}\left[-\sigma_1^{M-1}-\sigma_2^{M-1}\right] \tag{7.71c}$$

and there is no closed-form expression for $d\bar{\varepsilon}$ in terms of $d\varepsilon_1$ and $d\varepsilon_2$. The lack of an explicit function for $d\bar{\varepsilon}\ (d\varepsilon_i)$ may impede the testing and acceptance of

[22] W. F. Hosford, *7th Int. North Amer. Metalworking Res. Conf. Proc.* (Dearborn, Mich., SME, 1979): 191, and R. Logan and W. F. Hosford, *Int. J. Mech. Sci.*, 22 (1980): 419–430.

yield functions such as Eq. 7.70, but simulations have shown that there is virtually no computational penalty for finding $d\bar{\varepsilon}$ numerically.

The last nonquadratic yield function to be mentioned here is attributed to **Bassani:**[23]

$$f = |\sigma_1 + \sigma_2|^N + k|\sigma_1 - \sigma_2|^M \tag{7.72}$$

Equation 7.72 can be written in terms of effective stress as follows:

$$\bar{\sigma}^N + k\,\bar{\sigma}^M = |\sigma_1 + \sigma_2|^N + k|\sigma_1 - \sigma_2|^M \tag{7.73}$$

and the normality conditions can be derived directly, again assuming that $\sigma_1 > \sigma_2 > 0$:

$$d\varepsilon_1 = d\bar{\varepsilon}\left[N\,\bar{\sigma}^{N-1} + kM\,\bar{\sigma}^{M-1}\right]^{-1}\left[N\left(\sigma_1 + \sigma_2\right)^{N-1} + Mk\left(\sigma_1 - \sigma_2\right)^{M-1}\right] \tag{7.74a}$$

$$d\varepsilon_2 = d\bar{\varepsilon}\left[N\,\bar{\sigma}^{N-1} + kM\,\bar{\sigma}^{M-1}\right]^{-1}\left[N\left(\sigma_1 + \sigma_2\right)^{N-1} - Mk\left(\sigma_1 - \sigma_2\right)^{M-1}\right] \tag{7.74b}$$

$$d\varepsilon_3 = -d\bar{\varepsilon}\left[N\,\bar{\sigma}^{N-1} + Mk\,\bar{\sigma}^{M-1}\right]^{-1}\left[2N\left(\sigma_1 + \sigma_2\right)^{N-1}\right] \tag{7.74c}$$

Consideration of the strain ratios in a sheet tensile test reveals that k must take a particular value in Eqs. 7.72–7.74:

$$k = \frac{N(2r+1)}{M} \cdot \bar{\sigma}^{\frac{N-1}{M-1}} \tag{7.75}$$

so, either k or r must depend on effective stress or strain for this function to have meaning.

CHAPTER 7 - PROBLEMS

A. Proficiency Problems

1. What is the meaning of stress states lying outside of the yield surface?

---

[23] J. L. Bassani, *Int. J. Mech. Sci.*19 (1977): 651.

2. A researcher has found a way to measure the yield surface of sheet metal rapidly and automatically. He inserts a sheet into a biaxial testing machine and loads along proportional paths until, while he measures strains in the two directions, he obtains the 0.2% offset strength. Then he unloads, chooses a slightly different ratio and does the same thing, until he has generated many yield points. Criticize this procedure.

3. Critically evaluate the yield functions presented below in terms of isotropy, pressure-dependence, and existence of a Bauschinger effect. Demonstrate your results. Why would you choose to use each yield function?

   a. Hill's quadratic yield function (1948)[24]:

   $$f = F(\sigma_2 - \sigma_3)^2 + G(\sigma_1 - \sigma_3)^2 + H(\sigma_1 - \sigma_2)^2$$

   b. Hill's non quadratic yield function (1979)[25]:

   $$f = F|\sigma_2 - \sigma_3|^M + G|\sigma_1 - \sigma_3|^M + H|\sigma_1 - \sigma_2|^M$$

   c. Bourne and Hill[26] ($\sigma_3 = 0$, not principal axes):

   $$f = 3\sigma_x{}^3 - 6\sigma_x{}^2\sigma_y - 6\sigma_x\sigma_y{}^2 + 4\sigma_y{}^3 + \left(4\sigma_x + 21\sigma_y\right)\sigma_{xy}{}^2$$

   d. Drucker:[27]

   $f = (J_2')^3 - c(J_3')^2$, c = constant

   e. Edelman and Drucker:[28]

   $$f = \frac{1}{2} c_{ijkl} (\sigma_{ij}' - M\varepsilon_{ij}{}^P)(\sigma_{kl}' - M\varepsilon_{kl}{}^P),\ c_{ijkl},\ M = \text{constants};$$

   $\varepsilon_{ij}{}^P$ = plastic strain

   f. Gotoh[29] ($\sigma_3 = 0$):

   $$f = A_o(\sigma_x + \sigma_y)^4 + A_1\sigma_x{}^4 + A_2\sigma_x{}^3\sigma_y + A_3\sigma_x{}^2\sigma_y{}^2 + A_4\sigma_x\sigma_y{}^3 + A_5\sigma_y{}^4 + (A_6\sigma_x{}^2 + A_7\sigma_x\sigma_y + A_8\sigma_y{}^2)\sigma_{xy}{}^2 + A_9\sigma_{xy}{}^4,\quad A_i = \text{constants}$$

---

[24] R. Hill, *The Mathematical Theory of Plasticity*, (Clarendon Press, 1950), 318.
R. Hill, *Proc. Roy. Soc.* 198 (1949): 428.

[25] R. Hill, *Math. Proc. Camb. Philos. Soc.* 85 (1979), 179.
P.B. Mellor and A. Parmar, in D.P. Koistinen and N.M. Wang, eds., *Mechanics of Sheet Metal Forming* (New York: Plenum Press, 1978), p. 67.

[26] L. Bourne and R. Hill, *Philos. Mag.* 41 (1950): 671.

[27] D. C. Drucker, *J. Appl. Mech., Trans. ASME* 16 (1949): 349.

[28] F. Edelman and D.C. Drucker, *J. Franklin Inst.*251 (1951): 581.

[29] M. Gotoh, *Int. J. Mech. Sci.* 19 (1977): 505.

g. Bassani[30] ($\sigma_3 = 0$):

$$f = |\sigma_1 + \sigma_2|^N + (1+2r)\,k\,|\sigma_1 - \sigma_2|^M,$$

r, k, M = constants

h. Jones & Gillis[31] ($\sigma_3 = 0$):

$$f = c_{11}\,\sigma_x{}^2 + c_{11}\,\sigma_x\sigma_y + c_{13}\,\sigma_x\,\sigma_{xy} + c_{22}\,\sigma_y{}^2 + c_{23}\,\sigma_y\sigma_{xy} + c_{33}\sigma_{xy}{}^2,$$

c_{ij} = constants

i. Gupta:[32] $f = \sqrt{J_2{}'} - \alpha J_1,$

α = constant

4. Write the yield functions in Problem 3 in terms of the following definitions of effective stress:

   a. tensile test in the x_1 direction, $\bar{\sigma} = \sigma_1$

   b. tensile test in the x_2 direction, $\bar{\sigma} = \sigma_2$

   c. balanced biaxial test, $\sigma_1 = \sigma_2 = \bar{\sigma}$

   d. shear test, same principal axes, $\sigma_1 = \bar{\sigma}, \sigma_2 = \bar{\sigma}$

5. Derive the normality condition for each of the yield functions in Problem 3. For parts d, e, and i, assume the axes are orientedd for principal stress.

6. Assume that $\sigma_{i3} = 0$ and that we are working in principal axes for each yield functions a-d and g in Problem 3. Letting $\alpha = \sigma_2 / \sigma_1$ and $\beta = d\varepsilon_2 / d\varepsilon_1$, find expressions for $\alpha(\beta)$ and for $\beta(\alpha)$.

7. Construct a full set of useful equations for the following plane-stress yield functions (assume $\sigma_3 = 0$). Useful equations include the yield function in terms of $\bar{\sigma}$, the associated flow rule, the normality equations (forward and inverse), definition of $d\lambda$ in terms of $\bar{\sigma}$and $d\bar{\varepsilon}$, $d\bar{\varepsilon}$ in terms of $d\varepsilon_1$, $d\varepsilon_2$, and $d\varepsilon_3$, $\alpha(\beta)$, and $\beta(\alpha)$; where $\alpha = \sigma_2 / \sigma_1$ and $\beta = d\varepsilon_2 / d\varepsilon_1$, σ_3=0.

   a. von Mises:

   $$f = (\sigma_1 - \sigma_2)^2 + (\sigma_1 - \sigma_3)^2 + (\sigma_2 - \sigma_3)^2$$

   b. Hill quadratic:

   $$f = F\,(\sigma_2 - \sigma_3)^2 + G\,(\sigma_1 - \sigma_3)^2 + H\,(\sigma_1 - \sigma_2)^2$$

---

[30] J.L. Bassani, *Int. J. Mech. Sci.* 19 (1977): 651.

[31] Jones and Gillis, unpublished.

[32] Y.M. Gupta, *Acta Metall.* 25 (1977): 1509.

c. Hill normal anisotropic: modify function b such that the $\hat{x}_1$ and $\hat{x}_2$ axes are equivalent, and the strain ratio $d\varepsilon_2/d\varepsilon_1$ in an $\hat{x}_1$ tensile test is r ($\hat{x}_3$ is the sheet-thickness direction).

d. Hill nonquadratic (case IV):

$$f = (1+2r)\,|\sigma_1 - \sigma_2|^M + |\sigma_1 + \sigma_2|^M$$

(You may restrict your attention to one octant, where $\sigma_2 > \sigma_1 > 0$.)

e. Bassani 2-D:

$$f = |\sigma_1 + \sigma_2|^N + (1+2r)k\,|\sigma_1 - \sigma_2|^M$$

8. Verify that Eqs. 7.21 represent the von Mises yield function when balanced biaxial tension and pure shear are used to define the effective stress.

9. Consider the strain paths A, B, and C shown in the following figure. Each path begins at $\varepsilon_1 = \varepsilon_2 = 0$ and ends at $\varepsilon_1 = 2$, $\varepsilon_2 = 1$

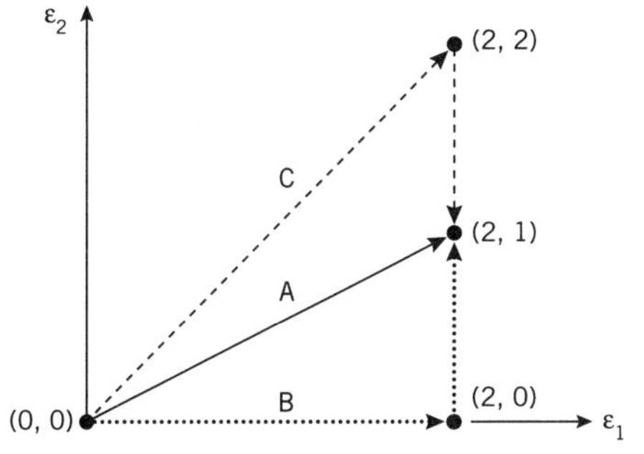

Find the effective strain for each path using the following yield functions:

a. von Mises
b. Hill : orthotropic, quadratic ($r_{RD} = 2$, $r_{TD} = 1.5$)
c. Hill : normal quadratic ($r = 1.75$)
d. Hill : nonquadratic ($r = 1.75$, $M = 2.5$)

10. Assume that the tensile stress-strain curve for a standard material is known to be

$$\bar{\sigma} = 500\bar{\varepsilon}^{0.25}\,(\text{MPa})$$

Consider a plane-strain tension test where x_1 = RD = principal tensile axis, x_2 = TD = zero strain direction, and $\sigma_3 = 0$. Use the yield functions in Problem 9 and find σ_1 as a function of ε_1 for this test. (First find the ratio σ_2/σ_1 for plane strain tension.) Does the work-hardening exponent depend on the form of anisotropy?

11. Repeat Problem 9 for a balanced biaxial test where $\sigma_1 = \sigma_2$, and $\sigma_3 = 0$. Also find the strain ratios for each case.

12. Plot the first quadrant of the yield functions in Problem 9.

13. By plotting, show how r affects Hill's normal quadratic yield function. Take values of $r = \frac{1}{2}$, r = 1 (von Mises), r = 2, r = 4 for illustration.

14. By plotting, show how M affects Hill's normal nonquadratic yield function. For r = 1, take values of M = 1.5, M = 2 (von Mises), M = 4, and M = 10.

15. Compare the form of Hill's normal nonquadratic yield function and Hosford's yield function by assuming that r = 2 and finding M in each case such that σ_1 (balanced biaxial tension) is equal to $1.1\,\bar{\sigma}_B$ (uniaxial tension).

B. DEPTH PROBLEMS

16. It has been proposed that the direction of the plastic-strain increment, $d\varepsilon$, is not always normal to the yield surface. Instead, the proposer suggests that $d\hat{\boldsymbol{\varepsilon}}$ is intermediate in direction between $d\hat{\boldsymbol{\varepsilon}}^{(n)}$, the normal direction, and $d\hat{\boldsymbol{\sigma}}$, the stress-increment direction [i.e., $d\hat{\boldsymbol{\varepsilon}} = \alpha\, d\hat{\boldsymbol{\varepsilon}}^{(n)} + (1-\alpha)\, d\hat{\boldsymbol{\sigma}}$, where $\alpha = 0-1$]. Criticize this model in terms of stability arguments.

17. It is convenient to introduce factors that can be used to multiply the effective stress or strain to obtain the stress or strain in a given state. For example,

$$\sigma_1^{BB} = F_\sigma^{(BB)}\,\bar{\sigma}$$

or

$$\varepsilon_1^{BB} = F_\varepsilon^{(BB)}\,\bar{\varepsilon}$$

might be used to find the in-plane strains and stresses in balanced biaxial tension (BB) for a given yield function at a given hardness of $\bar{\varepsilon}$, $\bar{\sigma}$.

a. Show that F_σ and F_ε are constants with respect to strain for a given yield function when isotropic hardening is obeyed.

b. Find the specific values of $F_\sigma^{(PS)}$ and $F_\varepsilon^{(BB)}$ in terms of r that relate plane-strain tension to uniaxial tension using Hill's normal quadratic yield criterion.

c. Repeat part b for Hill's normal nonquadratic yield function, finding F_σ and F_ε in terms of r and M.

d. What is the special relationship between $F_\sigma^{(ps)}$ and $F_\varepsilon^{(ps)}$? Why does this relationship always hold?

18. In view of results from Problem 17, what can you say about the complexity of yield function that would be required to account for the following observations? (Assume only normal anisotropy.)

    a. In uniaxial tension, $\sigma_1 = 500\ \varepsilon_1^{0.25}$
    In plane-strain tension, $\sigma_1 = 600\ \varepsilon_1^{0.25}$
    b. Same as part a, but r = 2 from the tensile test
    c. In uniaxial tension, $\sigma_1 = 500\ \varepsilon_1^{0.25}$
    In plane-strain tension, $\sigma_1 = 600\ \varepsilon_1^{0.35}$
    r = 2

19. Various authors have attempted to introduce work-hardening parameters, particularly to compare various strain states, or to compare material hardness when true strain is unknown. Here are three such quantities:

$$\left(\frac{d\bar{\sigma}}{d\bar{\varepsilon}}\right),\quad \left(\frac{d\bar{\sigma}}{\bar{\sigma}\,d\bar{\varepsilon}}\right) = \frac{d\ln\bar{\sigma}}{d\bar{\varepsilon}},\quad \left(\frac{\bar{\varepsilon}\,d\bar{\sigma}}{\bar{\sigma}\,d\bar{\varepsilon}}\right) = \frac{d\ln\bar{\sigma}}{d\ln\bar{\varepsilon}}$$

Based on your knowledge of isotropic hardening and results from Problems 17 and 18, which of these do you think is most suitable for comparing hardening in various strain states?

C. NUMERICAL PROBLEMS

20. Consider several closed strain paths represented in the following figure and find the effective strain at the end of each path by numerical integration of each law:

    a. von Mises
    b. Hill : orthotropic, quadratic $(r_{RD} = 2,\ r_{TD} = 1.5)$
    c. Hill : normal quadratic $(r = 1.75)$
    d. Hill : nonquadratic $(r = 1.75,\ M = 2.5)$

    Assume that all the strain is plastic.

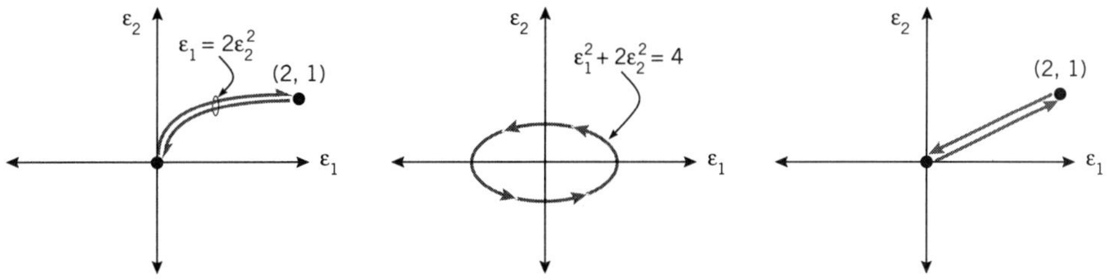

21. Repeat Problem 20 for the following yield functions, for which no explicit function of $d\bar{\varepsilon}$ in terms of $d\varepsilon_1$ exists.

    a. Hosford (r = 1.75, M = 6)

    b. Bassani (r = 1.75, N = 2, M = 4)

22. Consider the material triangle defined in Chapter 4, Problem 16. Find a matrix expression for the forces applied to each corner of the triangle in order to deform it plastically such that small displacements $\Delta\mathbf{a}$, $\Delta\mathbf{b}$, and $\Delta\mathbf{c}$ occur at each corner. Assume a unit thickness of material and start in terms of principal true strains ε_1 and ε_2 ($\ln \lambda_1$ and $\ln \lambda_2$). You may wish to use an equivalent work principle or make other approximations and assumptions about how the stress felt by the triangle relates to the forces applied at the corners. Consider the yield functions in Problem 9.

CHAPTER 8

Crystal-Based Plasticity[1]

In this chapter, we provide the basic framework necessary for deriving plasticity laws from crystallographic slip in metals. The ideas are by no means new, but recent advances in digital computation and automated texture analysis have made it possible to test predictions of these models against experiments that measure the evolution of texture and yield surface. Continuum plasticity measurements require unmanageable numbers of experiments in order to derive closed-form treatments, with the result that hardening was usually treated with very simple ideas such as isotropic hardening or kinematic hardening. More general evolutions simply could not be tested adequately.

The first sections of this chapter serve another purpose: to introduce **upper and lower bound methods** in plasticity. Like the slip-line field method outlined in Chapter 10, simple applications of upper and lower bound methods are usually limited to very simple geometries and thus must be implemented in a more elaborate numerical scheme for exact analysis of general forming operations. On the other hand, simple applications of these methods can provide physical insight that more elaborate numerical methods sometimes lack, and they provide a historical perspective of how many problems were solved before the advent of high-speed computation. In fact, the upper and lower bound methods will be used later in this chapter to develop important elementary concepts about **crystal slip**.

8.1 BOUNDS IN PLASTICITY

The upper and lower bound theorems in plasticity permit estimates of the critical load to plastically deform material. Each requires some estimate of either the deformation mode or stress state associated with the yield state, and each assumes that the material is rigid-perfectly plastic, has a convex yield surface,

[1] Chapter 8 was contributed by Peter M. Anderson, The Ohio State University.

and obeys normality. Despite the material restrictions, these theorems provide a powerful analytic tool on which numerical methods in plasticity are based. They will form the basis for estimates of macroscopic stresses required to yield crystals and polycrystals, as discussed in Sections 8.3 and 8.4. The bounds presented in this section are discussed in more detail by Calladine.[2]

According to the lower bound theorem, if one finds a distribution of stress that

1. satisfies the equations of equilibrium,
2. balances known external loads, and
3. does not violate the yield condition anywhere,

then the external loads required by the proposed stress field will be carried safely by the material. Equivalently, any stress field that meets the above conditions provides a **lower bound** to the actual collapse loads. Clearly, the best lower bound is produced by the actual stress field. Typically, that field is difficult to find because it not only must satisfy the three conditions above, it also must be related to a compatible deformation field. The utility of the lower bound theorem is that reasonable estimates of loads to yield a material or structure (i.e., collapse loads) can be determined from relatively simple, approximate stress fields that do not satisfy compatibility.

The lower bound theorem can be proven by considering a volume, *V*, of material with a surface, *S*, on which a distribution of force per area, or traction **T**, is acting. The actual collapse stress field, σ, and traction, **T**, must satisfy conditions (1) and (2) above, so that $\sigma_{ij,j} = 0$ everywhere in the volume; and the stress state just under any surface location must be related to the traction by $\sigma_{ij} n_j = \mathbf{T}_i$, where **n** is the outward normal to the surface. (Note: Unless noted otherwise, repeated indices imply summation over that index, from 1 to 3. For example, $T_i du_i = T_1 du_1 + T_2 du_2 + T_3 du_3$. Further, $T_{i,j} = \partial T_i / \partial x_j$.) Further, the incremental strain, dε, must be compatible, and this requires that $d\varepsilon_{ij} = \frac{1}{2}(du_{i,j} + du_{j,i})$ must hold at every point in the volume, where the displacement increment **du** is single-valued and has well-defined derivatives with respect to position. Under such conditions, the **principle of virtual work** states that the incremental work that is either dissipated plastically or stored elastically inside the body (the internal work increment, IW) must equal the external work increment (EW) done by tractions on the surface of the body:[3]

$$\int_V \sigma_{ij} d\varepsilon_{ij}\, dV = \int_S T_i du_i\, dS \tag{8.1}$$

[2] See C.R. Calladine, *Plasticity for Engineers* (Chichester, England: Halsted Press; 1985).

[3] See J. Chakrabarty, *Theory of Plasticity* (New York: McGraw-Hill, 1987) p. 40.

If we propose a trial collapse stress field, σ^*, and surface loading distribution, $\mathbf{T}^*$, which satisfy conditions (1) and (2) above, then the principle of virtual work also applies, where $d\varepsilon$ and $d\mathbf{u}$ remain the actual collapse values:

$$\int_V \sigma_{ij}^* d\varepsilon_{ij}\, dV = \int_S T_i^* du_i\, dS \tag{8.2}$$

If the difference between Eq. 8.1 and Eq. 8.2 is taken, then

$$\int_V (\sigma_{ij} - \sigma_{ij}^*) d\varepsilon_{ij}\, dV = \int_S (T_i - T_i^*) du_i\, dS \tag{8.3}$$

As long as σ^* does not violate yield, and the material is rigid-plastic and obeys convexity and normality, the volume integral in Eq. 8.3 cannot be negative. This feature is frequently called the **principle of maximum plastic work**,[4] and a motivation for it from descriptions of crystalline slip will be made in Section 8.3. However, on the continuum level, the yield surface geometry in Figure 8.1 shows that as long as σ_{ij}^* lies on or inside of the yield surface, the product $\sigma_{ij} d\varepsilon_{ij}$ is always greater than or equal to $\sigma_{ij}^* d\varepsilon_{ij}$ at each location in the body. According to Eq. 8.3, the corresponding external work increment must also be nonnegative. In the simple case where work is done through one load point displacement increment, $d\mathbf{u}$, on the body, the nonnegative condition becomes $F^* \le F$, where F^* and F are the lower bound and actual collapse load magnitudes in the direction $d\mathbf{u}$. The proof that an equilibrium trial stress distribution produces a lower bound to the actual collapse load is complete.

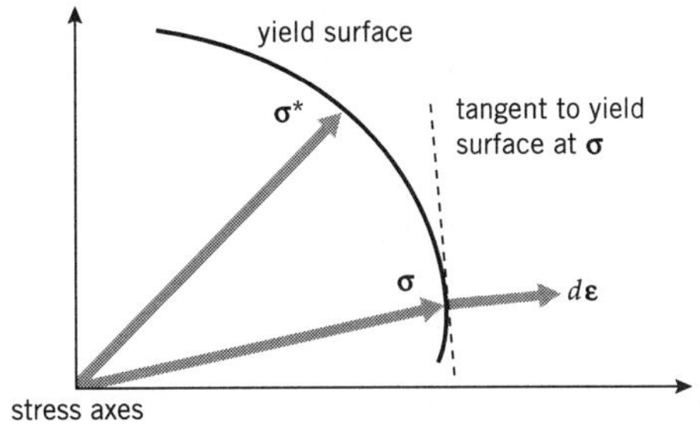

Figure 8.1 Section of a convex yield surface showing the principle of maximum plastic work.

According to the **upper bound theorem**, if a single-valued and differentiable trial displacement field, $\mathbf{u}^*$, is proposed that meets any imposed displacement boundary conditions, then a collapse load that is computed by equating the internal and external plastic work increments will be sufficiently large to

[4] R. Hill, *The Mathematical Theory of Plasticity* (Oxford: Oxford University Press, 1986): p. 66.

plastically deform the material. The upper bound theorem is proven by first using the principle of virtual work,

$$\int_V \sigma_{ij} d\varepsilon^*_{ij}\, dV = \int_S T_i du^*_i\, dS \tag{8.4}$$

where the proposed compatible strain increment is defined by $d\varepsilon_{ij}{}^* = \frac{1}{2}(du_{i,j}{}^* + du_{j,i}{}^*)$, and, as before, σ and **T** are the actual equilibrium fields during collapse. Again, if the material is rigid-perfectly plastic and the principle of maximum plastic work holds, then the stress field, $\sigma_{ij}{}^*$, which would be produced if $d\varepsilon_{ij}{}^*$ were applied, does the maximum plastic work,

$$\int_V \sigma^*_{ij} d\varepsilon^*_{ij} dV \geq \int_V \sigma_{ij} d\varepsilon^*_{ij} dV \tag{8.5}$$

When Eqs. 8.4 and 8.5 are combined, it is clear that

$$\int_V \sigma^*_{ij} d\varepsilon^*_{ij}\, dV \geq \int_S T_i du^*_i\, dS \tag{8.6}$$

The left-hand side of Eq. 8.6 is the internal work rate predicted by the proposed deformation field, and some examples to follow demonstrate that this is a straightforward quantity to calculate. Although $d\varepsilon^*$ is a compatible strain field, the corresponding stress field does not satisfy the equilibrium condition in general, so that $\sigma_{ij,j}{}^* \neq 0$. Further, we choose *not* to define the proposed tractions by the equilibrium condition, $\sigma_{ij}{}^* n_j = T_i{}^*$, but rather they are defined by setting the external plastic work increment equal to the internal work increment,

$$\int_S T^*_i du^*_i\, dS = \int_V \sigma^*_{ij} d\varepsilon^*_{ij}\, dV \tag{8.7}$$

As long as the proposed tractions are defined in this way, the combination of Eqs. 8.6 and 8.7 indicates that

$$\int_S T^*_i du^*_i\, dS \geq \int_S T_i du^*_i\, dS \tag{8.8}$$

In the case of a single imposed load point displacement, the condition becomes $F^* \geq F$, where F^* and F are the upper bound and actual collapse load acting in the same direction as $d\mathbf{u}^*$. The proof that a compatible trial displacement field produces an upper bound to a collapse load is complete.

8.2 APPLICATION OF BOUNDS

Biaxial Loading with a Single Slip Plane

The concept that crystals deform plastically by relative **slip** on **crystallographic planes** often provides trial deformation fields that make the upper bound approach simple to use. For example, Figure 8.2 shows a block of crystal containing a slip plane with a unit normal **m** and unit **slip direction s**.

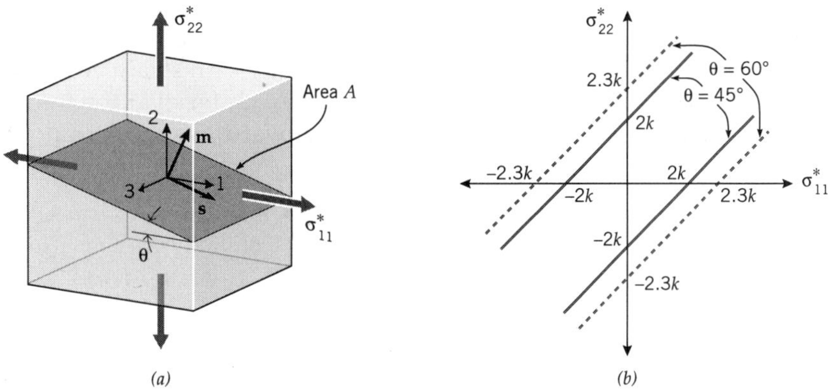

Figure 8.2 (*a*) A cube of material under applied stress σ_{11}* and σ_{22}*, containing a slip system oriented at angle θ and defined by a slip plane normal **m** and slip direction **s**. (*b*) Corresponding prediction of the yield surface, as a function of orientation angle θ.

The yield condition for this slip plane is that the resolved shear stress acting on the slip plane in the **s**-direction reaches a critical magnitude of ±k. The internal work increment is therefore IW = kAb, where A is the area of slipped plane and b is the magnitude of slip in the **s** direction. The external work increment is EW = σ_{11}*Abm$_1$s$_1$ + σ_{22}*Abm$_2$s$_2$. When the condition in Eq. 8.7, that EW = IW, is imposed, the collapse condition becomes

$$\sigma_{11}^* - \sigma_{22}^* = \frac{\pm k}{s_1 m_1} = \frac{\pm 2k}{\sin 2\theta} \tag{8.9}$$

and the result is plotted in Figure 8.2*b* for θ = 45° and 60°.

In this simple example, the lower bound theorem is easily applied also. The trial stress field, σ*, is uniform and has σ_{11}* and σ_{22}* as the only nonzero components. Because the stress state is uniform, it satisfies equilibrium, and, further, it balances known external loads because σ_{12}* = σ_{13}* = σ_{33}* = 0. The final requirement for the lower bound is that the field not violate yield anywhere. To meet this, the magnitude of the resolved shear stress must not exceed k. Obviously, the best lower bound is produced when the magnitude just equals k,

$$\sigma^*_{ij} m_j s_i = \pm k \tag{8.10}$$

and this may be rearranged to give Eq. 8.9. In fortunate situations such as this, where the upper and lower bound predictions coincide, the trial fields used correspond to the actual fields. In general, the trial fields are approximations to sufficiently complicated actual fields, so that the two types of bounds usually do not coincide.

Indentation

When a rigid, rectangular punch is pushed into a crystal as shown in Figure 8.3*a*, one imagines diffuse plasticity to occur, with strain concentrations near the edges of the indenter, and an uplifting of material at the free surfaces near the indenter. An increment, δ, in downward displacement of the punch

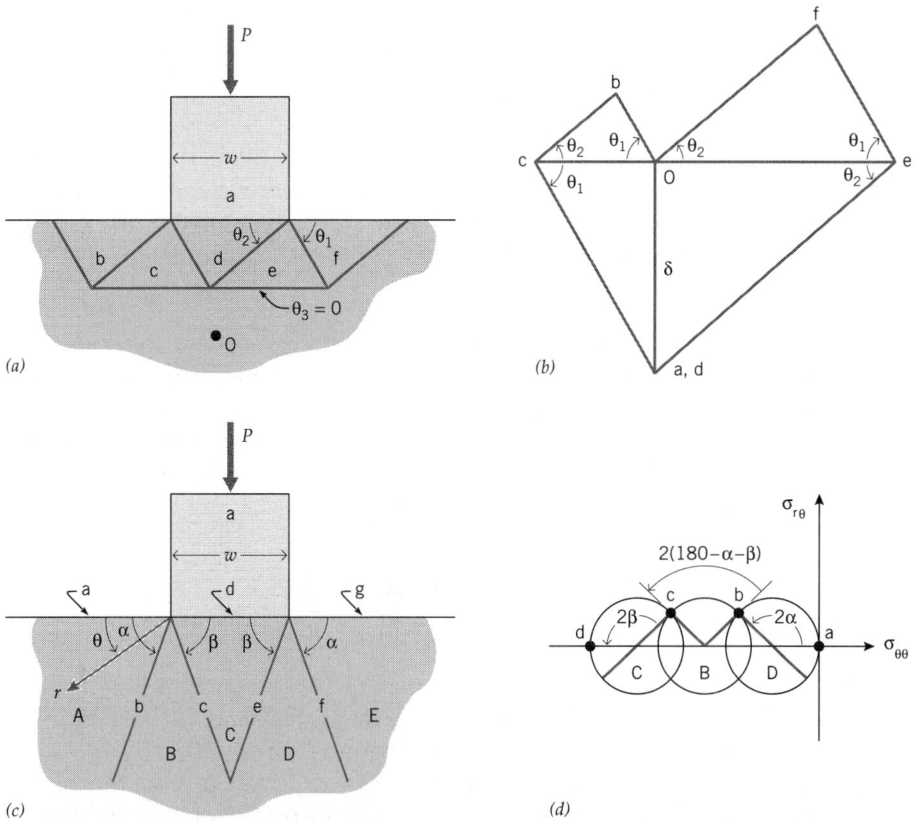

Figure 8.3 (*a*) Indenter geometry and proposed deformation field for an upper bound approach. (*b*) Corresponding hodograph for (*a*). (*c*) Proposed stress field for a lower bound approach to the indentation load. (*d*) Mohr's circle construction associated with (*c*).

requires a force, P, per unit depth, which can be estimated using simple upper and lower bound approaches. A simple deformation pattern of concentrated relative slip lines and rigid triangular sections is proposed in Figure 8.3*a*. Each type of slip line—i = 1, 2, or 3 here—is specified by an orientation angle, θ_i, and slip direction, s_i. Each slip line in the pattern may correspond to operation of either a single-crystal slip system or a linear combination of slip systems. Using the upper bound approach, an estimate, P*, of the collapse load is made by equating the external work increment, EW = +P*δ, to the internal work increment, IW. The internal work increment is determined by adding the work contributions from every sliding internal interface. For notation, the stationary part of the half space is labeled o, and the indenter and rigid triangular blocks are assigned other letters. The length of the sliding interface between blocks i and j is denoted by l_{ij}, and the relative slip distance between the blocks is δ_{ij}. Accordingly,

$$\begin{aligned} IW = k(&l_{bo}\delta_{bo} + l_{bc}\delta_{bc} + l_{co}\delta_{co} + l_{cd}\delta_{cd} \\ &+ l_{de}\delta_{de} + l_{eo}\delta_{eo} + l_{ef}\delta_{ef} + l_{fo}\delta_{fo}) \end{aligned} \tag{8.11}$$

The geometry in Figure 8.3*a* clearly indicates that

$$l_{bo} = l_{cd} = l_{ef} = \frac{w \sin\theta_2}{\sin(\theta_1 + \theta_2)}$$

$$l_{bc} = l_{de} = l_{fo} = \frac{w \sin\theta_1}{\sin(\theta_1 + \theta_2)}$$

and

$$l_{co} = l_{eo} = w$$

However, the corresponding relative displacements, δ_{ij}, in Eq. 8.11 must be determined. This is accomplished by ensuring that the proposed displacement field is single-valued. A **hodograph** as shown in Figure 8.3*b* is a convenient geometric way to impose this condition. Points labeled a, b, c, and so on in the hodograph are positioned so that, for example, the vector $\mathbf{v}_{oa}$ from o to a is the displacement of block a relative to block o. Clearly, $\mathbf{v}_{oa}$ is a downward vector of length δ, and since d does not move relative to a, points a and d in the hodograph coincide. Each subsequent point in the hodograph is determined by applying two relative displacement conditions. For example, point e is determined by noting that $\mathbf{v}_{de}$ must be parallel to $\mathbf{l}_{de}$, and $\mathbf{v}_{oe}$ must be parallel to $\mathbf{l}_{oe}$. The intersection of the two parallel lines constructed determines the position of point e. The δ_{ij} required in Eq. 8.11 are simply the lengths of vectors $\mathbf{v}_{ij}$ in the hodograph.

For the simple case where $\theta_1 = \theta_2 = 60°$,

$$IW = \frac{10kw\delta}{\sqrt{3}}$$

and the resulting estimate for the collapse load per unit depth is $P^* \approx 5.77$ kw. In closing, the upper bound estimate was based on a compatible displacement field involving localized slip planes between rigid blocks; however, the field is not an equilibrium one in that the sum of forces on each rigid block is not equal to zero.

A corresponding lower bound estimate of the indentation load is based on the stress state shown in Figure 8.3*c*.[2] In this case, the stress field consists of five sectors (A, B, C, D, E), each of which has a uniform stress state within it. The boundaries across which the stress state changes are labeled a, b, c, d, e, f, g. Since equilibrium must be satisfied in the lower bound approach, both normal and shear components of stress must be continuous across each boundary. Further, the yield condition cannot be violated anywhere. A convenient way to impose these conditions is invoke a Tresca yield criterion, and assume that the maximum and minimum principal stresses lie in the plane of the diagram. Therefore, the in-plane stress state in each sector can be described by a Mohr's circle with a radius that cannot exceed k, the shear stress required for yield. Further, the continuity of shear and normal components of stress across the boundaries will determine the relative positions of each Mohr's circle, producing the result shown in Figure 8.3*d*.

The most convenient coordinate system in which to express stress components is the cylindrical (r, θ) one shown in Figure 8.3*c*, so that $\sigma_{r\theta}$ and $\sigma_{\theta\theta}$ must be continuous across boundaries, as θ is increased from 0 to 180°. The corresponding estimate, P^*, of indentation load per unit depth is then $-\sigma_{\theta\theta}(r, \theta = 180°)w$. It is clear that the best lower bound is produced from an equilibrium stress field that makes $\sigma_{\theta\theta}(r, \theta = 180°)$ as negative as possible without violating yield anywhere. To accomplish this, propose that the stress state along boundary a is $\sigma_{\theta\theta} = 0$, $\sigma_{r\theta} = 0$, and $\sigma_{rr} = -2k$. The first two conditions are imposed by the free surface. Thus sector A is at yield in compression, and $\sigma_{\theta\theta}$ and $\sigma_{r\theta}$ along boundary a are given by the projections of point a in Figure 8.3*d* onto the horizontal and vertical axes, respectively. The corresponding values of $\sigma_{\theta\theta}$ and $\sigma_{r\theta}$ along boundary b are given by point b, which was obtained from a counterclockwise rotation, 2α. In order for $\sigma_{\theta\theta}$ and $\sigma_{r\theta}$ to be continuous across boundary b, Mohr's circle B for sector B must pass through point b. Continuity of $\sigma_{\theta\theta}$ and $\sigma_{r\theta}$ across boundary c is satisfied by the appropriate counterclockwise rotation, 2(180°–α–β), to produce point c, and then ensuring that Mohr's circle C passes through point c. Finally, $\sigma_{\theta\theta}(r, \theta = 180°)$ is simply the horizontal distance of point d from the origin.

The largest negative value, $\sigma_{\theta\theta}(r, \theta = 180°) = -2(1 + \sqrt{2})k$ occurs when all sectors are assumed to be yielding (the radius of each Mohr's circle is then k), and, further, α = β = 67.5°. The lower bound for *P* is then about 4.83kw, which is 6% below the exact answer of 5.14kw. The upper bound for *P* discussed previously is 5.77kw, which is about 12% above the exact answer.

Exercise 8.1 **Provide an upper bound to the density, ρ, that a vertical embankment of height h can sustain without yielding. The yield stress in shear of the embankment material is k.**

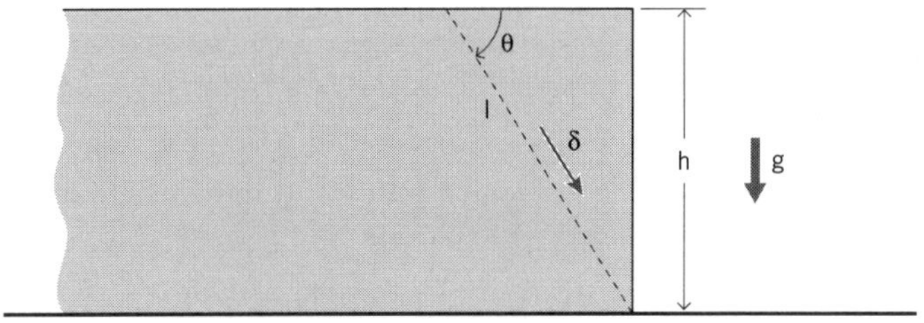

Solution: A proposed collapse mechanism is shown above, in which a triangular section shears away from the embankment along a line l, oriented at angle θ from the horizontal surface. The internal work associated with a relative displacement, δ, across line l is IW = lδk. The corresponding external work is provided by the gravitational force on the triangular block of material. In particular, EW = $\rho g l^2 \delta \cos\theta \sin^2\theta / 2$. When EW is equated to IW, the resulting expression is $\rho = 4k/(gh \sin^2\theta)$. This represents an upper bound to the actual ρ that may be sustained. The minimum upper bound, $\rho = 4k/gh$, for this collapse mechanism occurs when θ = 45°.

Exercise 8.2 **A two-phase sample of width l and height l is deformed in compression as shown in the following figure. Provide an upper bound to the collapse pressure, p, on the sample. The yield stress in shear for phase 1 is k_1, the corresponding value for phase 2 is k_2, and the interface between the crystals yields at a shear stress k_i**

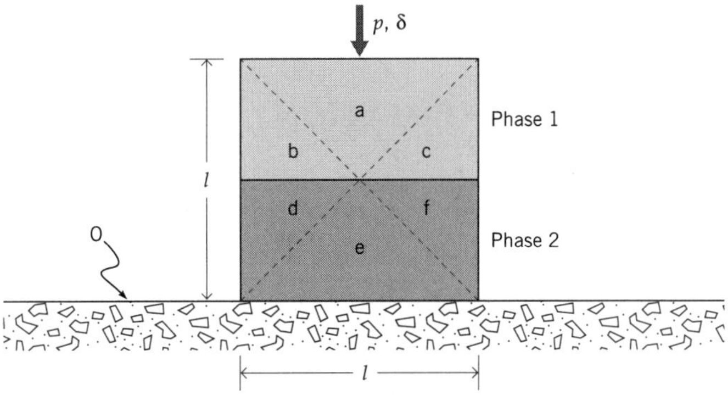

Solution: The external work per unit depth is EW = plδ, where δ is an increment in downward displacement of the top surface. Each of the rigid triangular sections in the proposed deformation mechanism is labeled as shown above. The corresponding internal work per unit depth is

$$IW = 2\left[l_{ac}\delta_{ac}k_1 + \left(\frac{1}{2}\right)\delta_{cf}k_i + l_{ef}\delta_{ef}k_2 \right]$$

A factor of 2 is used because a mode of deformation that is symmetric about the vertical axis through the center of the specimen is postulated. The geometry above indicates that

$$l_{ac} = l_{ef} = \frac{1}{\sqrt{2}}$$

The corresponding *hodograph* follows. It indicates that

$$\delta_{ef} + \delta_{ac} = \sqrt{2}\,\delta \qquad 8.11\text{-A}$$

$$\delta_{ef} - \delta_{ac} = \sqrt{2}\,\delta_{cf} \qquad 8.11\text{-B}$$

The above two relations may be used to replace δ_{ac} and δ_{ef} in the expression for IW, so that

$$IW = 1|\delta - \delta_{cf}|k_1 + 1|\delta_{cf}|k_i + 1|\delta + \delta_{cf}|k_2 \qquad 8.11\text{-C}$$

Accordingly, EW is set equal to IW. The corresponding estimate for p is a function of δ_{cf}, and the optimal choice of δ_{cf} is that which renders p a minimum. There are three possible minima corresponding to $\delta_{cf} = -\delta, 0,$ or $+\delta$, depending on the values of k_1, k_2, and k_i. Accordingly, the best upper bound is given by

$$p = \min[2k_2 = k_i, k_1 + k_2 = 2k_1 = k_i] \qquad 8.11\text{-D}$$

In the case of a homogeneous material for which $k_1 = k_2 = k_i = k$, the standard answer of $p = 2k$ is recovered.

8.3 CRYSTALLINE DEFORMATION

Slip Systems

The concept that some crystals deform at room temperature by incremental shearing on certain crystallographic planes has been recognized for almost 100 years. In particular, Ewing and Rosenhain[5] deformed polycrystalline lead samples and made several observations that demonstrate remarkable insight. The authors noted that there were slip bands throughout the material which, within a local region, appeared to be parallel and associated with the crystal structure. They noted that the amount of slip must be in multiples of lattice spacings in order for them to be observed. The implications were that deformation was inhomogeneous and concentrated on only certain crystallographic planes. They reasoned that existing slip made it harder for imperfections to activate further slip. Clearly, many qualitative, more macroscopic features of **crystalline** slip had been observed.

Dislocation theory has provided information about crystal slip that is centered on nucleation and incremental advance of crystal line defects that separate slipped from unslipped regions on crystal planes. From this theory have emerged estimates of the critical stress state to nucleate such defects, called **dislocations**, and the continued stress to move them across a slip plane past other crystal defects.[6] The relation between crystal structure and preferred **slip planes** and **slip directions** has also been examined, most notably by Peierls via analysis of two elastic half spaces joined by a nonlinear-shear force–shear-displacement relation.[7] In that analysis, the mean energy per unit length of dislocation line is proportional to μb^2, where b is the slip distance required to shift the lattice by a periodic unit and μ is the elastic shear modulus. The result suggests that the lowest energy dislocations have a small repeat distance b. The analysis also predicts that as the dislocation moves by a repeat distance, the energy fluctuates with position, so that a peak resolved shear stress of magnitude $\tau_p \approx \mu \exp(-d/b)$ is required for motion, where d is the spacing between slip planes.[8] The analysis predicts dense-packed planes to be favored, as they have the largest interplanar spacing and smallest repeat distance. These results are in qualitative agreement with observed slip systems in several common crystals, an excellent summary of which is given by Hirth.[9] Essentially, the {111}/<110>[10] slip plane/slip direction is the principal operative system in **face-centered cubic** structures, while for **body-centered cubic** structures such as iron, tungsten, vanadium, and molybdenum, {110}/<111> and {112}/<111> are predominant at room temperature, with some observations of {123}/<111>. For **hexagonal close-packed** structures such as zinc and magnesium, the predominant system is {0001}/<11–20>. For rock-salt type crystals, the predominant system is {110}/<110>.

Yield Surfaces Based on Critical Resolved Shear Stress

When a stress, σ, is applied to a crystal, a virtual force, f, to move a unit length of dislocation across the slip plane is produced. Peach and Koehler[11] developed the expression,

---

[5] A. Ewing and W. Rosenhain, *Philos. Trans. Roy. Soc.* A193 (1899): 353.

[6] For a comprehensive review, see J.P. Hirth, *Theory of Dislocations*, 2nd ed. (New York, John Wiley & Sons,1982).

[7] R.E. Peierls, *Proc. Roy. Soc.* 52 (1940): 23.

[8] [See Ref. [6], pp. 235-237.]

[9] [See Ref. [6], Tables 9-1 to 9-5.

[10] We use standard **Miller index** notation for slip planes and directions. A discussion of this convention is beyond the scope of this book. It is sufficient to note that in a cubic crystal, <abc> is equivalent to vector components a, b, c expressed in units of the unit cell edge dimensions, and {abc} is a plane normal to the same vector.

[11] M.O. Peach and J.S. Koehler, *Phys. Rev.* 80 (1950): 436.

$$f = b\tau, \quad \tau = s_i \sigma_{ij} m_j \tag{8.12}$$

where the unit slip direction is **s**, the slip magnitude is b, and the slip plane normal is m. The glide force, f, is proportional to the resolved shear stress, τ, that acts on the slip plane in the direction of b. Further, f acts normal to the dislocation line, and always acts to move the dislocation in the slip plane, so that a relative shear deformation of the same sign as τ is produced. In a more general sense, τ is the net shear stress at the dislocation, and it can be affected by nearby defects as well as the applied loading. In order for an existing dislocation to move a periodic distance, f must equal or exceed $\tau_p b$, the **Peierls resistance** caused by fluctuation in dislocation energy.

Most polycrystals contain dislocation densities that can range from 10^6 to 10^{12} cm of dislocation line in a cubic cm volume,[12] so that plasticity may not be limited by dislocation nucleation, but rather by the ability of existing dislocations to overcome obstacles that inhibit their motion, such as grain boundaries, impurities, and other dislocations. In any event, analyses to nucleate dislocations depend on more subtle details involving the shape of the nucleating loop and stress concentrations provided by defects such as cracks, grain boundaries, or other dislocations. Regardless of the critical event for onset of plasticity, the condition for yield of the crystal is that f in Eq. 8.12 reaches a critical value; or, equivalently, the resolved shear stress, τ, reaches a critical value, τ_c.

The concept of a critical resolved shear stress was proposed by Schmid[13] long before the Peierls model was developed. Schmid asserted that the resolved shear stress reaches a critical value, τ_c, for slip to occur. From experiments of tensile loading of zinc wires, Schmid reported a substantial variation with orientation of τ_c, the average values of which are 36 g/mm^2 at room temperature, and 126 g/mm^2 at -185°C. Given the concept of a **critical resolved shear stress**, a simple application is to calculate the yield surface of a f.c.c. crystal that is loaded with an arbitrary stress, σ. There are 12 possible slip systems, $(\mathbf{m}^{(\alpha)}, \mathbf{s}^{(\alpha)})$, $\alpha = 1$ to 12, to consider, so that 12 separate yield conditions follow:

$$\tau^{(\alpha)} \equiv \pm s_i^{(\alpha)} \sigma_{ij} m_j^{(\alpha)} \geq \tau_c^{(\alpha)}, \quad \alpha = 1, 12 \tag{8.13}$$

For the {111}/<110> class of **m**/**s** directions in f.c.c. materials, the 12 possible slip systems are listed in Table 8.1. The $\pm$ notation for **s** is included since slip on each system can occur in either of two opposing directions.

Since there are six independent components of stress, the yield surface can be expressed in six-dimensional stress space, where the components of the stress vector are defined as

---

[12] M.F. Ashby. In *Strengthening Methods in Crystals* A. Kelly and R.B. Nicholson, eds., (New York: Halsted, 1971).

[13] E. Schmid: *Proc. Int. Congr. Appl. Mech.* (Delft) (1924): 342.

$$\begin{bmatrix} \sigma_1 \\ \sigma_2 \\ \sigma_3 \\ \sigma_4 \\ \sigma_5 \\ \sigma_6 \end{bmatrix} = \begin{bmatrix} \sigma_{11} \\ \sigma_{22} \\ \sigma_{33} \\ \sigma_{23}(=\sigma_{32}) \\ \sigma_{13}(=\sigma_{31}) \\ \sigma_{12}(=\sigma_{21}) \end{bmatrix} \tag{8.14}$$

The yield condition, Eq. 8.13, may be restated as

$$\pm n_i^{(\alpha)} \sigma_i \geq \tau_c^{(\alpha)} \tag{8.15}$$

where

$$\begin{bmatrix} n_1 \\ n_2 \\ n_3 \\ n_4 \\ n_5 \\ n_6 \end{bmatrix}^{(\alpha)} = \begin{bmatrix} s_1 m_1 \\ s_2 m_2 \\ s_3 m_3 \\ (s_2 m_3 + s_3 m_2) \\ (s_1 m_3 + s_3 m_1) \\ (s_1 m_2 + s_2 m_1) \end{bmatrix}^{(\alpha)} \tag{8.16}$$

For application to the f.c.c. slip systems listed in Table 8.1, Eqs. 8.15 and 8.16 may be applied to each slip system, $\alpha = 1$ to 12. The yield condition then becomes

$$\pm \frac{1}{\sqrt{6}} \begin{bmatrix} 1 & -1 & 0 & -1 & 1 & 0 \\ 1 & 0 & -1 & -1 & 0 & 1 \\ 0 & 1 & -1 & 0 & -1 & 1 \\ -1 & 0 & 1 & 1 & 0 & 1 \\ 0 & 1 & -1 & 0 & 1 & -1 \\ -1 & 1 & 0 & 1 & 1 & 0 \\ -1 & 0 & 1 & -1 & 0 & 1 \\ 0 & -1 & 1 & 0 & 1 & 1 \\ 1 & -1 & 0 & 1 & 1 & 0 \\ 1 & -1 & 0 & 1 & -1 & 0 \\ 1 & 0 & -1 & 1 & 0 & 1 \\ 0 & 1 & -1 & 0 & 1 & 1 \end{bmatrix} \begin{bmatrix} \sigma_1 \\ \sigma_2 \\ \sigma_3 \\ \sigma_4 \\ \sigma_5 \\ \sigma_6 \end{bmatrix} \geq \begin{bmatrix} \tau_c \\ \tau_c \\ \tau_c \\ \tau_c \\ \tau_c \\ \tau_c \\ \tau_c \\ \tau_c \\ \tau_c \\ \tau_c \\ \tau_c \\ \tau_c \end{bmatrix} \tag{8.17}$$

where for now, all slip systems are assumed to operate at the same critical resolved shear stress. Therefore, Eq. 8.17 represents 12 conditions in which the components $n_i^{(\alpha)}$ are simply the components of row α of the 6×12 matrix in Eq.

TABLE 8.1 The twelve {111}/<110> f.c.c. slip systems

slip system	$\pm[s_1, s_2, s_3]$	$[m_1, m_2, m_3]$
1	$[1\ {-1}\ 0]/\sqrt{2}$	
2	$[1\ 0\ {-1}]/\sqrt{2}$	$[1\ 1\ 1]/\sqrt{3}$
3	$[0\ 1\ {-1}]/\sqrt{2}$	
4	$[1\ 0\ 1]/\sqrt{2}$	
5	$[0\ 1\ {-1}]/\sqrt{2}$	$[-1\ 1\ 1]/\sqrt{3}$
6	$[1\ 1\ 0]/\sqrt{2}$	
7	$[-1\ 0\ 1]/\sqrt{2}$	
8	$[0\ 1\ 1]/\sqrt{2}$	$[1\ {-1}\ 1]/\sqrt{3}$
9	$[1\ 1\ 0]/\sqrt{2}$	
10	$[1\ {-1}\ 0]/\sqrt{2}$	
11	$[1\ 0\ 1]/\sqrt{2}$	$[1\ 1\ {-1}]/\sqrt{3}$
12	$[0\ 1\ 1]/\sqrt{2}$	

8.17, including the $1/\sqrt{6}$ factor. Each yield condition for a slip plane generates two yield surface planes with outward normals $\pm n_i^{(\alpha)}$, and each plane is located a distance $\sqrt{3/2}\ \tau_c$ along that normal. Thus, the yield surface is the inner locus of the 24 planes. However, the surface is not closed, since all $n_i^{(\alpha)}$ have zero dot product with the hydrostatic direction, $[\sigma_i] = [1\ 1\ 1\ 0\ 0\ 0]$. The f.c.c. yield surface, like the von Mises yield surface, is a hypercylinder with the generator along the hydrostatic axis, so that hydrostatic loading cannot cause yield.

Unfortunately, six-dimensional plots cannot be displayed easily, but two-dimensional ones can. Figure 8.4 shows the resulting yield surface when all stress components but σ_{11} and σ_{22} are set to zero in Eq. 8.17. Equivalently, this projection is produced by slicing the complete yield surface along the plane containing the σ_{11} and σ_{22} axes. In this particular projection, there is a redundancy, in that each line of the yield surface is duplicated by four slip systems. For example, slip systems 6, 9, and 10 produce the same yield lines as slip system 1.

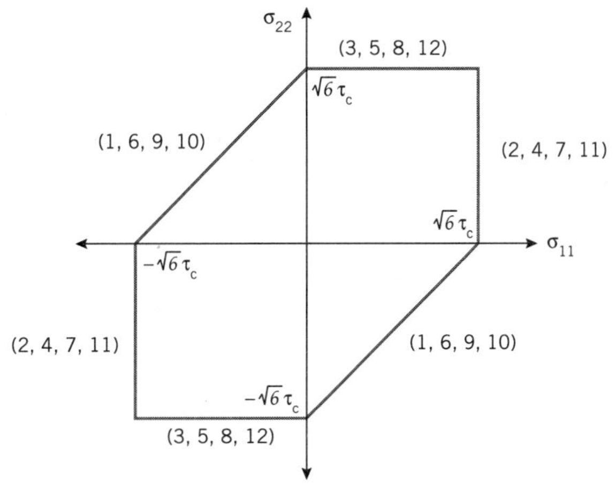

Figure 8.4 Yield surface for an f.c.c. crystal, projected onto the σ_{11}-σ_{22} stress plane. The critical shear stress on {111}/{110} slip systems is τ_c, and the numbers next to each facet of the surface indicate which slip systems in Table 8.1 are critically stressed.

Asaro and Rice[14] note that deviations to the Schmid concept of critical resolved shear stress can occur because of cross slip, where the resistance to slip may be a function of other stress components than τ. Thus the Schmid concept is regarded as an approximation.

Local and Macroscopic Strain Produced by Crystal Slip

When a slip plane with normal **m** produces a relative shear, such that the material on the positive m side has displaced by bs, (see Figure 8.2), then the resulting displacement field is

$$u_i(\mathbf{x}) = bs_i H(x_k m_k) \tag{8.18}$$

where the position, **x**, is measured from an origin located on the slip plane, and H is the **Heaviside step function**, which equals one when $x_k m_k > 0$ and equals zero otherwise. The corresponding displacement gradient is given by

$$\begin{aligned}\frac{\partial u_i}{\partial x_j} &= b\frac{dH(x_k m_k)}{d(x_q m_q)}\frac{\partial(x_p m_p)}{\partial x_j}s_i \\ &= b\delta(x_k m_k)s_i m_j\end{aligned} \tag{8.19}$$

[14] R.J. Asaro and J.R. Rice, *J. Mech. Phys. Solids* 25 (1977): 309.

The deformation gradient is Dirac singular on the slip plane, and only contributes components that are defined by **m** and **s**.

On a continuum level, one is not concerned about the local fluctuations from one slip plane to another, but rather, is concerned about the average deformation gradient over a representative volume. If the area of the slipped portion of the plane is A_s, and the volume of the crystal is V, then the average deformation gradient is

$$\overline{\frac{\partial u_i}{\partial x_j}} = \frac{1}{V}\int_V \frac{\partial u_i}{\partial x_j} dV = \frac{A_s b}{V} m_j s_i \tag{8.20}$$

This relation is particularly useful to understand how discrete slip on planes contributes to the average displacement gradient. To compute the average value of $u_{s,m}$ ($\equiv s_i u_{i,j} m_j$), we typically divide the amount of slip, b, by some characteristic distance, d, normal to the plane. In this case, Eq. 8.20 indicates that the appropriate distance is $d = V/A_s$, where V is the volume over which the deformation gradient is to be averaged. In this sense, a shear strain,

$$\gamma = \frac{b}{d} = \frac{A_s b}{V} \tag{8.21}$$

is defined to describe the amount of shear contributed by a slip plane.

The corresponding average strain is defined by

$$\begin{aligned} \varepsilon_{ij} &= (1/2)(\bar{u}_{i,j} + \bar{u}_{j,i}) \\ &= \frac{\gamma}{2}(m_j s_i + m_i s_j) \ . \end{aligned} \tag{8.22}$$

Equation 8.22 provides a useful measure of how discrete slip on a single plane contributes to the average strain of the crystal. During deformation, the average strain may increase by the incremental glide of existing dislocations, so that dA_s on a slip plane is increased while b remains fixed. On a more macroscopic level, one may not monitor the incremental slip of dislocations, but measure the accumulation of slip from b to 2b, and so forth, on an entire plane. In that case, the average strain increases by incrementing b while A_s remains fixed. In either case, an increment, $d\gamma$, is produced, and

$$d\varepsilon_{ij} = \frac{d\gamma}{2}(m_j s_i + m_i s_j) \tag{8.23}$$

An important concept is that crystal slip does not produce any distortion or rotation of **m** and **s** relative to the crystal basis. As long as the components $d\varepsilon_{ij}$ are referred to a coordinate system that is attached to the crystal basis, the components of **m** and **s** are treated as constant. Using Eq. 8.23 and the yield condition Eq. 8.13, it is straightforward to show that the macroscopic measure of plastic work increment equals the local plastic work done on the slip plane:

$$\sigma_{ij} d\varepsilon_{ij} V = \tau_c d\gamma V \tag{8.24}$$

When more than one slip system operates, the average strain increment is computed by the linear superposition of average strain increment from each system:

$$d\varepsilon_{ij} = \sum_{\alpha} \left[\frac{d\gamma^{(\alpha)}}{2} (m_j s_i + m_i s_j)^{(\alpha)} \right] \tag{8.25}$$

Exercise 8.3 **Idealize slip in a crystal in terms of the shearing of a deck of 101 cards. The 1- and 2-directions are along the edges of the cards, and the 3-direction is normal to the deck. Assume that the deck measures 9 cm × 6.5 cm × 1.5 cm in the 1-, 2-, and 3-directions, respectively, and that the cards are numbered from bottom to top from 1 to 101. Apply two types of relative slip. Displace all cards above each odd-numbered card by 0.01 cm in the 1-direction and displace all cards above each even-numbered card by 0.01 cm in the 2-direction. Therefore, a relative slip of 0.01 cm is imposed across each interface, but the direction of slip alternates between the 1 and 2-directions, depending on the interface. What is the average strain state produced?**

Solution: There are two slip systems present. There are 50 occurrences in which a slip of 0.01 cm in the 1-direction is activated on an area A = 9 cm × 6.5 cm (slip system 1), and 50 occurrences in which a slip of 0.01 cm in the 2-direction is activated on the same area A (slip system 2). Accordingly, if the volume V = 9 cm × 6.5 cm × 1.5 cm, then $d\gamma^{(1)} = 50 \times 0.01 \text{ cm} \times A/V = \frac{1}{3}$, and $d\gamma^{(2)} = 50 \times 0.01 \text{ cm} \times A/V = \frac{1}{3}$. Using Eq. 8.25, $d\varepsilon_{ij}$ $\frac{1}{6}[(m_j s_i + m_i s_j)^{(1)} + (m_j s_i + m_i s_j)^{(2)}]$, where the only nonzero components are $m_3^{(1)} = m_3^{(2)} = 1$, and $s_1^{(1)} = s_2^{(2)} = 1$. The resulting components of strain are

$$\left[d\varepsilon_{ij}\right] = \frac{1}{6} \begin{bmatrix} 0 & 0 & 1 \\ 0 & 0 & 1 \\ 1 & 1 & 0 \end{bmatrix} \tag{8.25-A}$$

A Plastic Potential for Crystals

The yield condition (Eq. 8.13) for a slip system (α) can be cast into the form, $\phi^{(\alpha)} = 0$, where

$$\phi^{(\alpha)} = \left[\frac{1}{2} (s_i m_j + s_j m_i)^{(\alpha)} \sigma_{ij} \right]^2 - \tau_c^2 \; [= 0 \text{ at yield}] \tag{8.26}$$

The term, $(s_i m_j + s_j m_i)/2$, was included to acknowledge that σ_{ij} always equals σ_{ji}. As discussed, the condition $\phi^{(\alpha)} = 0$ simply describes two parallel planes in stress space. When $\phi^{(\alpha)}$ is used as a plastic potential, the increment in strain during yield is assumed to be normal to the yield surface, so that

$$\begin{aligned} d\varepsilon_{ij}^{(\alpha)} &= d\lambda^{(\alpha)} \frac{\partial \phi^{(\alpha)}}{\partial \sigma_{ij}} \\ &= d\lambda^{(\alpha)} \tau_c \left(m_j s_i + m_i s_j\right)^{(\alpha)} \left[\text{when } \phi^{(\alpha)} = 0\right] \end{aligned} \tag{8.27}$$

The construction for $d\varepsilon_{ij}$ using the plastic potential yields the same plastic strain increment direction as the kinematic definition for $d\varepsilon_{ij}$ in Eq. 8.23. The symmetric feature, $d\varepsilon_{ij}^{(\alpha)} = d\varepsilon_{ji}^{(\alpha)}$ in Eq. 8.27 occurs since the components of σ_{ij} are symmetric, and hence $\partial\phi^{(\alpha)}/\partial\sigma_{ij} = \partial\phi^{(\alpha)}/\partial\sigma_{ji}$. The scalar, $d\lambda^{\alpha}$, represents the magnitude of slip to occur, and at the end of Section 8.3, a prescription to determine $d\lambda^{(\alpha)}$ will be given (see Eq. 8.57).

Maximum Plastic Work Due to Slip

The concept that plastic work is dissipated through discrete slip on critically stressed crystallographic planes can be used as the basis for a maximum plastic work principle. This principle states that for a given macroscopic strain increment, $d\varepsilon_{ij}$, the corresponding actual stress state, σ_{ij}, produces the maximum possible plastic work increment, $\sigma_{ij} d\varepsilon_{ij}$. Any other stress state, σ_{ij}^*, which loads an arbitrary number of slip planes up to but not exceeding the critical resolved shear stress, produces an amount of plastic work that is equal or smaller. Equivalently, the work to deform a crystal by a certain amount will be as large as physically possible.

To prove this, Bishop and Hill[15] first noted that σ_{ij} and σ_{ij}^* are two possible equilibrium stress fields, and the actual strain increment, $d\varepsilon_{ij}$, is compatible and involves activation of some subset, β, of the total available slip systems. Accordingly, the principle of virtual work applies, so that

$$\begin{aligned} \sigma_{ij} d\varepsilon_{ij} &= \sum \tau^{(\beta)} d\gamma^{(\beta)} \\ \overset{*}{\sigma_{ij}} d\varepsilon_{ij} &= \sum \tau^{(\beta)*} d\gamma^{(\beta)} \end{aligned} \tag{8.28}$$

Each of the conditions is an extension of Eq. 8.24 to multiple slip planes, and each states that the macroscopic plastic work is equal to the cumulative work on all of the active slip planes. The actual stress state loads each active slip plane to the required critical shear, so that $\tau^{(\beta)} = \tau_c^{(\beta)}$. However, for the other equilibrium

[15] J.F.W. Bishop and R. Hill, *Philos. Mag. Ser. 7* 42 (1957): 414.

stress state, one can ensure only that $\tau^{(\beta)*} \leq \tau_c^{(\beta)}$ if it does not exceed yield. Consequently, of all equilibrium stress states that do not violate yield, the actual stress state produces the maximum internal work. When Eqs. 8.28 are invoked, the same condition holds for the external work increments:

$$\sigma_{ij} d\varepsilon_{ij} \geq \sigma_{ij}^{*} d\varepsilon_{ij} \qquad (8.29)$$

The upper and lower bound constructions discussed at the outset of this chapter depend on the existence of the maximum work principle, and Eq. 8.26 provides an interpretation of it based on simple ideas about how plastic work is done through crystal slip.

Six-Dimensional Representation for Strain

Since there are only six independent components of stress, a convenient six-dimensional representation for stress state was introduced in Eq. 8.14. The representation provides a correspondence between components, σ_i, described in six-dimensional space and the real space components, σ_{ij}, representing the second rank tensor. In a similar manner, components of strain may be represented in six-dimensional space. The alternate representation for strain has the important feature that increments of work in each representation must be equal,

$$\sigma_k d\varepsilon_k = \sigma_{ij} d\varepsilon_{ij} \qquad 8.30$$

Equivalently, ε_i is defined to be the work conjugate of σ_i. The resulting correspondence between components in six-dimensional space and real space is given by the identity

$$\begin{bmatrix} d\varepsilon_1 \\ d\varepsilon_2 \\ d\varepsilon_3 \\ d\varepsilon_4 \\ d\varepsilon_5 \\ d\varepsilon_6 \end{bmatrix}^{(\alpha)} \equiv \begin{bmatrix} d\varepsilon_{11} \\ d\varepsilon_{22} \\ d\varepsilon_{33} \\ 2d\varepsilon_{23} \\ 2d\varepsilon_{13} \\ 2d\varepsilon_{12} \end{bmatrix}^{(\alpha)} \qquad 8.31$$

The factor of two associated with each component of shear strain occurs since each component of shear work (e.g., $\sigma_4 d\varepsilon_4$) represents two terms,(e.g., $\sigma_{23} d\varepsilon_{23} + \sigma_{32} d\varepsilon_{32}$) in real space.

All corresponding relations associated with stress and strain components can be cast into the convenient six-dimensional form. In particular, Eq. 8.23 between the strain increment and the local slip on a crystal plane may be expressed as

$$d\varepsilon_i^{(\alpha)} = d\gamma^{(\alpha)} n_i^{(\alpha)} \qquad 8.32$$

The components $n_i^{(\alpha)}$ depend on the components of slip direction and slip plane normal as defined in Eq. 8.16. Thus, $n_i^{(\alpha)}$ has several interpretations. According to Eq. 8.15, $n_i(\alpha)$ is the direction in stress space associated with reaching a critical resolved shear stress on slip system (α). According to Eq. 8.32, $n_i^{(\alpha)}$ is the direction of the strain increment contributed by operation of slip system (α). Finally, Eq. 8.27 may be expressed in six-dimensional space, to confirm that when a critical resolved shear stress on slip system (α) is reached, the normal to the yield surface at that point is given by $n_i^{(\alpha)}$. In particular, Eq. 8.17 indicates that for the f.c.c. slip systems discussed earlier, all of the $n_i^{(\alpha)}$ are perpendicular to the [1 1 1 0 0 0] direction. Therefore, hydrostatic stress will not contribute to the resolved shear stress on any slip system, nor will it produce any dilatation. Equivalently, all portions of the yield surface are parallel to the [1 1 1 0 0 0] direction.

Determining Active Slip Systems and the Principle of Minimum Shear

For some crystals, there may be several possible combinations of slip systems that can be used to accomplish a certain macroscopic strain increment. This is generally the case for f.c.c. materials, in which there are 12 slip systems, or equivalently, twelve vectors $n_i^{(\alpha)}$.

For crystals with more than five distinct strain directions $\mathbf{n}^{(\alpha)}$, there is more than one combination of the $d\gamma^{(\alpha)}$ to produce some arbitrary strain increment, $d\varepsilon$. Taylor[16] proposed that the actual combination of slip, $d\gamma^{(\beta)}$, to occur produces the minimum amount of slip, compared to other combinations $d\gamma^{(\beta)*}$:

$$\sum \left| d\gamma^{(\beta)} \right| \le \sum \left| d\gamma^{(\beta)*} \right| \qquad (8.33)$$

To prove **Taylor's assertion**, Bishop and Hill[13] observed that the actual combination $d\gamma^{(\beta)}$ and any other kinematically equivalent combination $d\gamma^{(\beta)*}$ produce the same $d\varepsilon_{ij}$ and therefore the same external work increment, $\sigma_{ij} d\varepsilon_{ij}$. If σ_{ij} is the actual equilibrium stress state and $\tau^{(\beta)}$ is the corresponding resolved shear stress on slip system β, then according to the principle of virtual work, the internal work increments must also be equal:

$$\sum \tau^{(\beta)} \left| d\gamma^{(\beta)} \right| = \sum \tau^{(\beta)} \left| d\gamma^{(\beta)*} \right| \qquad (8.34)$$

The left-hand sum over β involves only critically stressed planes on which $\tau^{(\beta)} = \tau_c^{(\beta)}$, but the right-hand sum over β may involve noncritically stressed

[16] G.I. Taylor, *J. Inst. Metals* 62 (1938): 307P.

planes on which $\tau^{(\beta)} \le \tau_c^{(\beta)}$. It follows that

$$\sum \tau^{(\beta)} \left| d\gamma^{(\beta)} \right| = \sum \tau^{(\beta)} \left| d\gamma^{(\beta)*} \right| \le \sum \tau_c^{(\beta)} \left| d\gamma^{(\beta)*} \right| \tag{8.35}$$

For the simple case where τ_c is uniform on each slip plane, the minimum slip principle Eq. 8.33 follows.

Exercise 8.4 **Consider a f.c.c. crystal in which the incremental deformation is plane strain compression, $[d\varepsilon_i] = d\varepsilon[1\ -1\ 0\ 0\ 0\ 0]$, and in which all slip planes operate at a critical resolved shear stress, τ_c. Show that the minimum slip principle does not uniquely specify the magnitude of slip on each system.**

Solution: One possible construction is to have equal operation of slip systems (1) and (10), as listed in Table 8.1. According to Eq. 8.32,

$$[d\varepsilon_i^{(1)}] = d\gamma^{(1)}[n_i^{(1)}] = d\gamma^{(1)}[1\ -\ 1\ 0\ -1\ 1\ 0]/\sqrt{6}$$
$$[d\varepsilon_i^{(10)}] = d\gamma^{(10)}[n_i^{(10)}] = d\gamma^{(10)}[1\ -1\ 0\ 1\ -1\ 0]/\sqrt{6},$$

where the components, $n_i^{(\alpha)}$, are provided by entries in the corresponding rows 1 and 10 of the 12×6 matrix in Eq. 8.17. If $d\gamma^{(1)} = d\gamma^{(10)} = \sqrt{3/2}\ d\varepsilon$, then the total strain increment in the crystal is

$$[d\varepsilon_i] = d\varepsilon[1\ -1\ 0\ 0\ 0\ 0]$$

The cumulative slip = $\sum d\gamma^{(\alpha)} = \sqrt{6}$.

The combination, $d\gamma^{(9)} = -d\gamma^{(6)} = \sqrt{3/2}\, d\varepsilon$, also produces the same total strain increment with the same cumulative slip. Accordingly, any linear combination of the two pairs (1, 10) and (6, 9) will accomplish the same. Thus the principle of minimum slip does not uniquely specify the linear combination of slip systems to operate. The portions of the yield surface labeled with the notation (1, 6, 9, 10) also indicate a degree of nonuniqueness among these slip systems.

In general, the deformation of individual crystals is constrained sufficiently so that more than one slip system is operative. Even in less-constrained situations, different slip systems will be activated during the course of deformation. To describe hardening behavior in a more quantitative way, an increment, $d\gamma^{(\beta)}$, in strain on slip system β is assumed to produce an increase in the critical shear stress for all slip systems according to

$$d\tau_c^{(\alpha)} = \sum_{\beta} h_{\alpha\beta} d\gamma^{(\beta)} \tag{8.36}$$

where $h_{\alpha\beta}$ with equal indices (e.g., h_{11}) is called the **self-hardening coefficient**. With different indices, it is called the **latent-hardening coefficient**. Asaro[17] summarizes particular dislocation configurations that contribute to each type of hardening. For example, it is clear that continued operation of Orowan sources can contribute to self hardening, while cross slipping provides an important contribution to latent hardening.

Investigations of self-hardening and latent-hardening effects, beginning with early work by Taylor and Elam,[14,18] generally show that the latent-hardening coefficients are comparable in value to self-hardening coefficients. In fact, more recent investigations, summarized by Asaro,[6] suggest that latent-hardening coefficients relating non-coplanar systems may be larger, particularly for materials with higher stacking-fault energies. The results of Taylor and Elam are discussed in more detail later.

Lattice Rotation and Geometric Softening Under Uniaxial Loading

When crystal slip occurs, the lattice structure is generally maintained and $\mathbf{m}^{(\alpha)}$ and $\mathbf{s}^{(\alpha)}$ remain orthogonal to within elastic distortions. However, a material line scribed on the crystal, such as the one connecting the load points A and B of the tensile specimen shown in Figure 8.5*a*, does rotate with respect to the crystal basis. Using the expression for average deformation gradient, Eqs. 8.20 and 8.21, the change, $d\mathbf{T}$, in a unit vector $\mathbf{T}$ along the tensile axis of the specimen is

$$\begin{aligned} dT_i &= d\bar{u}_{i,j}T_j \\ &= d\gamma(m \cdot T)s_i \end{aligned} \tag{8.37}$$

That is, the increment in $\mathbf{T}$ must be parallel to the slip vector $\mathbf{s}$. The amount, $d\theta$, and unit axis $\mathbf{r}$ about which $\mathbf{T}$ rotates is

$$\begin{aligned} d\theta\,\mathbf{r} &= \mathbf{T} \times d\mathbf{T} \\ &= d\gamma(\mathbf{m} \cdot \mathbf{T})(\mathbf{T} \times \mathrm{s}) \end{aligned} \tag{8.38}$$

The change in angle β shown in Figure 8.5*a* is found by projecting the rotation in Eq. 8.38 onto the axis, $(\mathbf{s} \times \mathbf{T})/\,|\,\mathbf{s} \times \mathbf{T}\,|$,

---

[17] R.J. Asaro, *Advances in Appl. Mech.* 23 (1983): 1.

[18] G.I. Taylor and C.F. Elam, *Proc. Roy Soc. London Ser A*, 108 (1925): 28.

$$d\beta = d\theta\, \mathbf{r} \cdot \frac{\mathbf{s} \times \mathbf{T}}{|\mathbf{s} \times \mathbf{T}|}$$
$$= -d\gamma(\mathbf{m} \cdot \mathbf{T})|\mathbf{s} \times \mathbf{T}| \tag{8.39}$$

In the special case where **T** lies in the **s**-**m** plane, then $\mathbf{m} \cdot \mathbf{T} = |\mathbf{s} \times \mathbf{T}| = \sin\beta$, so that from Eq. 8.39, $d\beta = -d\gamma \sin^2\beta$. Integration from some starting orientation β_o yields

$$\beta = \cot^{-1}[\cot\beta_o + \gamma] \tag{8.40}$$

When the loading is tensile, so that γ is increasing, Eq. 8.40 shows that the **s** and **T** axes approach one another and eventually coincide when $\gamma \rightarrow \infty$. For compressive deformation, so that γ is negative, Eq. 8.40 indicates that **T** rotates away from **s** in Figure 8.5*a*, along a path of increasing β. In fact, under compressive loading, **T** is again rotating *toward* the direction of slip, which in this case is the minus-**s** direction in Figure 8.5*a*.

A more standard measure of strain during the test is ε_{TT}, the tensile strain along the loading direction. The incremental strain, $d\varepsilon_{TT}$, along the current tensile axis equals the incremental change, $dT = d\mathbf{T} \cdot \mathbf{T}$, in the length of the unit vector **T**. Using Eq. 8.37 and specializing to the case where **m**, **s**, and **T** are coplanar, $d\varepsilon_{TT} = -\cot\beta\, d\beta$. This incremental relation may be integrated to give

$$\varepsilon_{TT} = \ln(\sin\beta_o / \sin\beta) \tag{8.41}$$

Figure 8.5*c* shows how β depends on ε_{TT}. Here, β decreases with ε_{TT} as for γ. In the limit of large tensile strain, β approaches zero, so that the tensile axis and slip direction coincide.

The observation that β changes with strain is important for two reasons. First, the result suggests that any single active slip system will eventually rotate to decrease the resolved shear stress, and thus require higher applied stress to operate it. For the applied uniaxial stress along **T** considered here, the condition for yield Eq. 8.13 becomes

$$\sigma = \tau_c / [(\mathbf{m} \cdot \mathbf{T})(\mathbf{s} \cdot \mathbf{T})]$$
$$= 2\tau_c / \sin 2\beta \quad \text{when } \mathbf{m}, \mathbf{s}, \mathbf{T} \text{ are coplanar.} \tag{8.42}$$

This result is also plotted in Figure 8.5*c*. Clearly, the applied stress is minimum at $\beta = 45°$, and it increases as β approaches 0° or 90°. Accordingly, if $\beta_o \leq 45°$ and tensile loading is applied, or if $\beta_o \geq 45°$ and compression is applied, then the magnitude of applied σ must increase monotonically with deformation, even when there is no hardening. Conversely, if $\beta_o > 45°$ and tension is applied, or if $\beta_o < 45°$ and compression is applied, σ decreases with deformation until $\beta = 45°$.

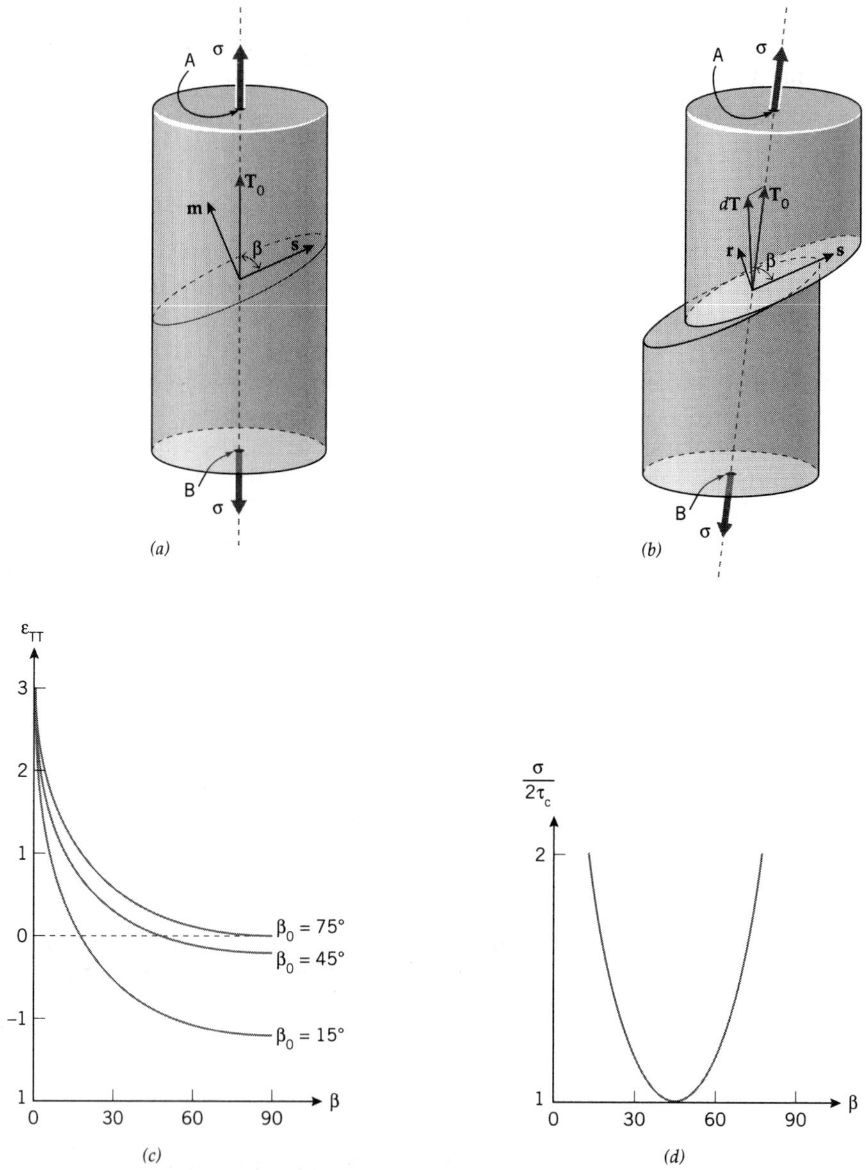

Figure 8.5 *(a)* Undeformed geometry of a cylindrical crystal loaded in tension along the A-B axis. (*b*) Deformed geometry where $\mathbf{r} = \mathbf{T} \times \mathbf{s}$ is the axis of rotation of tensile axis $\mathbf{T}$ relative to the crystal. (*c*) Prediction of tensile strain, ε_{TT}, and Schmid factor, $\sigma/2\tau$, as a function of orientation angle β. σ is the applied tensile stress and τ is the resolved shear stress on the slip plane.

The geometrical hardening and softening caused by changes in β are important in initiating instability in crystal deformation. When self hardening of the form Eq. 8.36 is considered, the initial slope, $d\sigma/d\varepsilon_{TT}$, of the stress-strain curve is obtained by differentiating Eq. 8.42:

$$\frac{d\sigma}{d\varepsilon_{TT}} = \left[\frac{2h}{\sin^2 2\beta} + \frac{\sigma \cos 2\beta}{\cos^2 \beta}\right] \tag{8.43}$$

The first term is positive when self hardening is present ($h > 0$), and the second term produces geometric softening when $\sigma \cos 2\beta < 0$. Asaro[19] and Harren and colleagues[20] emphasize that instability can occur even when hardening is present ($h > 0$), because geometric softening may be present.

Observations of Latent and Self Hardening (Taylor)

An important issue is the amount of hardening induced on one system due to slip on another system. Taylor and Elam[16] performed an elegant series of tests in which aluminum crystals were elongated in tension, and the rotation of the tensile axis relative to the crystal basis was recorded as a function of strain. The authors presented their results in the form of a **stereographic diagram** as shown in Figure 8.6. In this figure, the origin of the f.c.c. crystallographic basis is imagined to be at the center of a sphere, and the points at which various crystallographic axes pierce through the sphere are projected onto the circle shown. In this case, the orientation normal to the plane of the figure is [1 0 0], and the other two cube directions, [0 1 0] and [0 0 1], are oriented at ± 45°. The 45° lines are traces of directions that are normal to either [0 1 0] or [0 0 1]; the horizontal and vertical lines are traces of directions that are normal to [0 1 -1] and [0 1 1], respectively.

In the experiments conducted, the tensile axis **T** was oriented in the [0 1 1]-[1 0 0]-[0 1 0] stereographic triangle, as shown in Figure 8.7*a*. According to Eq. 8.13, the resolved shear stress on any slip system due to a direct tensile stress σ along the T-axis is $\sigma(\mathbf{s} \cdot \mathbf{T})(\mathbf{m} \cdot \mathbf{T})$. The quantity $(\mathbf{s} \cdot \mathbf{T})(\mathbf{m} \cdot \mathbf{T})$ measures the ratio, τ/σ, and is called the **Schmid factor** for this geometry. If **T** is in the [0 1 0]-[1 1 1]-[0 1 1] stereographic triangle, the (–1 1 1)/[1 1 0] slip system has the largest resolved shear stress, and this system was observed to be the first to operate. As such, it is called the ***primary*** slip system. According to Eq. 8.38, **T** will rotate toward **s** about the axis $\mathbf{r} = \mathbf{T} \times \mathbf{s}$, and the trajectory is indicated by the dashed arc C. In the process of rotation toward [1 1 0], the Schmid factor on the primary system will decrease. In fact, if **T** rotates close enough to [1 1 0], the slip system with the largest Schmid factor changes to (1 1 -1)/[0 1 1]. This slip system, which eventually becomes favored due to rotation, is called the ***conjugate*** system.

---

[19] R.J. Asaro, *Acta Metall* 27 (1979): 445.

[20] S.V. Harren, H.E. Deve, and R.J. Asaro, *Acta Metall.* 36(9) (1988): 2435.

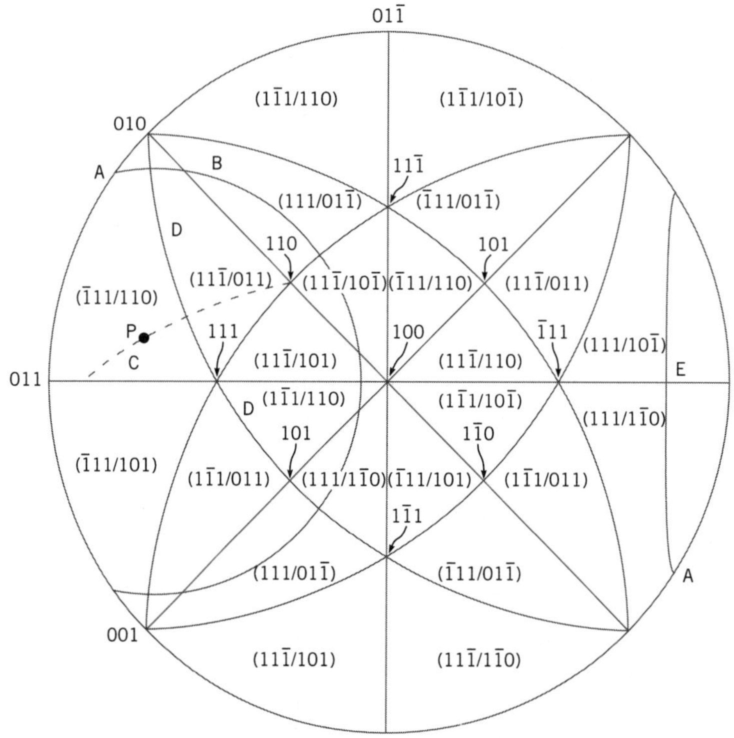

Figure 8.6 Stereographic triangle produced from Figure 1 of Taylor and Elam.[14]

The value of **T** for which the primary and conjugate systems have equal Schmid factors is found by satisfying $(\mathbf{m}^{(p)} \cdot \mathbf{T})(\mathbf{s}^{(p)} \cdot \mathbf{T}) = (\mathbf{m}^{(c)} \cdot \mathbf{T})(\mathbf{s}^{(c)} \cdot \mathbf{T})$, where p and c denote primary and conjugate, respectively. One solution of this equation is that **T** must be normal to [1 0 –1], and the arc D shows values of **T** that satisfy this. The simplicity of the result is that if **T** moves along arc C into the conjugate slip region, then operation of the primary system has apparently caused more hardening in the conjugate than the primary system. Conversely, if **T** begins to deviate from arc C before reaching the conjugate region, then the conjugate system has not hardened as much as the primary system. Two types of deviations are obvious. If **T** begins to move toward [0 1 1], then the conjugate system has become dominant. In comparison, if both systems operate equally, then **T** rotates toward the new slip direction, $\mathbf{s} = \mathbf{s}^{(p)} + \mathbf{s}^{(c)}$. This is simply the [1 2 1] direction located along arc D.

Figures 8.7*a-c* show Figures 3, 4, and 5 from Taylor and Elam.[16] The numbers above the trajectory and associated tick marks display the tensile strain and location of **T**, as predicted by Eq. 8.39 with (–1 1 1)/[1 1 0] as the only operating slip system. The numbers below the trajectory and corresponding points show the measured values from X-ray crystallography. In Figure 8.7*a*, predictions and measurements agree remarkably well until a strain of 0.56. At larger strain,

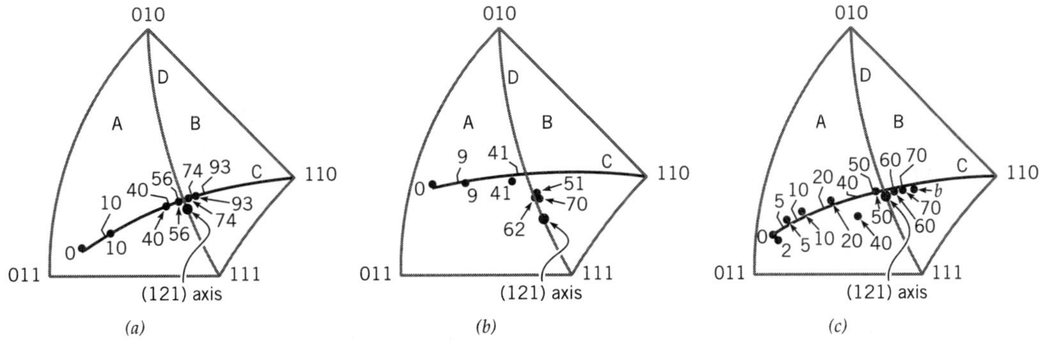

Figure 8.7 Annotated Figures 3, 4, and 5 from Taylor and Elam.[14] The dark solid line is the arc along which the tensile axis **T** rotates, and the boundary between stereographic triangles A and B is the arc along which the Schmid factor for the primary and conjugate slip systems are equal.

conjugate slip appears to intervene where the conjugate Schmid factor is largest. However, **T** continues into the conjugate region, apparently due to substantial hardening of the conjugate system by primary slip. Figure 8.7*b* shows a representative case where equal slip intervenes and **T** rotates along the equal Schmid arc toward [1 2 1]. Figure 8.7*c* displays an intermediate case where there is a modest deviation from the primary trajectory in the vicinity of [1 2 1]. The authors comment that trajectories associated with double slip could not be studied extensively, because fracture soon followed the initiation of double slip. However, these experiments provide evidence that latent hardening can be comparable or greater than self hardening.

Material Rotation Relative to the Crystal Basis

In a previous section, we considered the rotation of a tensile axis relative to the crystal basis, due to operation of a slip system. The resulting expression Eq. 8.39 for rotation depends on the particular axis, **T**, chosen. This definition becomes ambiguous for multiple-axis loading. A unique, more objective measure of rotation is to express the increment, d**T**, due to slip as

$$dT_i = \left[d\varepsilon_{ij} + d\omega_{ij}\right]T_j \, . \tag{8.44}$$

where $d\varepsilon_{ij}$ is recognized as an increment in strain due to crystal slip, and its connection to crystal slip is given in Eq. 8.25. The quantity $d\omega_{ij}$ is defined by

$$d\omega_{ij} = \sum_{\alpha}\left[\frac{d\gamma^{(\alpha)}}{2}(m_j s_i - m_i s_j)^{(\alpha)}\right] \tag{8.45}$$

and is recognized as an increment in material rotation, as observed in a coordinate system attached to the crystal basis. In particular, Figure 8.8 shows two orthogonal material axes, **T** and **Q**, the ends of which displace relative to point O by the same amounts, $d\mathbf{T} = d\mathbf{Q}$, by activation of the **m/s** slip system. The displacements $d\mathbf{T}$ and $d\mathbf{Q}$ are decomposed into the components $dT^e = \mathbf{T} \cdot d\varepsilon \cdot \mathbf{T}$, $dQ^e = \mathbf{Q} \cdot d\varepsilon \cdot \mathbf{Q}$ associated with elongation of **T** and **Q**, the components $dT^s = \mathbf{Q}{\cdot}d\varepsilon{\cdot}\mathbf{T}$, $dQ^s = \mathbf{T}{\cdot}d\varepsilon{\cdot}\mathbf{Q}$ associated with shear deformation, and the components $\mathbf{dT^r} = \mathbf{Q}{\cdot}\mathbf{d\omega}{\cdot}\mathbf{T}$, $\mathbf{dQ^r} = \mathbf{T}{\cdot}\mathbf{d\omega}{\cdot}\mathbf{Q}$ associated with material rotation. Since $d\varepsilon_{ij}$ is symmetric and $d\omega_{ij}$ is antisymmetric, it follows that $dT^s = dQ^s$ and that $dT^r = -dQ^r$. The first condition imposes that **T** and **Q** shear to equally increase or decrease the right angle initially made by **T** and **Q**. The second condition imposes that the sense and magnitude of rotation of **T** and **Q** are equal.

In comparison, the prescription for rotation in Eq. 8.39 does not satisfy these conditions. In particular, Eq. 8.39 is based on attributing the entire displacement, $dT^s + dT^r$, to rotation of **T**. The prescription Eq. 8.45 for rotation is adopted in most current work, because it does not depend on the choice of a particular axis, **T** or **Q**. As shown in Figure 8.8, a simple interpretation of $d\omega$ is that the product $\mathbf{P} \cdot d\omega \cdot \mathbf{Q}$ gives the incremental material rotation about the axis $\mathbf{P} \times \mathbf{Q}$.

Implementing the Crystal-Constitutive Relation

The implementation of the crystal-constitutive relation is straightforward when a hardening slip process such as Eq. 8.36 is used. Integration of the constitutive relation will furnish a macroscopic stress-strain relation. For notation, a *reference* basis ($\mathbf{e}^{(1)*}$, $\mathbf{e}^{(2)*}$, $\mathbf{e}^{(3)*}$) is chosen to rotate with the material, as do vectors **T** and **Q** in Figure 8.8. The basis ($\mathbf{e}^{(1)}$, $\mathbf{e}^{(2)}$, $\mathbf{e}^{(3)}$) is attached to the *lattice* of slip planes and directions, as are **m** and **s** in Figure 8.8. For example, if ($\mathbf{e}^{(1)}$, $\mathbf{e}^{(2)}$, $\mathbf{e}^{(3)}$) are chosen to correspond to the cube directions of a f.c.c. lattice, then the components $m_i^{(\alpha)}$ and $s_i^{(\alpha)}$ take on values as shown in Table 8.1 and do not change with plastic deformation. However, the components $m_i^{*(\alpha)}$ and $s_i^{*(\alpha)}$, referred to the reference basis, do evolve with plastic deformation. Regardless, **m** and **s** must always remain orthogonal and not change length to within elastic deformation, so that $m_i^{(\alpha)} s_i^{(\alpha)} = m_i^{(\alpha)*} s_i^{(\alpha)*} = 0$.

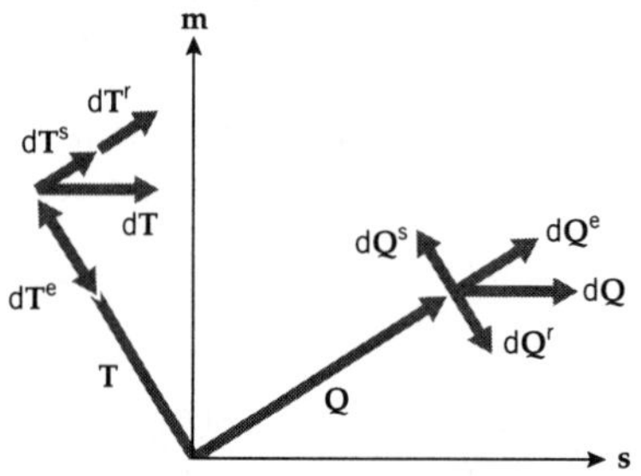

Figure 8.8 Schematic of the symmetric and antisymmetric partitions of the deformation gradient, as measured by the displacements of vectors **T** and **Q** due to slip in the direction **s**.

The orientation of the reference basis relative to the lattice basis is defined by

$$A_{ij} = e^{(i)*} \cdot e^{(j)} \tag{8.46}$$

so that the components $m_i^{(\alpha)*}$ and $s_i^{(\alpha)*}$ of each potentially active slip system, as referred to the reference (*) basis, are

$$\begin{aligned} m_i^{(\alpha)*} &= A_{ij} m_j^{(\alpha)} \\ s_i^{(\alpha)*} &= A_{ij} s_j^{(\alpha)} \end{aligned} \tag{8.47}$$

If a macroscopic stress with components σ_{ij}^*, as measured in the reference basis, is applied, the resolved shear stress on each slip system is calculated as

$$\tau^{(\alpha)} = s_i^{(\alpha)*} \sigma_{ij}^* m_j^{(\alpha)*} \tag{8.48}$$

and initially, it is assumed that $\tau^{(\alpha)} \le \tau_c^{(\alpha)}$ for all slip systems. Accordingly, if the yield condition is not met, $d\gamma^{(\alpha)} = 0$ for all slip systems, and, from Eq. 8.36, there is no hardening of any slip system. The behavior is purely elastic.

Yield occurs when σ_{ij}^* is sufficiently high so that $\tau^{(\alpha)} = \tau_c^{(\alpha)}$ occurs on one or more slip systems. Under such conditions, the hardening, $d\tau_c^{(\alpha)}$, on those slip systems must equal any subsequent increment, $d\tau^{(\alpha)}$, in order to ensure that the stress state remains on or inside the yield surface. On those slip systems for which $\tau^{(\alpha)} \le \tau_c^{(\alpha)}$, the strain increment, $d\gamma^{(\alpha)}$, is set to zero. Thus either $d\tau_c^{(\alpha)}$ or $d\gamma^{(\alpha)}$ is known on each slip system, and Eq. 8.36 permits the remaining $d\tau_c^{(\alpha)}$ and $d\gamma^{(\alpha)}$ to be determined. In some cases, the solution is not unique, and in Section 8.4., a rate-dependent formulation is discussed that removes ambiguities in the solution of $d\gamma^{(\alpha)}$.

A change in resolved shear stress on slip system (α) is given by differentiating Eq. 8.48,

$$d\tau^{(\alpha)} = ds_i^{*(\alpha)} \sigma_{ij}^* m_j^{*(\alpha)} + s_i^{*(\alpha)} d\sigma_{ij}^* m_j^{*(\alpha)} + s_i^{*(\alpha)} \sigma_{ij}^* dm_j^{*(\alpha)} \qquad 8.49$$

where the quantities $ds_i^{*(\alpha)}$ and $dm_i^{*(\alpha)}$ indicate that during plastic deformation, slip planes change orientation with respect to directions along which loading is prescribed, such as **T** in Fig. 8.8. In general, the increment in slip direction and slip plane normal for slip system (α) is described by

$$ds_i^{*(\alpha)} = d\omega_{ij}^* s_j^{*(\alpha)}$$
$$dm_i^{*(\alpha)} = d\omega_{ij}^* m_j^{*(\alpha)} \qquad 8.50$$

where $d\omega_{ij}^*$ describes the incremental rotation of the *lattice* basis, relative to the *reference* basis.

There are several possible definitions for the reference basis. A common convention is to attach the reference basis to the material, as are **P** and **Q** in Fig. 8.8. Accordingly, the rotation of the lattice basis relative to the reference basis is given by an expression opposite in sign to Eq. 8.45, and with components $()^*$ referred to the reference basis,

$$d\omega_{ij}^* = \sum_\beta W_{ij}^{*(\beta)} d\gamma^{(\beta)} \qquad 8.51$$

where

$$W_{ij}^{*(\beta)} = \frac{m_i^{*(\beta)} s_j^{*(\beta)} - m_j^{*(\beta)} s_i^{*(\beta)}}{2} \qquad 8.52$$

The summation is over all active slip systems (β), for which $\tau^{(\beta)} = \tau_c^{(\beta)}$.

The condition that permits the $d\gamma^{(\beta)}$ to be determined is that $d\tau^{(\beta)}$ in Eq. 8.49 must equal the change in critical shear stress due to slip,

$$d\tau^{(\beta)} = \sum_\beta h_{\beta\eta} d\gamma^{(\eta)} \qquad 8.53$$

This condition provides *n* equations to determine the $d\gamma^{(\eta)}$ on the *n* active slip systems. If $d\tau^{(\beta)}$ in Eqs. 8.53 and 8.49 are set equal, and if Eqs. 8.50-8.52 are used to replace $ds_i^{*(\alpha)}$ and $dm_i^{*(\alpha)}$, then

$$d\gamma^{(\beta)} = k_{\beta\eta}^{-1}\left(s_i^{*(\eta)} d\sigma_{ij}^* m_i^{*(\eta)}\right) \qquad 8.54$$

where

$$k_{\beta\eta} = h_{\beta\eta} - \left(s_k^{*(\beta)} \sigma_{ij}^* m_j^{*(\beta)} + s_j^{*(\beta)} \sigma_{ij}^* m_k^{*(\beta)}\right) W_{ik}^{*(\eta)} \qquad 8.55$$

If $k_{\beta\eta}$ is not positive definite, then it cannot be inverted, indicating that there is more than one possible combination of $d\gamma^{(\beta)}$. Clearly, the positive definite requirement depends sensitively on the hardening matrix, $h_{\alpha\beta}$, and orientation of the applied stress state to the lattice.

Once the $d\gamma^{(\beta)}$ are determined for a given increment, $d\sigma_{ij}^*$, the increment in slip plane (lattice) orientation, $d\omega_{ij}^*$, may be updated according to Eq. 8.51, and the strain increment is computed according to

$$d\varepsilon_{ij}^* = \sum_{\beta} d\gamma^{(\beta)} \frac{\left(m_j^{*(\beta)} s_i^{*(\beta)} + m_i^{*(\beta)} s_j^{*(\beta)}\right)}{2} \qquad 8.56$$

This definition of ε_{ij}^* may be derived from the Principle of Virtual Work statement that $\sigma_{ij}^* d\varepsilon_{ij}^* = \sum \tau^{(\beta)} d\gamma^{(\beta)}$.

The increments, $ds_i^{*(\alpha)}$ and $dm_i^{*(\alpha)}$, in slip directions and slip plane normals are given by Eq. 8.50, and Eq. 8.36 provides the increase in $\tau_c^{(\alpha)}$, for all slip systems (α), regardless of whether active or not during the stress increment. All slip systems should satisfy $\tau^{(\alpha)} \leq \tau$, with all active slip systems satisying $\tau^{(\alpha)} = \tau_c^{(\alpha)}$. The process may be continued by applying repeated increments in stress.

In some cases, it is more realistic and convenient to apply an increment, $d\varepsilon_{ij}^*$, in deformation, and determine the corresponding increment, $d\sigma_{ij}^*$ in stress. In such cases, $d\gamma^{(\beta)}$ in Eq. 8.56 may be replaced by the expression for $d\gamma^{(\beta)}$ in Eq. 8.54. The result is a set of equations that permit the $d\sigma_{ij}^*$ to be determined,

$$d\varepsilon_{ij}^* = \frac{1}{2}\left(m_j^{*(\beta)} s_i^{*(\beta)} + m_i^{*(\beta)} s_j^{*(\beta)}\right) k_{\beta\eta}^{-1}\left(s_k^{*(\eta)} d\sigma_{kl}^* m_l^{*(\eta)}\right) \qquad 8.57$$

where β and η sum over the active slip systems. In general, one prescribes five components of $d\varepsilon_{ij}^*$, and inversion of Eq. 8.57 permits five components of stress increment, $d\sigma_{ij}^*$, to be determined. Equivalently, conditions on a mixed combination of components of $d\varepsilon_{ij}^*$ and $d\sigma_{ij}^*$ may be specified , and the unknown counterparts may then be determined using Eq. 8.57. In general, $k_{\beta\eta}$ must be invertable, in order for a unique relation between stress and strain increments to exist. More elaborate treatments which include the effect of elastic distortion of the lattice are summarized, for example, by Asaro[17].

In some cases, it is useful to measure or prescribe the applied stress or deformation relative to a basis that is different than the material basis depicted in Fig. 8.8. For example, the simple tension test depicted in Fig. 8.5(*a*, *b*) is described most simply by applying successive increments, $d\sigma$, along the **T** axis, and measuring the corresponding increment, $d\varepsilon_{TT}$, along the **T** axis. Thus, the most convenient reference basis is one which contains the axis **T**. Earlier, we obtained the rotation of **T** relative to the lattice, found in Eq. 8.38. Therefore, the rotation of the lattice relative to a reference basis containing **T** is the negative of that in Eq. 8.38. The prescription for $d\omega_{ij}^*$ in Eq. 8.51 still holds, but instead,

$$W_{ij}^{*(\beta)} = m_k^{*(\beta)} T_k^* \left(T_i^* s_j^{*(\beta)} - T_j^* s_i^{*(\beta)}\right) \qquad 8.58$$

Thus, the formulation, Eqs. 8.46 to 8.57, described above is still applicable, but the prescription for $W_{ij}^{*(\beta)}$ is specified by Eq. 8.58 rather Eq. 8.52. Clearly, the reference basis in which applied stress and strain components are measured may be arbitrarily changed, with the result that $W_{ij}^{*(\beta)}$ must also be changed.

Exercise 8.5 **Consider a single crystal with tensile axis along the T-direction as depicted in Fig. 8.5 (a,b). Assume that only one slip system with normal m and slip direction s is activated, and that T lies m-s plane. Use the implementation of the crystal constitutive relation above to determine the slope, dσ/dε, of the tensile stress-strain relation.**

Solution: For convenience, we choose a Cartesian reference (*) basis in which **T** corresponds to $\mathbf{e}^{*(1)}$, and $\mathbf{e}^{*(2)}$ lies in the $\mathbf{m}^{(1)}$-$\mathbf{s}^{(1)}$ plane of slip system (1), which is assumed to be the only active slip system. Accordingly, $\{T^*_1, T^*_2, T^*_3\} = \{1, 0, 0\}$, $\{s^*_1, s^*_2, s^*_3\} = \{\cos\beta, \sin\beta, 0\}$, and $\{m^*_1, m^*_2, m^*_3\} = \{\sin\beta, -\cos\beta, 0\}$. The applied stress corresponds to σ^*_{11}, and the corresponding tensile strain is ε^*_{11}. Accordingly, i and j are both set equal to 1 in Eq. 8.57, and further, the only contribution from the internal sum over indices k and l occurs when k and l are both equal to 1. The result is

$$d\varepsilon = (m^*_1 s^*_1)^2 k_{11}^{-1} d\sigma$$

or

$$\frac{d\sigma}{d\varepsilon} = \frac{k_{11}}{(m^*_1 s^*_1)^2} = \frac{k_{11}}{\cos^2\beta\sin^2\beta}$$

From Eq. 8.55,

$$k_{11} = h_{11} - (s^*_1 m^*_2 + s^*_2 m^*_1)\sigma W^*_{12} = h_{11} + \sigma\cos 2\beta\sin^2\beta.$$

When k_{11} is substituted into the expression for dσ/dε, the result is identical to the independent derivation, Eq. 8.43, of the slope of the stress-strain curve.

$$\frac{d\sigma}{d\varepsilon} = \frac{4h_{11}}{\sin^2 2\beta} + \frac{\sigma\cos 2\beta}{\cos^2\beta}$$

8.4 POLYCRYSTALLINE DEFORMATION

Constraints Due to Neighboring Grains

From a macroscopic point of view, grain boundaries are modeled as generally planar regions across which there is a change in crystal orientation so that the $\mathbf{s}^{(\alpha)}$, $\mathbf{m}^{(\alpha)}$ change orientation. Typically, the displacements across the boundary are modeled as continuous, so that the distributions of incremental displacement $d\mathbf{u}^{(1)}(\mathbf{x})$ and $d\mathbf{u}^{(2)}(\mathbf{x})$ in the adjoining crystals 1 and 2 have a common distribution $\mathbf{du}^{(b)}$ in the plane of the boundary. If **S** and **T** are two orthogonal unit vectors in the plane of the boundary, and **N** is a unit normal to the boundary, then the quantities

$$S_j \frac{\partial(du_i)}{\partial x_j}$$

$$T_j \frac{\partial(du_i)}{\partial x_j} \qquad (8.59)$$

must be continuous. Therefore, when an arbitrary deformation gradient is imposed on one side of the boundary, the continuity relations Eq. 8.59 impose six conditions in the adjoining grain.

Equilibrium conditions also impose that the increment in force per area, or traction, exerted by grain 1 on grain 2 is equal and opposite to that exerted by 2 on 1. Therefore, the quantity

$$d\sigma_{ij} N_j \qquad (8.60)$$

must be continuous across the boundary where N is a unit vector normal to the boundary. This type of continuity condition was imposed across constant stress sectors, using a Mohr's circle approach, in the construction of a lower bound for indentation in Section 8.2. If increments in stress state and slip configuration are imposed on one side of a boundary, then Eqs. 8.59 and 8.60 impose continuity of six components of deformation gradient and three components of stress increment. Thus, it is clear that deformation in neighboring grains imposes several constraints. A particular example of slip in one crystal inducing slip in an adjoining crystal is discussed by Hirth[21], and criteria for such activation have been discussed by Shen and colleagues.[22]

Maximum Work and Minimum Slip Principles for Polycrystals

Bishop and Hill[13] also demonstrated that the maximum work and minimum slip principles are valid for polycrystals, provided that the macroscopic stress and strain increments are properly defined. In particular, consider a polycrystal in which the local strain increment, $d\varepsilon_{ij}$, varies with position throughout the volume V. The macroscopic strain increment is defined as the average of the local strain increment,

[21] See, for example, Ref [6], Figure 9-19 and related discussion.

[22] Z. Shen, R. H. Wagoner, and W. A. T. Clark, *Acta Metall.* 36 (1988): 3231–3242.

$$dE_{ij} = \frac{1}{V}\int_V d\varepsilon_{ij} dV \tag{8.61}$$

Since the strain increment is compatible with a single-valued displacement increment,

$$dE_{ij} = \frac{1}{V}\int_V \frac{1}{2}\left(\frac{\partial(du_i)}{\partial x_j} + \frac{\partial(du_j)}{\partial x_i}\right) dV \tag{8.62}$$

Provided that du_i and the spatial derivatives of it are defined throughout the cell, the **Green-Gauss theorem** may be used to change the volume integral in Eq. 8.62 to an integral over the surface S that encloses V, where **n** is the outward normal to S:

$$dE_{ij} = \frac{1}{2V}\int_S \left(du_i n_j + du_j n_i\right) dS \tag{8.63}$$

This definition of average strain increment is appropriate even when there are cavities within the polycrystal, provided that no work is done on the surfaces of those cavities.

In order to define the macroscopic stress, the macroscopic work increment on the polycrystal is calculated by integrating the internal work increment over *V*, noting that σ_{ij} is symmetric and that local equilibrium must be satisfied, and then converting the volume integral to a surface integral by the Green-Gauss theorem:

$$\begin{aligned} dW &= \frac{1}{2}\int_V \sigma_{ij}\left(\frac{\partial(du_i)}{\partial x_j} + \frac{\partial(du_j)}{\partial x_i}\right) dV \\ &= \int_V \frac{\partial(\sigma_{ij} du_i)}{\partial x_j} dV \quad (\text{since } \sigma_{ij} = \sigma_{ji} \text{ and } \sigma_{ij,j} = 0) \\ &= \int_S \sigma_{ij} n_j du_i dS \quad (\text{Green - Gauss theorem}). \end{aligned} \tag{8.64}$$

The above relations hold even when there are deforming internal surfaces or sliding boundaries, as long as the internal surfaces are traction-free and no internal work is dissipated inside the boundary that is not already accounted for by the internal work product, $\sigma_{ij} d\varepsilon_{ij}$.

To develop a maximum work principle on the macroscopic level, the macroscopic stress, Σ_{ij}, must be defined as a work conjugate to dE_{ij}, so that dW

$= \Sigma_{ij} dE_{ij}$. Also, to define macroscopic quantities in an unambiguous way, Bishop and Hill discuss the concept that a local quantity such as stress or displacement increment should not have a value that is correlated with position. As a particular example, consider a polycrystal that is large enough so that when deformed, the surface, S, of the polycrystal deforms according to a homogeneous macroscopic deformation:

$$du_i(\mathbf{x}) = \overline{\frac{\partial(du_i)}{\partial x_j}} x_j \quad \text{on S} \tag{8.65}$$

even though the local internal displacement increment is quite inhomogeneous. When Eq. 8.65 is used to replace du_i in the last relation in Eq. 8.64, and dW is equated to $\Sigma_{ij} dE_{ij}$, then the definition of macroscopic stress becomes

$$\Sigma_{ij} = \frac{1}{V} \int_S \sigma_{ik} x_j n_k dV = \frac{1}{V} \int_V \sigma_{ij} dV \tag{8.66}$$

Bishop and Hill make two important points from this development. First, suppose that Σ_{ij} and dE_{ij} are the actual macroscopic stress and increment in strain, and σ_{ij} and $d\varepsilon_{ij}$ are the corresponding distributions in the polycrystal. If Σ^*_{ij} and σ^*_{ij} correspond to some other equilibrium distribution that does not violate yield, then

$$\left(\Sigma_{ij} - \Sigma^*_{ij}\right) dE_{ij} = \int_V \left(\sigma_{ij} - \sigma^*_{ij}\right) d\varepsilon_{ij} \, dV \geq 0 \tag{8.67}$$

The equality holds because internal and external work increments must be equal, and the inequality holds because the maximum work principle Eq. 8.29 holds on the microscopic level. Therefore, the parallel yield surface construction in macroscopic variables can be made where, for example, σ_{ij} and $d\varepsilon_{ij}$ in Figure 8.1 are replaced by Σ_{ij} and dE_{ij}, respectively.

The second point concerns two continuous displacement increments, where one is the actual increment, du_i, corresponding to a local stress distribution σ_{ij}, and the other, du^*_i, produces the same surface displacements and thus the same macroscopic strain increment according to Eq. 8.63. The macroscopic stress does the same amount of work increment through each of these displacement fields. Therefore, the internal work increments must be equal:

$$\int_V \Sigma_{ij} \left(dE_{ij} - dE^*_{ij}\right) dV = \int_V \sigma_{ij} \left(d\varepsilon_{ij} - d\varepsilon^*_{ij}\right) dV = 0 \tag{8.68}$$

Accordingly, the same argument to demonstrate the principle of minimum slip on the individual crystal level, Eq. 8.35, can be used to show that

$$\int_V \tau_c |d\gamma^*|\, dV \leq \int_V \tau_c |d\gamma^*|\, dV \tag{8.69}$$

where the integrals represent the incremental work due to slip over all crystals and slip systems. In general, τ_c can vary from one slip system and crystal to another, but in the limit of uniform τ_c for all slip systems, Eq. 8.69 shows that of all the combinations of slip that produce the macrostrain increment dE_{ij}, the actual one produces the minimum amount of internal slip.

Estimates of Polycrystalline Yield

In 1928, Sachs investigated the validity of the Schmid concept of critical resolved shear stress to activate slip, by comparing critical values yield of copper, nickel, and iron polycrystals loaded in either pure tension or torsion.[23] In particular, the geometric factor, S_σ, for pure tension was defined as $(\sigma/2) = S_\sigma\tau$, where σ is the applied tension and τ is the resolved shear stress on a (1 1 1)/[1 -1 0] type slip system. Although a valid lower bound for a polycrystal would be to compute $\sigma = S_\sigma 2\tau_c$, where the minimum value of S_σ over all orientations is used, Sachs computed the average $\bar{S}_\sigma \approx 1.11$ over various orientations to produce the estimate, $\bar{\sigma}_{\text{yield}} \approx 1.11(2\tau_c)$. The corresponding procedure for pure torsion produced $\bar{\tau}_{\text{yield}} \approx 1.29\,\tau_c$. If τ_c remained the same in tension and torsion, then $2\bar{\tau}_{\text{yield}}/\bar{\sigma}_{\text{yield}} = 1.29/1.11 \approx 1.15$. Sachs noted that the measured value, $2\bar{\tau}_{\text{yield}}/\bar{\sigma}_{\text{yield}} \approx 1.12$, was in acceptable agreement with the predicted value of 1.15. However, this averaging technique does not meet the criteria for a lower bound to polycrystalline yield. In particular, an applied tension of $1.11(2\tau_c)$ or an applied shear of $1.29\tau_c$ would violate the condition that the proposed equilibrium stress field not exceed yield anywhere in the polycrystal.

In his May 1938 Institute of Metals lecture, Taylor provided a simple description of the principle of virtual work, and proposed an upper bound procedure to determine the yield locus of an f.c.c. polycrystal.[14] He noted that earlier work by Cox and Sopwith estimated polycrystalline yield based on a collection of single crystals of different orientation, each of which was constrained to elongate by the same amount in the tensile direction. However, compatibility was not satisfied in the transverse directions. Based on his observations of aluminum polycrystals loaded in compression, Taylor asserted that, qualitatively, individual grains deformed in a self-similar manner to the macroscopic polycrystal. Therefore, the macroscopic deformation increment, $d\varepsilon_{11} = d\varepsilon$, $d\varepsilon_{22} = d\varepsilon_{33} = -d\varepsilon/2$, was imposed, and the tensile direction was varied within the [1 1 1]-[1 0 1]-[1 0 0] stereographic triangle to ensure a uniform sampling of

[23] G. Sachs, *Z. d. Vereines Deut. Ing.* 72 (1928): 734.

crystal orientation. To calculate the tensile yield stress, σ_y, for a given orientation, the fractions of slip, $f^{(\alpha)} = d\gamma^{(\alpha)}/d\varepsilon$, on 5 of the 12 slip systems were chosen to produce the strain increment $(d\varepsilon, -d\varepsilon/2, -d\varepsilon/2)$. An upper bound, σ_y^*, to the actual yield stress was determined by equating the external work, $\sigma_y^* d\varepsilon$, to the internal work, so that

$$\sigma_y^* = \frac{\tau_c}{d\varepsilon}\sum_{\alpha=1}^{5}\left|d\gamma^{(\alpha)}\right| = \tau_c\sum_{\alpha=1}^{5}\left|f^{(\alpha)}\right| \tag{8.70}$$

The combination of 5 slip systems was varied to render the cumulative shear a minimum, or, equivalently, to render σ_y^* a minimum.

Figure 8.9 shows a stereographic triangle from that work, in which the number reported next to each (+) location is the minimum $\Sigma\,|\,d\gamma^{(\alpha)}\,|$ necessary to produce a strain $d\varepsilon = 272$, in arbitrary units. Based on these results, σ_y^*/τ_c ranges from $667/272 \approx 2.45$ when the tensile axis is near [1 0 0], to $1000/272 \approx 3.68$ when the tensile axis is near [1 0 1]. The average σ_y^*/τ_c over all orientations is ≈ 3.06.

In subsequent work, Bishop and Hill[24] used an upper bound approach that yields approximately the same average, $\sigma_y^*/\tau_c \approx 3.06$. The upper bound approach is stated as

$$\Sigma_{ij}^* dE_{ij}^* \equiv \frac{1}{V}\int \sigma_{ij}^* d\varepsilon_{ij}^* dV \geq \frac{1}{V}\int \sigma_{ij} d\varepsilon_{ij}^* dV \equiv \Sigma_{ij} dE_{ij}^* \tag{8.71}$$

where the imposed macroscopic strain increment, dE_{ij}^*, has a more general form, $[d\varepsilon, -\lambda d\varepsilon, (1-\lambda)d\varepsilon]$, than used by Taylor. The yield surface of the polycrystal is acknowledged to be isotropic, and the range $0.5 \leq \lambda \leq 1$ is imposed in order to determine a 60° sector of the polycrystal yield surface, as projected in the deviatoric stress plane. Like Taylor, the proposed local deformation, $d\varepsilon_{ij}^*$, is taken to be uniform and equal to dE_{ij}^*. The internal work is computed by considering among twelve f.c.c. slip systems, the various combinations, $d\gamma^{*(\beta)}$, that will accomplish the strain increment, dE_{ij}^*, and setting the corresponding internal work equal to $\Sigma_{ij}^* dE_{ij}^*$. According to Eq. 8.71, this approach provides an upper bound, Σ^*, to the *perpendicular* distance from the origin to the portion of the yield surface with normal dE_{ij}^*,

$$\Sigma^* = \frac{\Sigma_{ij}^* dE_{ij}^*}{\left(dE_{kl}^* dE_{kl}^*\right)^{1/2}} = \frac{\sum_{\beta}\tau_c d\gamma^{*(\beta)}}{\left(dE_{kl}^* dE_{kl}^*\right)^{1/2}} \qquad 8.72$$

---

[24] J.F.W. Bishop and R. Hill, *Philos. Mag Ser. 7* 42 (1951): 1298.

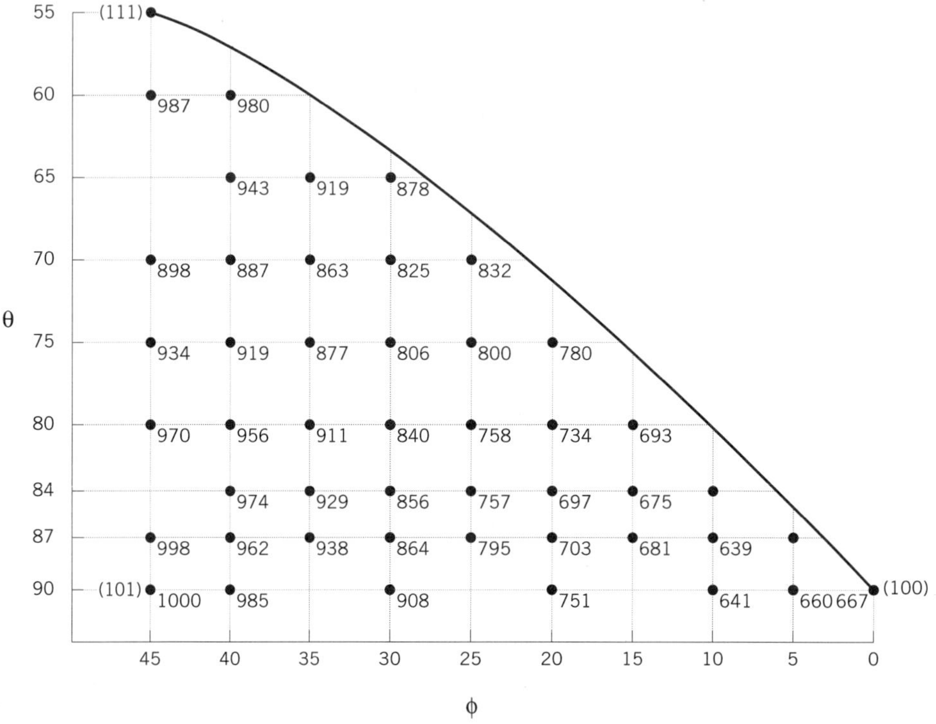

Figure 8.9 Stereographic triangle projection showing, for discrete positions of the tensile axis, the cumulative slip required on f.c.c. slip systems to produce a tensile strain of 272 (in arbitrary units). Taken from Figure 13 of Taylor.[15]

Clearly, of all the combinations of $d\gamma^{*(\beta)}$ which produce the strain increment dE^*_{ij}, that which has the minimum internal work increment furnishes the minimum, or best, upper bound.

Figure 8.10 shows the predicted yield surface, as projected in the deviatoric plane. It is based on an average of Σ^* in Eq. 8.72 over all possible crystal orientations. The surface is positioned between the corresponding Tresca and von Mises surfaces, and passes through the values for pure tension predicted by Taylor. Although the predicted surface is an upper bound, it appears to predict yield values that are lower than those measured experimentally. A possible source of discrepancy stems from the accuracy with which polycrystalline yield can be determined experimentally, given that yield occurs incrementally, from one crystal to another.

Assessment of the Taylor Model

The implementation of crystal-constitutive relations into polycrystalline geometries via a finite element scheme is an important application of crystal-based plasticity. In particular, such approaches predict the evolution of crystallographic texture, or preferred crystal orientation, during polycrystalline deformation. The general outcome is that polycrystalline models that adopt the

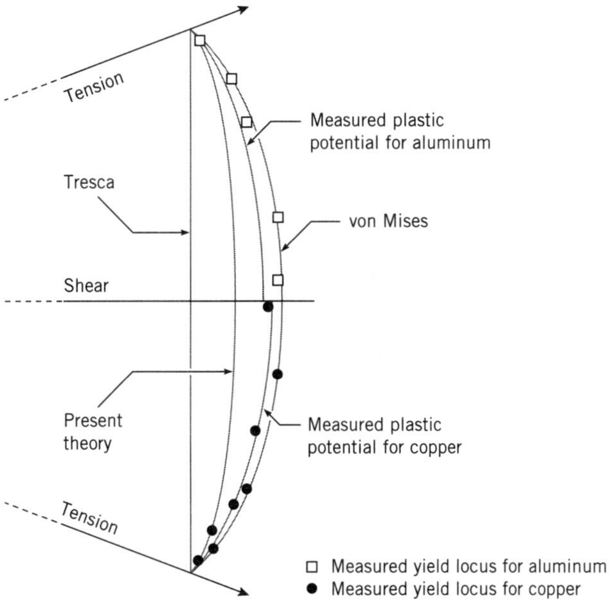

Figure 8.10 Comparison of the predicted yield locus in the deviatoric plane with corresponding Tresca and von Mises predictions. Taken from Bishop and Hill.[22]

Taylor assumption are quite successful in predicting the average texture evolution in a polycrystal. Of course, the onset of inhomogeneous deformation and shear instability in polycrystals is a striking deviation from the Taylor assumption, and the chapter will conclude with a summary of such observations.

The minimum slip calculations performed by Taylor provide predictions of the crystal rotation under simple tension, because for each crystal orientation, those calculations furnished a combination of $d\gamma^{(\alpha)}$. Figure 8.11 is the projection of the (111)-(101)-(100) triangle as shown earlier, and the vector at each nodal point indicates the displacement of a tensile axis initially situated at that nodal point, after a tensile strain of 0.0237. Two vectors emanating from the node indicate that the minimum slip criterion did not furnish a unique combination of $d\gamma^{(\alpha)}$, and that any linear combination of the two vectors is suitable. However, the texture development is evident, in that tensile axes in region G and many positions within region EC will rotate toward (111), those in the vicinity of (100) will rotate toward a cube axis, and those near (101) will rotate to either (111) or (101).

Figure 8.12*d*, left-hand side, shows a stereographic projection, in which darker shaded regions indicate more frequently observed orientations of the tensile axis after tensile straining to 0.37.[25] The crystallographic axis normal to

[25] C.A. Bronkhorst, S.R. Kalidindi, and L. Anand, *Modeling the Deformation of Crystalline Solids* (Warrendale, PA: 1991), TMS, T.C. Lowe, A.D. Rollett, P.S. Follansbee, and G.S. Daehn, eds., p. 211.

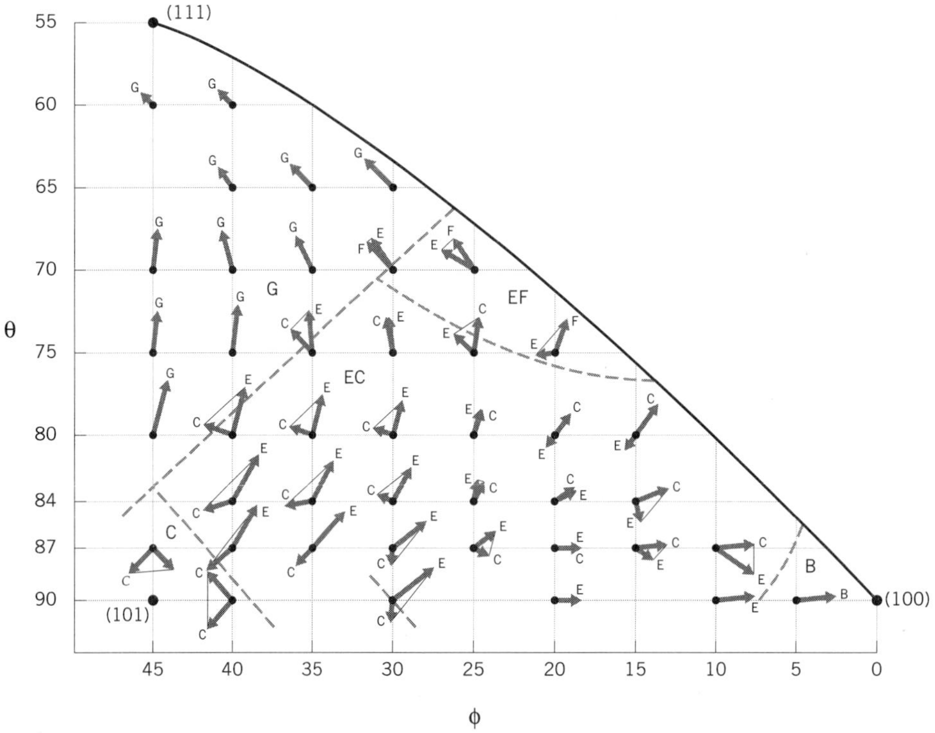

Figure 8.11 Predicted rotation of the tensile axis as a function of tensile axis orientation in a stereographic triangle. Taken from Taylor.[15]

the diagram is [1 1 1], and the X_1 and X_2 directions in the figure are of [1 0 – 1] and [1 – 2 1] type. The [1 1 1] orientation is represented by the dark-centered dot, while the more diffuse dark ring includes the cube axis directions. The [1 0 – 1] directions are located on the outer edge of the projection, where the density is low. Taylor noted that in compression, the vectors drawn in Fig. 8.11 would reverse, so that the compression axis rotated away from [1 1 1] and [1 0 0], but toward [1 0 1] directions. The corresponding measurements for simple compression, shown in Figure 8.12*b*, left-hand side, confirm the predictions. The results for tension and compression discussed here are representative of aluminum, copper, and nickel textures, for example, but differ from textures observed in brass, austenitic stainless steel, and silver. Hosford[26] provides a good summary of observed textures in these systems, and reviews rolling textures for

[26] W.F. Hosford, *The Mechanics of Crystals and Textured Polycrystals* (Oxford: Oxford University Press, 1993).

b.c.c. and h.c.p. polycrystals. Havner also provides an extensive description of experimental observations, coupled with analysis of those results.[26a]

A more recent modification to the Taylor model by Asaro and Needleman[27] is to include strain-rate hardening effects by adopting a slip system constitutive relation in which the slip rate, $\dot{\gamma}^{(\alpha)}$, depends on the current resolved shear stress, $\tau^{(\alpha)}$, according to:

$$\dot{\gamma}^{(\alpha)} = \dot{\gamma}_0^{(\alpha)} \left(\frac{\tau^{(\alpha)}}{\tau_0^{(\alpha)}} \right)^m \qquad 8.73$$

In this constitutive relation, $\tau_0^{(\alpha)}$ is defined as a reference shear stress at which the slip rate on slip system (α) is equal to $\gamma_0^{(\alpha)}$. The strain rate sensitivity is represented by m. For example, values of m on the order of 10^2 are typical of a pure f.c.c. metal, and indicate that $\tau^{(\alpha)}$ approximately equals $\tau_0^{(\alpha)}$ for plastic slip, independent of the slip rate. Conversely, relatively low values of m on the order of 10^1 imply that $\tau^{(\alpha)}$ to produce plastic slip depends more sensitively on the slip rate. In principle, the reference quantities, $\dot{\gamma}_0^{(\alpha)}$ and $\tau_0^{(\alpha)}$, depend on temperature and evolve with deformation. One possible prescription is that $\tau_0^{(\alpha)}$ evolves with slip according to

$$d\tau_0^{(\alpha)} = \sum_\beta h_{\alpha\beta} d\gamma^{(\beta)} \qquad 8.74$$

where again, the matrix $h_{\alpha\beta}$ denotes the amount of self and latent hardening present. A more extensive discussion of rate-dependent slip relations is presented by Asaro[17].

An important feature is that rate-independent formulations avoid ambiguities in determining $\dot{\gamma}^{(\alpha)}$, as present in a rate-independent constitutive relation such as Eq. 8.36. In particular, the $\tau^{(\alpha)}$ are computed according to Eq. 8.48, and the slip rate is computed according to Eq. 8.73. In this sense, all of the $d\gamma^{(\alpha)}$ are nonzero, indicating that all slip systems contribute to the deformation. However, the $d\gamma^{(\alpha)}$ may vary by orders of magnitude, particularly in the rate-independent limit of large m. Accordingly, Eq. 8.56 provides the corresponding increment, $d\overset{*}{\varepsilon}_{ij}$, strain, and Eq. 8.50 provides the corresponding increment in slip system directions $m_i^{*(\alpha)}$ and $s_i^{*(\alpha)}$, measured in the reference basis. This simplicity is very attractive compared to a rate-independent formulation.

[26a]K.S. Havner, *Finite Plastic Deformation of Crystalline Solids* (Cambridge: Cambridge University Press, 1992).

[27] R.J. Asaro and A. Needleman, *Acta metall.* 33(6) (1985): 923.

[28] S. Harren, T.C. Lowe, R.J. Asaro, and A. Needleman, *Phil Trans. R. Soc. Lond. A* 328 (1989): 443.

The observation from rate-dependent work is that increasing either strain hardening or strain-rate hardening generally increases the strain to fracture. Further, increased strain rate sensitivity is observed to decrease the amount of crystal lattice rotation, relative to a material reference frame,[28] and presumably, this decreases the geometric instabilities attributed to lattice rotation. In summary, Harren and Asaro[29] comment that the Taylor model shows modest overestimates of yield compared to more exact finite element calculations that permit inhomogeneous deformation of individual crystals. In particular, the differences in tension/compression curves predicted by Taylor and finite element approaches are about 2%, and for shear deformation, the predictions differ by about 4%. The Taylor model also appears to overestimate the axial extension under torsional loading.[25]

A comparison of texture predictions for aluminum, copper, and nickel show that the Taylor model predictions are generally reasonable; but the texture bands, for example, as shown on the right-hand side of Figure 8.12, appear to be too sharp. Harren and Asaro[26] offer the explanation that localized deformation, which the Taylor model cannot admit as a deformation field, is the source of the discrepancy. More exact finite element studies[26, 30, 31] demonstrate that although the Taylor model predictions are good for average grain orientation at lower strains, large local grain interaction and localized shear bands can produce dominant modes of deformation at larger strains, so that the Taylor model provides poor predictions of individual grain orientations. Nevertheless, the Taylor approach appears to have utility in modeling texture evolution in multipass rolling simulations,[32] for example, as well as offering a more efficient numerical treatment of texture evolution. Recent hybrid finite element formulations such as this now model considerably larger polycrystals as a continuum in which the deformation is inhomogenous on the macroscopic level, but with a local (pointwise) constitutive relation that is determined by a Taylor model.

CHAPTER 8 - PROBLEMS

A. Proficency Problems

1. In an effort to refine the upper bound estimate to density, ρ, that a vertical embankment of height h can sustain, you propose a trial displacement field that is different than shown in Exercise 8.1. The field is shown below, where the boundary between the slipping and stationary parts of the

---

[29] S.V. Harren and R.J. Asaro, *J. Mech. Phys. Solids* 37(2) (1989): 191.

[30] R. Becker, *Modeling the Deformation of Crystalline Solids* (Warrendale, PA: 1991), TMS, T.C. Lowe, A.D. Rollett, P.S. Follansbee, and G.S. Daehn, eds., p. 249.

[31] R. Becker and S. Panchanadecswaran, Alcoa Technical Report, April, 1994.

[32] K.K. Mathur and P.R. Dawson, *International Journal Plasticity*, 5 (1989): 67.

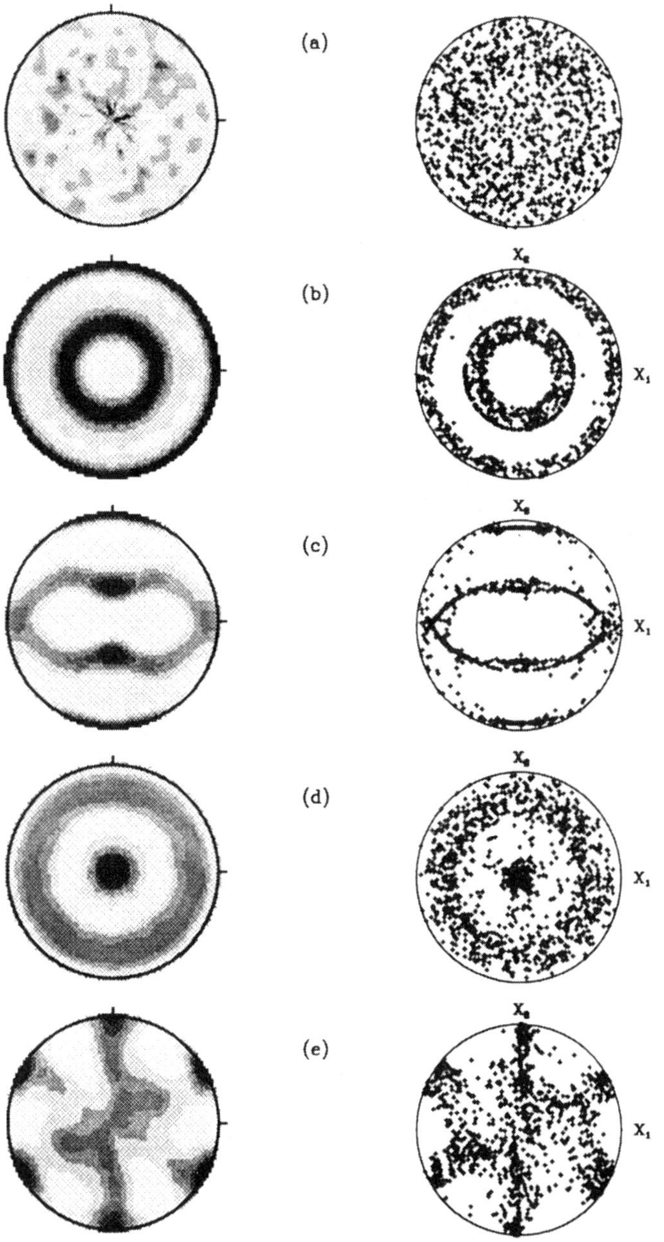

Figure 8.12 Experimentally determined (left-hand side) and computed (right-hand side) {111} pole figures showing texture development as a function of type of loading. The predicted pole figures are based on a Taylor model of a polycrystal. Reproduced from Bronkhorst et al.[22]

embankment is a quarter circle of radius R, from $\theta = 0$ to $\pi/2$. Calculate the new upper bound to the density.

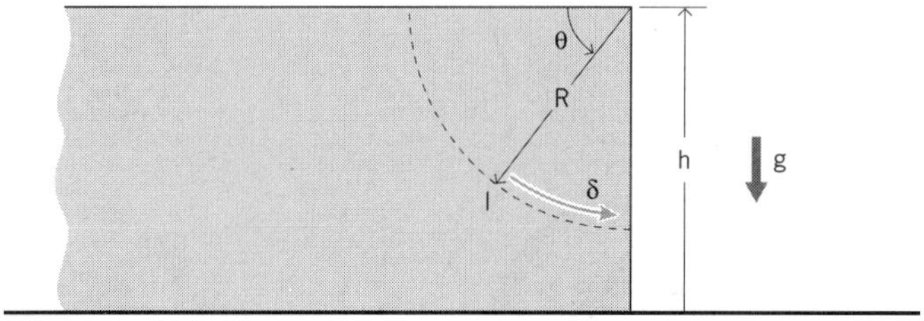

2. Draw a free-body diagram of the triangular block d in Fig. 8.3(a). Show that the forces that act on the block, corresponding to the assumed displacement field, do not satisfy equilibrium for general θ_1, θ_2.

3. Construct a lower bound for the indentation load, P, per unit depth, based on a three-sector stress field. Vary the angle θ to obtain the best lower bound.

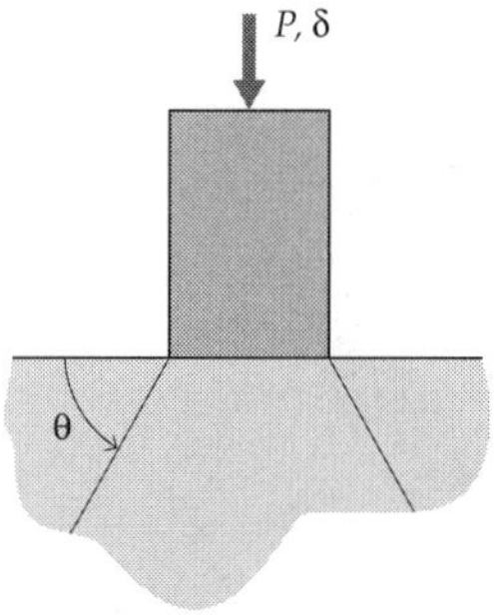

4. Using the yield condition, Eq. 8.13, show why hydrostatic loading, $\sigma_{11} = \sigma_{22} = \sigma_{33} = \sigma$, cannot cause yield on any slip system, regardless of the orientation of the slip system to the loading axes.

5. Experiments that apply hydrostatic loading to engineering materials have shown that materials such as steel, aluminum, copper, or silicon will yield, although the magnitude of hydrostatic loading to cause yield is many times that required for simple tension or compression. Discuss why materials yield in hydrostatic loading.

6. A single crystal is indented with a square knife edge as shown in the following diagram. The candidate slip systems are oriented at discrete angles θ_1, θ_2, and θ_3.
   a. Construct a deformation field out of rigid, sliding triangles, produce the corresponding hodograph, and determine an upper bound to the indentation load.
   b. If the crystal had only two slip planes, could you construct a deformation field? Support your answer with some sketches. What would the indentation load P be under such a case?

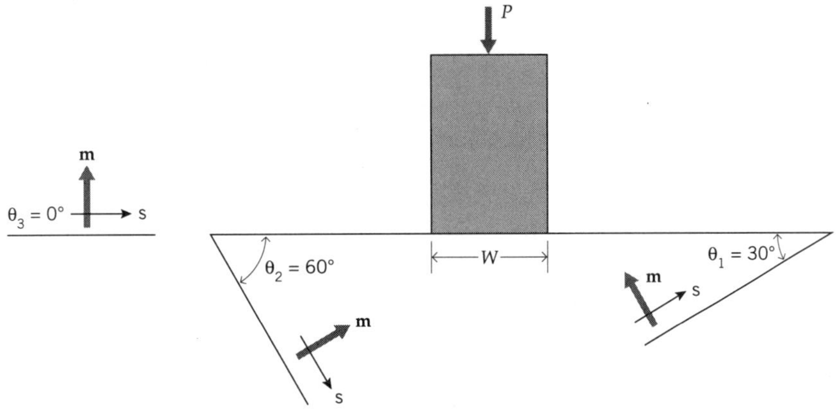

7. In the upper bound analysis of a block of dimensions l_1, l_2, and l_3, which is sheared by an amount b along a direction **s**, on a plane with normal **m** (see Fig. 8.2), the internal dissipation is

$$IW^* = k\,Ab$$

and the external dissipation is

$$EW^* = \sigma_{11}^*\,Abm_1s_1 + \sigma_{22}^*\,Abm_2s_2$$

where σ_{11}^* and σ_{22}^* are uniform stresses, and A is the area of slipped plane.

   a. Sketch the resulting upper bound load to the collapse surfaces in σ_{11}^*–σ_{22}^* space, assuming the inclination angle, θ, shown in Figure 8.2 equals 30°.
   b. Although the internal strain is concentrated at the slip plane, define an appropriate average, or macroscopic strain state for the block, after a slip of b has occurred on the plane.

c. Do your macroscopic strains in Part *b* satisfy normality with the collapse surface in Part *a*? Explain why. If they do not, propose a definition of macroscopic strain that will satisfy normality.

8. Construct the projection of the f.c.c. yield surface onto the σ_{11}–σ_{12} stress plane. Compare your answer to the prediction based on a Tresca yield criterion, $\sigma_1 - \sigma_3 = \pm 2k$, where σ_1 and σ_3 are the maximum and minimum principal stresses, respectively.

9. Imagine that you can load a f.c.c. crystal along any crystallographic direction. Assume that the potential slip systems are of the type {111}/<110>, as listed in Table 8.1. Find a crystallographic direction along which the tensile stress to yield is a minimum, and report the minimum value of tensile stress in terms of τ_c, the critical resolved shear stress to activate slip.

10. A biaxial stress state is applied to a single crystal of f.c.c. material. However, the crystal is oriented so that one applied stress, $\sigma_{1'1'}$, is along the [111] crystallographic direction, and the other applied stress, $\sigma_{2'2'}$, is along the [1-1 0] crystallographic direction. Assume that all slip planes of the type {111}/<110> (see Table 8.1) slip when the critical resolved shear stress reaches τ_c. Construct a projection of the yield surface onto the $\sigma_{1'1'}$-$\sigma_{2'2'}$ stress plane. Comment on the difference between this projection and the one shown in Fig. 8.4.

11. A sphere with radius r = 1 cm is a single f.c.c. crystal. Slip systems 1, 6, and 12 listed in Table 8.1 are activated, so that a slip of 1 μm in the corresponding +**s** directions listed occurs on slip planes that pass right through the center of the sphere. Calculate the resulting macroscopic strain in the crystal.

12. Consider the portion of the yield surface in Figure 8.4 that is contributed by slip systems 2, 4, 7, and 11. Using Eq. 8.25, show that operation of any individual or linear combination of these slip systems produces a strain increment that is perpendicular to the portion of the yield surface shown in the figure.

13. Equation 8.24 states that external and internal plastic work increments are equal. Show that this equation may be converted to the notation

$$\sigma_i d\varepsilon_i = \Sigma(\tau_c d\gamma)^{(\alpha)}$$

where the components $[\sigma_i]$ are defined by Eq. 8.14, and the components $[d\varepsilon_i]$ are defined by the left-hand identity in Eq. 8.31.

14. Use Eq. 8.25 to show that no combination of slip systems can produce dilatation.

15. Of the 12 slip systems listed in Table 8.1, choose slip systems 1, 2, and 6. Determine whether these 3 slip systems are independent.

16. Suppose you load in tension along the [3 8 6] direction in a f.c.c. material in which the candidate slip systems are of the type listed in Table 8.1. The crystal begins to yield at a tensile stress of 10MPa. Assume that all slip systems require the same critical resolved shear stress, τ_c, for activation. What slip system will be activated first? What is the value of τ_c? To what crystallographic direction will the tensile axis rotate? What is the axis about which the tensile axis rotates?

17. Derive the results

$$\dot{\beta} = -\frac{\dot{b}}{H_o \sin\beta_o}\sin^2\beta$$

$$\dot{\beta} = -\dot{\bar{\varepsilon}}_{TT}\frac{\sin\beta}{\cos\beta}$$

by analyzing the 2-D slip-plane geometry in the following diagram, and where $\dot{b}$ and $\dot{\bar{\varepsilon}}_{TT}$ are the increment in slip on the plane and average strain increment parallel to the current tensile axis, respectively. The other parameters are labeled in the sketch. *Do not* use the general formulation associated with the discussion of Figure 8.5*a* and *b*, but derive the expression from a trigonometric analysis of the 2-D geometry shown.

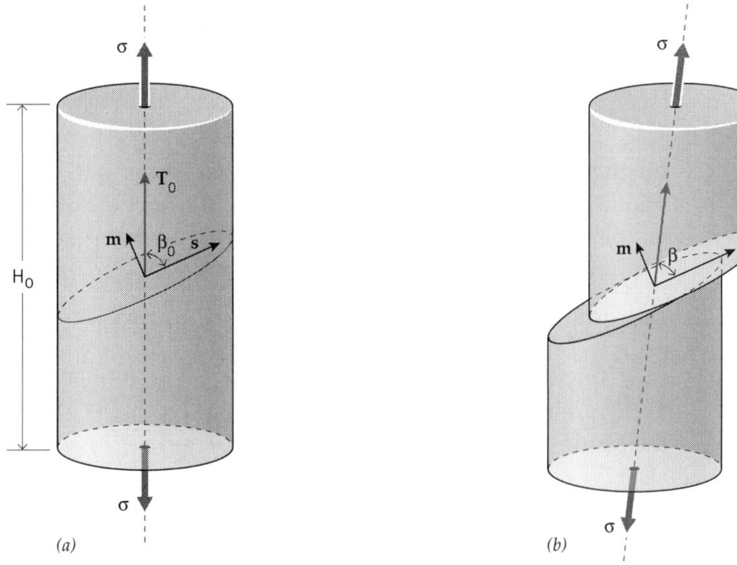

18. Eq. 8.56 defines the increment, $d\varepsilon_{ij}^*$, in components of strain that are expressed in the reference coordinate system. Derive this relation, beginning with a statement of the Principle of Virtual Work that $\sigma_{ij}^* d\varepsilon_{ij}^* = \sum \tau^{(\beta)} d\gamma^{(\beta)}$.

19. Suppose you have a geometry similar to that shown in Problem 17, but with two competing slip systems, ($\mathbf{m}^{(1)}$, $\mathbf{s}^{(1)}$) and ($\mathbf{m}^{(2)}$, $\mathbf{s}^{(2)}$). β_o is the initial angle between $s^{(1)}$ and the initial tensile axis direction, T_o, and γ_o is the corresponding initial angle for slip system (2). Suppose you have documented the angle β as a function of macroscopic strain, ε_{TT}, along the $\mathbf{T}$ (*not* $\mathbf{T}_o$) axis. Explain how you could determine the relative amount of slip, $b^{(1)}$ and $b^{(2)}$, on each system. Do you have enough information?

20. Consider a material in which the critical resolved shear stress to shear a plane with normal $\mathbf{m}^{(\alpha)}$ depends on the stress normal to the plane, according to

$$\tau_c^{(\alpha)} = \tau_o - \mu_f \left(m_i^{(\alpha)} \sigma_{ij} m_j^{(\alpha)} \right)$$

where τ_o is the critical resolved shear with zero normal load, and μ_f is a coefficient of friction. Such behavior is consistent with a hard sphere atomic picture of materials, in which the activation barrier to slide particles past one another increases with confining pressure on the material. Show that when a macroscopic stress with nonzero components σ_{11} and σ_{22} is applied, the yield condition for a slip system (α) with slip direction $\mathbf{s}^{(\alpha)}$ and slip plane normal $\mathbf{m}^{(\alpha)}$ is

$$\left(\pm s_1^{(\alpha)} + \mu_f m_1^{(\alpha)} \right) m_1^{(\alpha)} \sigma_{11} + \left(\pm s_2^{(\alpha)} + \mu_f m_2^{(\alpha)} \right) m_2^{(\alpha)} \sigma_{22} \geq \tau_o$$

where $\pm$ must be interpreted as either + for both terms ($+s_1^{(\alpha)}$ and $+s_2^{(\alpha)}$) or - for both terms ($-s_1^{(\alpha)}$ and $-s_2^{(\alpha)}$).

Further, produce a projection of the yield surface onto the σ_{11}-σ_{22} plane, assuming that σ_{11} and σ_{22} are applied along the cube directions of a f.c.c. single crystal, and that $\mu_f = 0.1$. Compare your result to the yield surface projection in Fig. 8.4.

21. Consider the previous example where the critical resolved shear stress for a slip plane depends on the stress normal to the plane. Construct a yield function for this case that reduces to Eq. 8.26 when $\mu_f = 0$. Find the normal to the yield surface. Compare your result to $d\varepsilon_{ij}^{(\alpha)}$ predicted by Eq. 8.23. Is the strain increment parallel to the normal to the yield surface?

22. Assume that you have a polycrystalline wire with a bamboo-type structure, in which each grain α occupies a slice of the wire as shown in the following diagram. Also, each grain has a single slip system described by slip-plane normal $\mathbf{m}^{(\alpha)}$ and slip direction $\mathbf{s}^{(\alpha)}$. The critical resolved shear stress to operate any slip system is 10 MPa, and the amount of slip on any plane is assumed to be negligible compared to the diameter, D, of the wire. A single angle $\beta^{(\alpha)} = \mathbf{T} \cdot \mathbf{s}^{(\alpha)}$ describes the orientation of the slip system relative to the wire axis **T**, along which there is an applied stress, σ. Initially, the probability, P, of finding a grain with angle $\beta^{(\alpha)} = \gamma$ is the same, regardless of the value of γ, which ranges from 0 to 90°. Sketch the probability of finding a grain with angle $\beta^{(\alpha)} = \gamma$, as a function of γ, after loading the wire to four different values of tension: $\sigma = 0$, 20, 30, and 40 MPa. To obtain your result, use a lower bound approach and make the approximation that the stress state in each grain is one of simple tension, σ. What aspect of compatibility is not satisfied here?

23. Outline the components of a computer program which will calculate the uniaxial stress-strain response of a f.c.c. single crystal loaded in tension. The tensile axis **T** is a material one, similar to the case depicted in Fig. 8.5, and initially, **T** is parallel to [1 12 11]. The sample is loaded from 0 to 40MPa over a period of 300s, so that the stress rate, $d\sigma_{TT}/dt$ is constant. Use a rate-dependent constitutive relation as described in Eqs. 8.73 and 8.74 with the following parameters:

$$\dot{\gamma}_o^{(\alpha)} = 10^{-3}/s$$

$$m = 20$$

$$\tau_o^{(\alpha)} = 10\text{MPa (initially, prior to any slip)}$$

$$\text{Case A:} \quad h_{\alpha\beta} = \begin{cases} 0\text{MPa if } \alpha = \beta \\ 5\text{MPa if } \alpha \neq \beta \end{cases}$$

$$\text{Case B:} \quad h_{\alpha\beta} = \begin{cases} 5\text{MPa if } \alpha = \beta \\ 0\text{MPa if } \alpha \neq \beta \end{cases}$$

Note that Case A simulates a larger latent hardening situation, and Case B simulates a larger self hardening situation.

Produce for each case the tensile stress-tensile strain curves along **T**, from $\varepsilon_{TT}= 0$ to approximately 0.8. Note the direction of **T** in each case when $\varepsilon_{TT} = 0.8$. For each case, discuss the dominant slip system(s) when $\varepsilon_{TT} = 0$ and when $\varepsilon_{TT} = 0.8$.

24. Consider a polycrystal with a random orientation of crystals. Produce a lower bound to yield in tension.

CHAPTER 9

Friction

Friction between a workpiece and tools or dies dominates the strain patterns and performance of many forming operations, and yet is often the least quantified of all phenomena involved in forming. This status is a result of two conditions:

- Friction laws depend sensitively on a host of variables that are not themselves usually well characterized, and are quite diverse for various forming operations.
- Until recently, forming analysis was qualitative, and there was little motivation to derive quantitative laws for friction under large-strain forming conditions.

These two conditions interact because friction conditions depend on the forming operation characteristics itself (and vice-versa), so it is very difficult to separate the effects into two classes of behavior.

In this chapter, we will briefly review the principal empirical laws of friction usually applied to forming operations, and their application to analytical models. A taste of the complexity of these laws will be presented along with a few examples of tests to measure friction. While the field of friction and lubrication is an active one that holds great promise for metal-forming analysis, we will not devote time to fundamental studies of the phenomenon itself.

9.1 BASIC CONCEPTS OF FRICTION

The basic idea of friction, presented in most undergraduate courses in physics, is shown in Figure 9.1*a*. A block of a given mass sits on a flat table such that a **normal force** of $\mathbf{f}_N$ is felt between the table and the block. ($\mathbf{f}_N$ is simply the weight of the block in the simplest case, or an additional external normal force may be applied). A rising **tangential force,** $\mathbf{f}_T$, is applied until the block starts to move.

The tangential force is taken to be the **friction force**, $\mathbf{f}_F$, which initiates motion. This limiting reaction force is assumed to be a *property of the particular block and table.*

Figure 9.1*b* is identical to Figure 9.1*a* except that the table is inclined by an angle θ to horizontal, which automatically decomposes the weight of the block, **w**, into $\mathbf{f}_N$ and $\mathbf{f}_T$: $\mathbf{w} = \mathbf{f}_N + \mathbf{f}_T$, where $|\mathbf{f}_N| = |\mathbf{w}| \cos\theta$ and $|\mathbf{f}_T| = |\mathbf{w}| \sin\theta$. The angle of the table can be changed until the block just begins to move, thus establishing the friction force and normal force applicable to the critical friction condition.

Therefore, in the simplest picture of friction, we observe that for a given table and block (for a given $\mathbf{f}_N$), we can find two friction conditions:

for $|\mathbf{f}_T| < |\mathbf{f}_F|$, the reaction force is equal to $\mathbf{f}_T$ (characteristic of applied force)

for $|\mathbf{f}_T| = |\mathbf{f}_F|$, the reaction force is equal to $\mathbf{f}_F$ (characteristic of block and table)

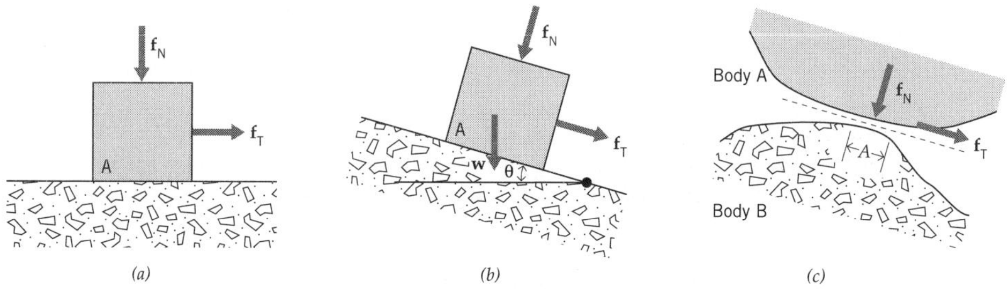

Figure 9.1 Conception of friction forces.

A few other observations may be made using the simple block-and-table experiment:

- As $\mathbf{f}_N$ goes up, so does $\mathbf{f}_T$ ($\mathbf{f}_T$ is zero for $\mathbf{f}_N$ equal to zero).
- Once $\mathbf{f}_T$ reaches $\mathbf{f}_F$, the block accelerates.
- For the same block and table, the surface conditions (roughness, lubrication) have a major contribution to determining $\mathbf{f}_F$.
- If we were to look carefully at the contact between the block and table, we would find that only local areas actually come into contact.

Now, let's extend our consideration to more general contact, Figure 9.1*c*, using the lessons learned from the block and table. The situation is the same once contact normal and contact tangent directions ($\hat{\mathbf{N}}$ and $\hat{\mathbf{T}}$, e.g.) have been defined. The general contact force, $\mathbf{f}_C$, may be decomposed into normal and tangential components in the usual way:

$$\mathbf{f}_C = \mathbf{f}_N + \mathbf{f}_T = f_N \hat{\mathbf{N}} + f_T \hat{\mathbf{T}}, \tag{9.1}$$

where $f_N = |\mathbf{f}_T|$ and $f_T = |\mathbf{f}_T|$, and, as before, f_T may be less than or equal to some limiting value f_F, depending on the normal force, the surfaces, and the materials. In three dimensions, it is necessary to be more precise about the direction $\hat{\mathbf{T}}$ because the two bodies make contact at a tangent plane, and thus any direction lying in the plane could be considered a tangent direction. In fact, the resisting frictional force always opposes either (a) the applied tangential force at the point (in the case of no relative motion), or (b) the relative motion of one contacting body relative to the other. For example, if we wish to calculate the friction force restraining the motion of body A relative to body B (imagine body B fixed in space for convenience), the friction force is applied opposite to the component of the relative motion, which lies in the tangent plane:

$$\hat{\mathbf{T}} = -\frac{\hat{\mathbf{N}} \times (d\mathbf{u}_A - d\mathbf{u}_B) \times \hat{\mathbf{N}}}{\left|\hat{\mathbf{N}} \times (d\mathbf{u}_A - d\mathbf{u}_B) \times \hat{\mathbf{N}}\right|} \tag{9.2}$$

where $\hat{\mathbf{N}}$ is a normal to the contact plane oriented away from body B and into body A.

It is possible to define average macroscopic **contact stress** σ_C, **normal stress** σ_N, **tangential stress** τ_T, and **friction stress** τ_F by normalizing the forces to the **apparent contact area**, A:

$$\boldsymbol{\sigma}_C \equiv \mathbf{f}_C / A, \quad \boldsymbol{\sigma}_N \equiv \mathbf{f}_N / A, \quad \boldsymbol{\tau}_T \equiv \mathbf{f}_T / A, \quad \boldsymbol{\tau}_F \equiv \mathbf{f}_F / A, \tag{9.3}$$

The negative of the normal stress, σ_N, is also called the **contact pressure**. Analogous to the continuum approach to material behavior, we can consider an ideal definition of these contact stresses in terms of a limiting procedure:

$$\boldsymbol{\sigma}_C \equiv \lim_{A \to 0} \mathbf{f}_C / A \tag{9.4}$$

However, note that the limit in Eq. 9.3, will not exist for A less than some size that corresponds approximately to the spacing of the true contact regions. This limitation is similar to the continuum limitation in real materials, as the length scale approaches the important physical length scale (e.g., atomic spacing, dislocation spacing, or grain size).

As shown in Figure 9.2, the **actual contact area** may be much less than the apparent contact area, perhaps only 5% to 10%, so the true local stresses at local areas of contact may be much greater than the macroscopic ones that appear in Eq. 9.2. In standard friction terminologies, the high points of surface roughness—those most likely to come in contact with the other surface—are called **asperities**. Adjacent regions that are not in contact may have virtually no true contact stress. However, we shall restrict our attention to the apparent contact area and **apparent stresses** because these are the ones that are used in phenomenological measurements and macroscopic simulations of metal forming. The microscopic picture is more suitable for fundamental studies of friction.

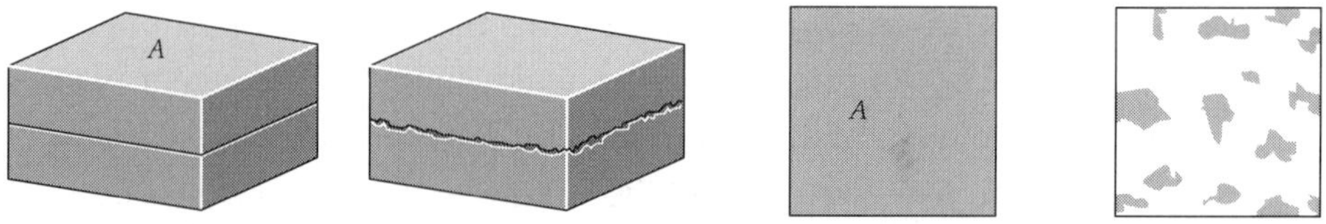

(a) (b) (c) (d)

Figure 9.2 Macroscopic and microscopic views of contact areas.

9.2 PARAMETERS AFFECTING FRICTION FORCES

It is instructive to consider the kinds of variables that can affect friction forces, even though we will focus only on a **phenomenological approach** to friction in which friction forces are measured under conditions close to those in forming operations of interest.

From our simple picture of friction from a physics textbook, it is possible to identify immediately some of the effects.

- *Contact pressure.* Simple experiments show that the friction force increases as the contact pressure or normal force increases, at least at low pressures.
- *Sliding speed.* Most simple experiments show that the **dynamic friction** force (i.e., the dissipative force opposing sliding motion) is less than the **static friction** force (i.e., the critical force to start sliding).
- *Materials.* The elastic and plastic properties of the materials in contact will certainly affect how difficult it is to slide the two materials across one another. At low true contact pressures, the asperities are more likely to interact elastically. Also, the materials may have different fracture characteristics that lead to changes in the surface topology and debris (from **spalling**) on the friction surface.
- *Surface roughness.* The shape and density of asperities may have a major effect on friction, whether lubricated or dry. However, even the measurement of surface roughness is difficult and its quantification is not standard, except for a few simple parameters.
- *Lubrication and debris.* The presence of other material at the contact surfaces is important. The properties of the lubricant (rheology, compressibility, temperature sensitivity, and so on) enter into consideration. The distribution of lubricant depends intimately on surface conditions, pressure, and sliding speed, and knowledge of all of these depends on the forming operation analysis.
- *Temperature.* The local temperature at contact points and in lubricants will depend on thermal conductivity and plastic work dissipation.
- *Concurrent deformation.* This effect is particularly germane for metal forming because the workpiece is usually deforming plastically while the friction forces are operating. The deformation changes the surface

roughness, eliminates most elastic effects in the workpiece, and opens up new surface by the action of dislocation slip.

With all of these variables influencing friction, it is apparent why it is necessary to make measurements under conditions as close as possible to the real ones in the forming operations. For purposes of simulation, it is necessary to adopt simple laws that allow measurement of a limited number of coefficients. These two principles, **simplicity** and **similitude**, are sometimes contrary, but they guide the design of most applied friction tests.

9.3 COULOMB'S LAW

The most common form of friction law is known variously as **Coulomb's law,** or **Amonton's law**. Simply put, it states that the magnitude of the friction force is proportional to the magnitude of the normal force, or, equivalently, that the friction stress is proportional to the normal stress (or pressure):

$$f_F = \mu\, f_N \qquad \text{or} \qquad \tau_F = \mu\, \sigma_N \tag{9.5}$$

where, as before, the friction force or friction stress are simply limiting values of the tangential force or stress that may be present because of frictional effects. μ is called the **friction coefficient**, or **Coulomb constant**, with the basic idea that it is a constant with respect to contact pressure or stress for a given situation.

Coulomb's law, Eq. 9.5, is conceptually correct at very small pressures, in that the limiting tangential force approaches zero as the pressure tends toward zero. In fact, this law is usually found applicable at low contact pressures relative to material strength, before there is a great deal of deformation. However, at higher contact pressures, material properties should enter into consideration.

9.4 STICKING FRICTION AND MODIFIED STICKING FRICTION

While Coulomb's law has the proper form suggested by simple friction experiments for nondeforming bodies (which are by definition lightly loaded), it drastically overestimates friction at high contact pressures. A simple thought experiment shows that friction at the interface cannot exceed the shear strength of the material, because the material can slide at this shear strength even if the interface is bound tightly (welded). This extreme condition is called **sticking friction**, and may be represented generally as

$$\tau_F = \tau_{max} \tag{9.6}$$

where τ_{max} is a property of the softer material alone. Figure 9.3 illustrates conceptually the transition from very low contact pressure to pressures near the shear strength of the material.The term τ_{max} in Eq. 9.6 is the shear strength of the

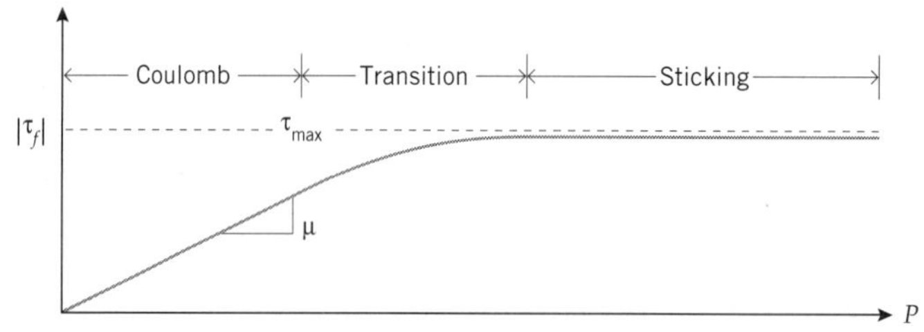

Figure 9.3 Conceptual view of friction as a function of contact pressure.

softer material. For a Tresca material, the shear strength is equal to one-half of the material strength in uniaxial tension:

$$\textbf{\textit{Tresca yield criterion:}}\ \tau_{max} = \frac{\bar{\sigma}}{2} \tag{9.7}$$

whereas for a von Mises material the shear strength can easily be found:

$$\textbf{\textit{von Mises yield criterion:}}\ \tau_{max} = \frac{\bar{\sigma}}{\sqrt{3}} \tag{9.8}$$

See, for example, Exercise 9.1 for derivation of τ_{max} for von Mises or other yield criteria.

A modification to sticking friction is often introduced to account for the fact that frictional forces are seldom as high as the shear strength of a material, but are in a high-contact-pressure regime where they are only weakly dependent on pressure. This generalization may be written as follows:

$$\textbf{\textit{Modified sticking friction:}}\ \tau_F = m\,\tau_{max} \tag{9.9}$$

where m is a constant known as the **friction factor.**[1] As before, τ_{max} can be determined from Tresca or von Mises yield functions, or from other descriptions of the plastic flow stress in shear. Eq. 9.9 defines what is sometimes refered to as the **Tresca friction law**.

A family of friction laws can be constructed based on the idea presented in Figure 9.3, where the friction force depends on contact pressure in an arbitrary way, such that the slope is nearly constant at small pressures, and that it approaches a constant force (τ_F or perhaps m τ_F) at high pressures:

$$\tau_F = f(P) = f(\sigma_N) \tag{9.10}$$

[1] m is typically used for both the friction factor and the material strain-rate sensitivity, so the usage must be determined by context.

Exercise 9.1 **Find the shear flow stress of a material that obeys either the von Mises yield criterion or Hill quadratic normal anisotropic yield criterion.**

Assume that a state of pure shear stress producing plastic deformation exists, so that $\tau_{12} = \tau_{max} = \tau$, with all other stress components equal to zero:

$$\boldsymbol{\sigma} \Leftrightarrow [\sigma] = \begin{bmatrix} 0 & \tau & 0 \\ \tau & 0 & 0 \\ 0 & 0 & 0 \end{bmatrix} \tag{9.1-1}$$

In order to find the tensile equivalent stress, $\bar{\sigma}$, we must transform the components of $\boldsymbol{\sigma}$ to the principal coordinate system, and then substitute into the yield function an expression written in terms of $\bar{\sigma}$. The required rotation is by 90° about $\mathbf{x}_3$, which may be seen most simply by inspection or by reference to Mohr's circle. The principal stress components (the eigenvalues) may also be obtained as shown in Chapter 2:

$$|\boldsymbol{\sigma} - \lambda \mathbf{I}| = 0 \tag{9.1-2}$$

$$\tau^2 - \lambda^2 = 0, \qquad \lambda_1 = \tau, \qquad \lambda_2 = -\tau \tag{9.1-3}$$

Thus the stress components in the principal axes are

$$[\sigma'] = \begin{bmatrix} \tau & 0 & 0 \\ 0 & -\tau & 0 \\ 0 & 0 & 0 \end{bmatrix} \tag{9.1-4}$$

By substitution into the appropriate yield functions written in principal components (see Chapter 7), we obtain the desired shear flow stress in terms of the tensile equivalent stress:

$$2\,\bar{\sigma}^2 = \left[(\tau + \tau)^2 + (\tau)^2 + (\tau)^2\right] \tag{9.1-5}$$

$$\tau = \frac{\bar{\sigma}}{\sqrt{3}} \tag{9.1-6}$$

as shown in Eq. 9.8. For Hill's quadratic normal anisotropic yield (Eq. 7.63), the equivalent relationships are

$$\bar{\sigma}^2 = \left(\tau^2 + \tau^2 + \frac{2r}{1+r} \tau^2 \right) \tag{9.1-7}$$

where we have assumed that $\hat{x}_3$ is normal to the sheet, but that we seek τ_{12}. Solving Eq. 9.1-7 obtains

$$\tau_{max} = \sqrt{\frac{1+r}{2(1+2r)}}\, \bar{\sigma} \tag{9.1-8}$$

If r = 1, we recover the isotropic result, Eq. 9.8:

$$\tau_{max} = \frac{\bar{\sigma}}{\sqrt{3}} \tag{9.1-9}$$

The foregoing solution can be simplified considerably by starting from the general form of the von Mises yield function, Eq. 7.14*b*.

9.5 VISCOPLASTIC FRICTION LAW

It is often convenient for modeling purposes to consider friction in terms of the relative velocity between two surfaces; for example, a tool and workpiece in forming operations. The basic idea arises from consideration of two rigid, smooth surfaces separated by a layer of **viscoplastic lubricant**, as shown in Figure 9.4.

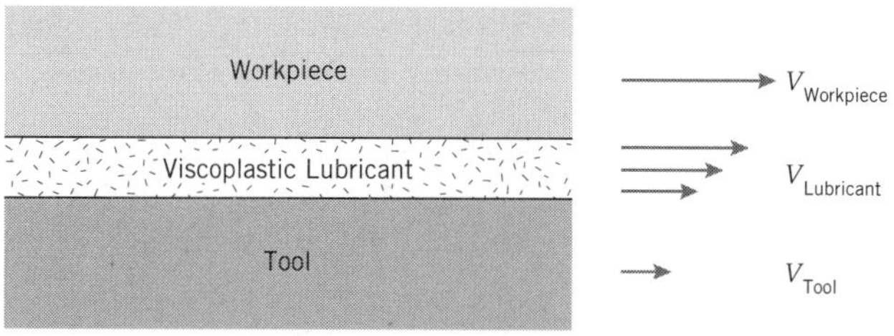

Figure 9.4 Conceptual view of viscoplastic friction condition.

The friction law may be written following simple viscoplastic behavior:

$$\tau_F = \alpha \left| \mathbf{v}_{WP} - \mathbf{v}_{TOOL} \right|^m \tag{9.11}$$

where m can be viewed as the strain-rate sensitivity index of the intermediate material.

When the normal stress (or contact pressure) is small, the friction stress will again be overestimated by Eq. 9.11. A possible modification to improve the fit at low contact pressures based on the Coulomb approach can be made:

$$\tau_F = \alpha |\sigma_N| |v_{WP} - v_{TOOL}|^m \tag{9.12}$$

9.6 REGULARIZATION OF FRICTION LAWS FOR NUMERICAL APPLICATION

Contact and friction conditions introduce several very nonlinear behaviors that make life difficult for numerical modelers. For example, a surface point on a workpiece experiences no boundary condition as long as it is infinitesimally removed from the tool surface, but once it makes contact, its position is subject to a **unilateral contact condition.** Put simply, the point may lie on the surface or outside it, but may not penetrate into the tool. This abrupt change makes smooth convergence of solutions difficult and time-consuming.

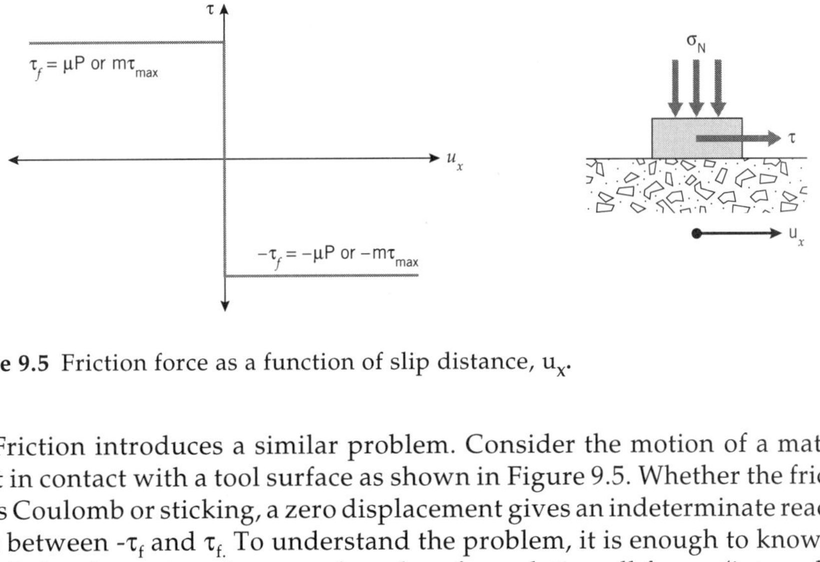

Figure 9.5 Friction force as a function of slip distance, u_x.

Friction introduces a similar problem. Consider the motion of a material point in contact with a tool surface as shown in Figure 9.5. Whether the friction law is Coulomb or sticking, a zero displacement gives an indeterminate reaction force between -τ_f and τ_f. To understand the problem, it is enough to know that most finite element analyses are based on formulating all forces (internal and external, the latter of which include contact and friction) at nodes in terms of nodal displacements (analogous to u_x in Figure 9.5). The resulting set of equations is solved using the derivatives of these forces with respect to nodal displacements. However, Figure 9.5 shows that at $u_x = 0$, neither the force nor its derivative are known, so there will in general be no solution. It is possible to overcome this difficulty in a way inspired by the physical situation: that is, when the external force demanded at a node is less that τ_f, then a true sticking boundary condition is applied to the node (i.e., u_x is set identically to zero).

However, this change of boundary conditions (so-called **stick-slip** condition) can introduce instabilities, particularly when the procedure is coupled with contact pressure (which depends on internal forces) via a Coulomb law, and coupled with the unilateral contact condition, which is usually present. The result, at best, is greatly increased computation times, and, at worst, instability and nonconvergence, or errors in accuracy.

In order to avoid these numerical difficulties, an **ever-slipping** condition is introduced by **regularizing** the friction law. Under these conditions, a contact point (or contact node in finite element analysis) is always considered to slip relative to the tool surface, except at zero applied tangential force. For reasonable accuracy, it is necessary to insure that the slippage for τ less than τ_f is very small, less than some constant δ, for example.

Figure 9.6 shows two typical forms used to regularize the friction law for use in numerical analysis. (We have shown both nodal displacement magnitude u_x and nodal speed v_x, which are roughly equivalent in a small time step, where $u_x \approx v_x \Delta t$. The choice depends on the formulation used in the numerical analysis.) In the first approach,[2] a linear relationship between u_x and τ_f is employed. This has the benefit of simplicity, but there are two small drawbacks, not usually serious: the slope of the curve changes abruptly at $u_x = \delta$, which may cause convergence difficulties in some programs; and the friction function must be written for two domains, such that a logical check is required at each node at each time to decide which domain the nodal displacement falls into. Figure 9.6*b* shows an alternative approach[3] that has a much smoother relationship for all nodal displacements. The conceptual disadvantages to this method are the need to evaluate hyperbolic tangent when the nodal displacement is less than δ, and the very high slope near $u_x = 0$, which may make convergence difficult.

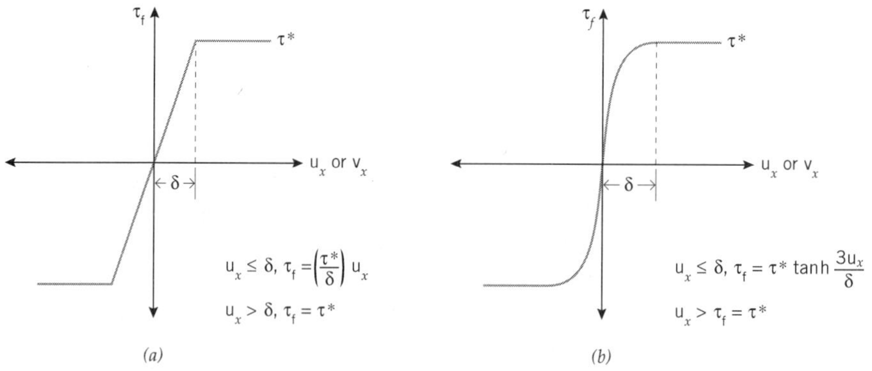

Figure 9.6 Regularization of friction forces by imposition of ever-slipping condition.

---

[2] See, for example, Y. Germain, K. Chung, and R. H. Wagoner, *Int. J. Mech. Sci.*, 31, 1989: 1-24.

[3] M. J. Saran and R. H. Wagoner: *ASME Trans. — J. Appl. Mech.*, 58, 1991: 499-506.

As long as δ is chosen sufficiently small relative to the average tangential displacements in each time step, the error introduced by these approximations is not significant. However, care must be taken that errors are not introduced for small time steps, especially in programs that have automatic time step selection.

9.7 THE ROPE FORMULA

Whereas the classic test defining ideas of friction is shown in Figures 9.1*a* and *b*, the sliding block is not very similar to the situation in many forming operations. For example, the sliding block does not involve the macroscopic deformation inherent to forming.

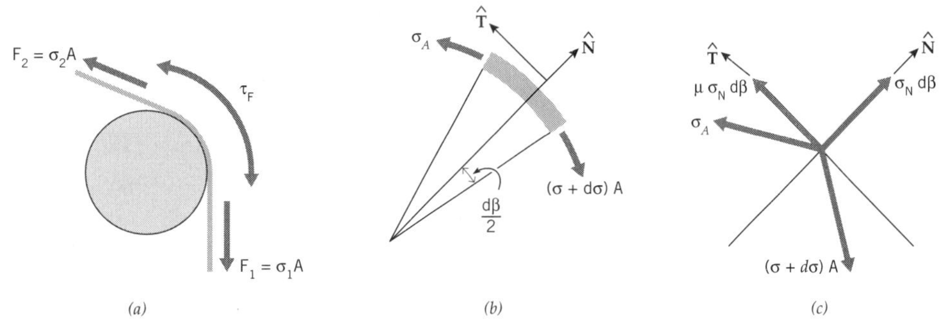

Figure 9.7 Derivation of differential equation leading to the rope formula.

For generally curved contact and friction, like that shown in Figure 9.1*c*, different approaches are preferred. The basic tool for use with friction of thin deforming sheets is known as the **rope formula** because the conditions and results are characteristic of a flexible rope sliding over a fixed surface. The situation is shown in Figure 9.7*a*, where the following assumptions are made.

- The sheet or rope has no significant bending strength; i.e., it is a membrane.
- All points on the rope or sheet are sliding relative to the fixed surface according to the Coulomb friction law.
- The cross-sectional area is assumed to remain constant.[4]

With these assumptions, we consider a differential element of the rope or sheet, Figure 9.7*b*, and perform force balances in the normal $\hat{\mathbf{N}}$ and tangential $\hat{\mathbf{T}}$ directions as shown in Figure 9.7*c*:

$$0 = \Sigma F_{\hat{T}} = \sigma A - (\sigma + d\sigma) A + \mu \, \sigma_N \, d\beta \tag{9.13}$$

[4] This assumption affects only the relationship between membrane stress and force, but the formula may be derived in terms of force without reference to cross-sectional area.

$$0 = \Sigma F_{\hat{N}} = \sigma_N \, d\beta - \sigma A \frac{d\beta}{2} - (\sigma + d\sigma) \, A \frac{d\beta}{2} \tag{9.14}$$

Solving Eqs. 9.13 and 9.14 obtains

$$\mu\beta = \frac{d\sigma}{\sigma} \tag{9.15}$$

which may be integrated over the entire **contact angle**:

$$\mu \int_0^\beta d\beta = \int_{\sigma_2}^{\sigma_1} \frac{d\sigma}{\sigma} \tag{9.16}$$

and the standard rope formula is obtained:

$$\mu = \frac{1}{\beta} \ln \frac{\sigma_2}{\sigma_1} \quad \text{or} \quad \mu = \frac{1}{\beta} \ln \frac{F_2}{F_1} \tag{9.17}$$

The shape of the tool surface does not affect the calculation of friction coefficient, only the contact angle, also known as the **wrap angle** or **angle of wrap**.

> ***Note***: *Equation 9.17 is often used in the reverse sense in sheet-forming analysis. That is, a standard friction coefficient may be used in conjunction with a known geometry to estimate the forces (and thus the stresses and strains via the constitutive equation) in terms of the wrap angle:*

$$F_2 = F_1 \exp(\beta\mu) \tag{9.18}$$

9.8 EXAMPLES OF SHEET FRICTION TESTS

As mentioned earlier, the basic principles guiding the selection of a friction test are simplicity (to separate frictional effects from others) and similitude (to assure that the many variables are similar to the actual operation of interest). In addition, there are many practical considerations involving tooling cost, complexity of the machinery needed, size of the specimens, and so on. In order to illustrate the trade-offs, we consider a few typical tests that have been proposed and used to determine the friction in sheet metal pressing or forming operations.

Figure 9.8 shows several typical geometries for **pinch-type friction tests** of sheet metal. These tests may be one-sided (Figure 9.8*a*) or two-sided[5] (Figures 9.8*b* and *c*), and may have flat, cylindrical, or inclined contact geometries. A **back force** may be applied to introduce a degree of tensile

[5] S. Nakamura, M.Yoshida, A. Nishimoto, *Proc. 15th Biennial Congr. International Deep Drawing Research Group* (May 1998): 77-83.

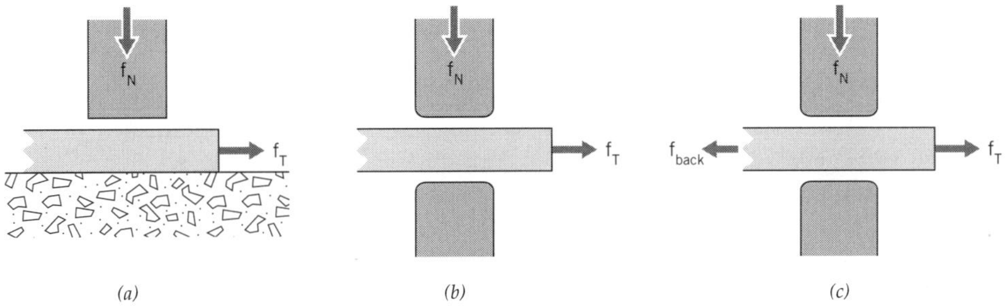

Figure 9.8 Pinch-type friction tests for metal sheets.

deformation that can be controlled independently from the normal force, f_N. Pinch-type devices have the advantage of simple interpretation:[6]

$$\mu = \frac{1}{2}\frac{f_T}{f_N} \quad \text{or} \quad m = \frac{1}{2}\frac{f_T}{\tau_{max} A} \tag{9.19}$$

However, the sticking case is complicated by the unknown contact area if curved dies are used. The disadvantages of such tests are serious. The geometry and deformation patterns induced in a pinch-type apparatus are not similar to the tensile deformation and sliding in press dies, where through-thickness compression is nearly always absent. Pinch-type tests are much more similar to conditions in rolling, drawing, or extrusion operations.

Tests with the most similitude to sheet forming are based on tensile deformation around a radius, and these make use of the rope formula or direct measurement of normal forces. The device most immediately suggested from the rope formula involves drawing over a single radius, as shown in Figure 9.9. In the simplest case, a back force, F_1, is applied, and F_2 is measured to move the sheet. In more complicated arrangements, the wrap angle may be varied as shown in Figure 9.9*b*, and computer control may be used to vary F_1, F_2 and θ throughout the test to obtain combinations of stretch and wrap and their histories. In addition, the radius of the tool can be changed and a roller may be substituted to investigate bending and other non-friction effects.

Devices based on Figure 9.9 offer a great deal of flexibility at the expense of very complicated and specialized equipment. Because of the number of variables present, it is necessary to have a good characterization of the forming operation of interest. However, interpolation is simple in view of the direct measurement of F_1 and F_2, and use of the rope formula.

[6] The factor of $\frac{1}{2}$ is required because the friction force is generated by two contact surfaces.

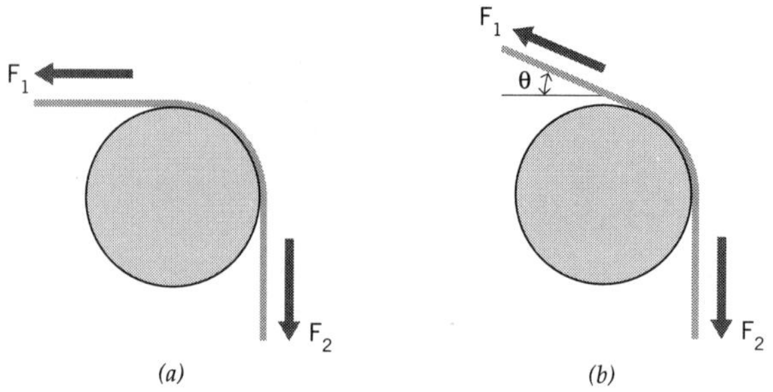

Figure 9.9 Single-bead sheet friction tests.

The strip friction test, Figure 9.10, is similar to the single-bead tests except that it may be performed using a standard tensile testing machine. The analysis of the data available is more complicated because the force acting on the center leg is unknown and must be inferred from the strain measure there, e_2. In fact, once the constitutive equation is known (from separate tensile tests, for example), the two strain measurements (e_1 is assumed to be equal on the two vestical legs) are used to obtain the corresponding forces, and the rope formula is then used to obtain μ.

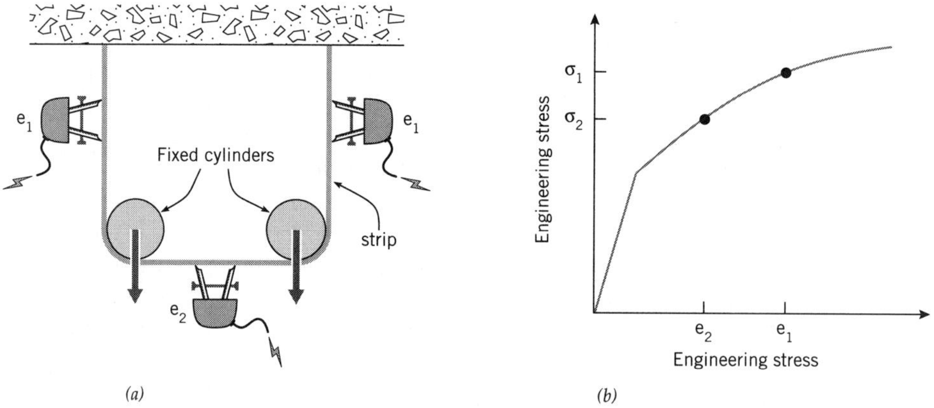

Figure 9.10 (*a*)Strip friction test and (*b*)data interpretation.

While eliminating the need for expensive, complex, and specialized machinery, the strip test has several undesirable characteristics. In terms of similitude, the most serious is the fixed wrap angle, which must be established by arbitrary bending of the flat sheet during installation. Perhaps more serious is the need to know the constitutive equation well (including strain and strain-rate effects) in order to obtain accurate forces. Finally, multichannel data acquisition is needed to record the various strains and forces.

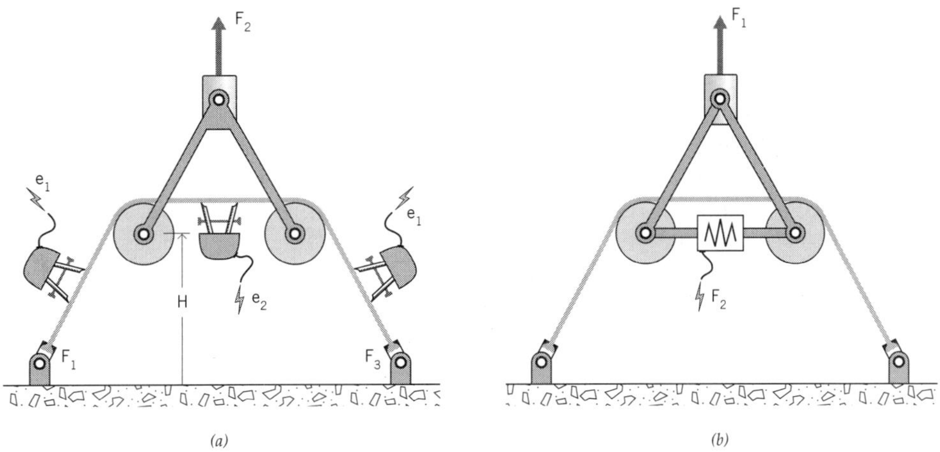

Fig. 9.11 OSU Friction Test and proposed modification

Modifications of the strip friction test address some of these issues while retaining the use of standard tensile testing machinery. **The OSU friction test,** Figure 9.11*a*, used swiveling grips with embedded load cells to allow starting the test from a flat strip and developing the wrap angle in a manner similar to that experienced in a press-forming operation. The data analysis is similar to that for the strip friction test, and the need to know the constitutive equation accurately is the same. The cylindrical tools may be replaced by rollers to investigate the nonfrictional contributions to differences between F_1 and F_2.[7]

Figure 9.11*b* shows a modified OSU friction test that incorporates a load cell between the rollers in order to measure F_2 directly. With this added capability, the data interpretation is direct, and the need for extensometer and measured constitutive equations is eliminated.

The rope formula may be used directly with production testing to estimate the friction coefficient. For example, consider a cross section through a long side of sheet- forming dies, as shown in Figure 9.12*a*. After a part is formed,[8] the strains in the sidewall and bottom may be used with the constitutive equations to obtain an apparent **friction coefficient** in the usual way. In this case, the similitude is ideal—the real forming operation is used to measure friction. The disadvantages lie in the expense and size of production tools and uncertainties introduced by bending, constitutive equation, and strain state,[9] none of which

---

7 For example, our results show that bending is a significant factor in the apparent friction coefficient for r/t ratios less than 5 or 6, where r = radius of curvature and t = sheet thickness.

8 It may also be possible and desirable to stop the forming operation at intermediate punch heights to evaluate friction.

9 Variation in strain state is quite significant when trying to evaluate forces from strains. For example, if the sidewall is in plane strain while the bottom section is uniaxial tension, the ratio of forces determined by assuming uniaxial tension will be in error by a large factor, a factor which may be larger than that corresponding to the effect of friction. See Exercise 9.2 for details.

can be completely controlled in production. These problems may be safely ignored if only an apparent friction coefficient is needed—for example, if the purpose is to rank several lubricants for use with a given sheet metal. Since the geometry and sheet metal are the same, the rankings by apparent friction coefficient should be consistent with a ranking by absolute coefficient.

Figure 9.12*b* shows a simplified friction test used at The Ohio State University based on a small set of dies to simulate production press forming. A small press or double-action testing machine may be used. Besides the use of cheaper tooling and equipment, the test offers several advantages. If very narrow or very wide specimens are used, consistent plane strain or uniaxial tension may be obtained, thus eliminating the large uncertainties attributable to strain state.[10] Different tools may be produced cheaply in order to vary the punch radius so as to reduce bending uncertainties to insignificance.

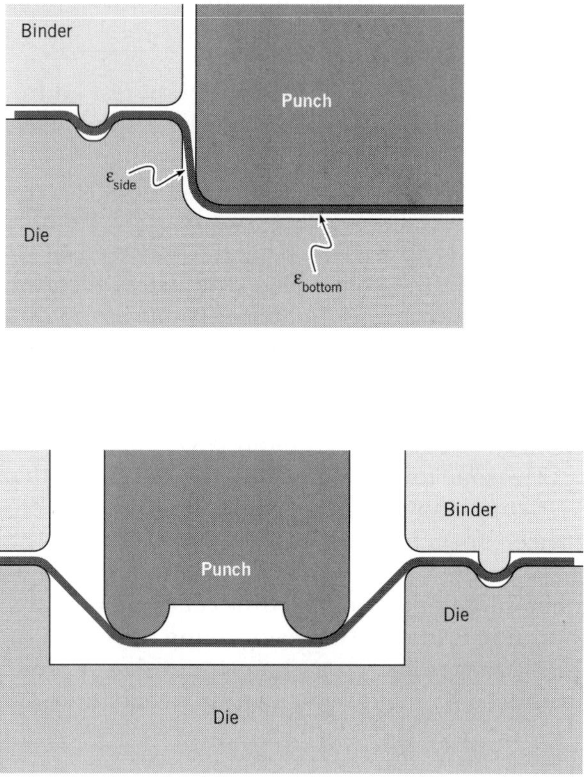

(b)

Figure 9.12 Friction measurement from production tools and similar testing device.

[10] Conceptually, the plane-strain condition is closer to most forming operations, but very wide specimens are required in order to ignore edge effects, and wide specimens require expensive tooling and machinery, and careful alignment.

Exercise 9.2 **Estimate the magnitude of the error in applying the rope formula to measured strains in a deformed strip if the strain state varies along the length.**

Let's assume a von Mises material and take an extreme case, where the bottom of the punch is in uniaxial tension and the sidewall is in the plane-strain tension, as shown. We'll take the wall angle to be $\pi/4$.

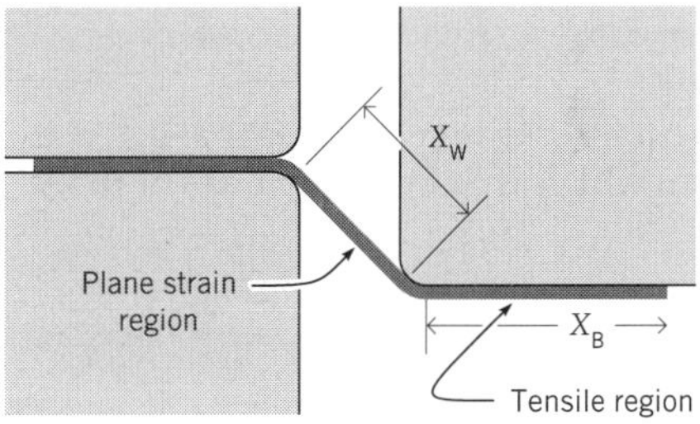

If y is the distance into the paper, then this approximation would hold for $X_w << y << X_B$. For the purposes of this calculation, we'll assume that the friction coefficient is zero, such that $F_{wall} = F_{Bottom}$, and we'll compute the strains (along directions X_w and X_B) required for equilibrium. Then we'll compute the apparent friction coefficient calculated from these strains, assuming a uniaxial tensile state using a tensile constitutive equation such as the following:

$$\bar{\sigma} = K\,\bar{\varepsilon}^{0.25} \tag{9.2-1}$$

First, we compute the axial strains for equilibrium. For the bottom section in uniaxial tension the result is obtained directly from Eq. 9.2-1, because $\sigma_B = \bar{\sigma}$, $\varepsilon_B = \bar{\varepsilon}$:

$$\sigma_B = K\,\varepsilon_B^{\,0.25} \tag{9.2-2}$$

For the sidewall section in plane-strain tension, we use our knowledge of the normality condition (see Chapter 7) to find that $\sigma_1 = 2\sigma_2$ ($\sigma_3 = 0$) for the plane-strain state where $\varepsilon_1 = -\varepsilon_3$, $\varepsilon_2 = 0$. We substitute these expressions into equations of von Mises effective stress and strain to obtain

$$\bar{\sigma} = \frac{1}{\sqrt{2}}\left[\left(\sigma_w - \frac{1}{2}\sigma_w\right)^2 + \left(\frac{1}{2}\sigma_w\right)^2 + (\sigma_w)^2\right]^{\frac{1}{2}} \tag{9.2-3}$$

or

$$\sigma_w = \frac{2}{\sqrt{3}}\,\bar{\sigma} \tag{9.2-4}$$

$$\bar{\varepsilon} = \frac{2}{3}\left[\varepsilon_1^{\,2} + 0^2 + \varepsilon_1^{\,2}\right]^{\frac{1}{2}} \tag{9.2-5}$$

or

$$\bar{\varepsilon} = \frac{2}{\sqrt{3}} \varepsilon_W \tag{9.2-6}$$

such that

$$\sigma_W = \frac{2}{\sqrt{3}} K \left[\frac{2}{\sqrt{3}} \varepsilon_W \right]^{0.25} \tag{9.2-7}$$

The equilibrium condition requires that $\sigma_{wall} = \sigma_{bottom}$ (assuming constant thickness and width), from which we obtain the ratio of the strains in the two sections:

$$(K\varepsilon_B{}^{0.25}) = \frac{2}{\sqrt{3}} K \left(\frac{2}{\sqrt{3}} \varepsilon_W \right)^{0.25} \tag{9.2-8}$$

or

$$\frac{\varepsilon_B}{\varepsilon_W} = \left(\frac{2}{\sqrt{3}} \right)^5 = 2.05 \tag{9.2-9}$$

or

$$\varepsilon_B \approx 2\,\varepsilon_W \tag{9.2-10}$$

Because these are the strains that will be measured, we can now substitute into the tensile constitutive equation to find the apparent friction coefficient caused by failing to take into account the difference in strain state:

$$\mu_{app} = \frac{1}{\beta} \ln \frac{\sigma_W{}^{app}}{\sigma_B{}^{app}} = \frac{4}{\pi} \ln \frac{(\varepsilon_W)^{0.25}}{(2\varepsilon_W)^{0.25}} \tag{9.2-11}$$

$\sigma_B{}^{app}$

$$\mu_{app} = \frac{1}{\pi} \ln \left(\frac{1}{2} \right) = -0.22 \tag{9.2-12}$$

This is not only in error, it has a physically impossible sign. To put the magnitude of this error into perspective, we note that a typical range of friction coefficients for sheet forming is 0.1 to 0.3. Therefore, the uncertainty introduced by unknown strain state must be avoided if any significant result is to be obtained.

9.9 EXAMPLES OF BULK FRICTION TESTS

Bulk forming usually requires large compressive loads and may involve large sliding distances (extrusion, drawing) or only minimal sliding distances (upsetting, heading). While sheet forming is usually carried out at room temperature, bulk forming may be done over a wide range of temperatures, sometimes approaching the melting point of the alloy. Because the conditions are diverse and can be quite different from those found in sheet-forming applications, different tests are generally used. As in sheet-forming tests, the main aspect to look for is how well the test simulates the forming condition in production, and whether the test is easily interpreted.

Ring Compression Test [11]

This test is widely used for evaluating the friction coefficient for forging operations. In this test, a flat ring is compressed plastically between dies. If friction at the interface is zero, both the inner and outer diameters expand. With increasing friction, the rate of increase of the internal diameter decreases, and for high values of friction, the internal diameter actually decreases. The change in the internal diameter as a function of the change in thickness of the specimen has been theoretically calculated for a variety of friction conditions.

Several theoretical analyses have been performed.[12,13,14] In one of these, a computer program has been developed for mathematically simulating the metal flow in ring compression with bulging.[2] Thus ring dimensions for various reductions in height and shear factors, m, can be determined. The shear factor, m, can be determined by comparing the measured dimensions of the specimen to the values in the calibrating figure. Figure 9.13 shows the geometry of the test, and Figure 9.14 shows the computed curves against which the experimental results are compared.

The advantages of the ring test are:

1. No load or strain measurements are needed to evaluate the friction coefficient, so standard presses may be used.
2. The test is simulative, easy to perform, and can be used for a variety of strain rates and temperatures.

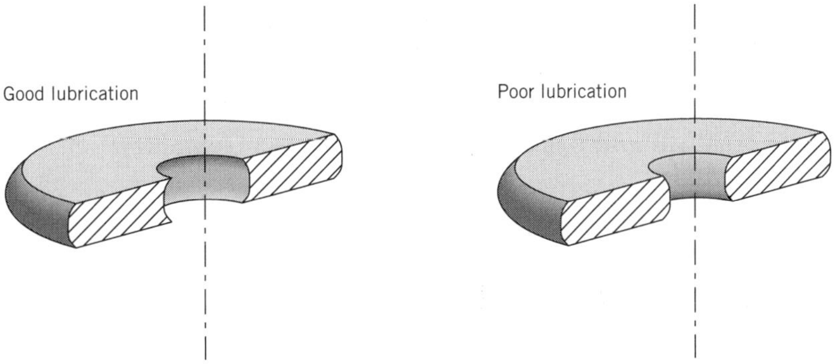

Figure 9.13 Geometry of the ring test.[11]

[11] S. *Kalpakjian, Manufacturing Processes for Engineering Materials* (Reading, Mass: Addison-Wesley, 1984) pp. 203-205.

[12] C. H. Lee and T. *Altan Trans. ASME. J. Engr. Industry* 94(1972):775.

[13] B. Avitzur, *Metalforming: Processes and Analyses* (New York:McGraw-Hill) 1988.

[14] J. B. Hawkyard and W. Johnson, *Int. J. Mech. Sci.* 1967:9–163.

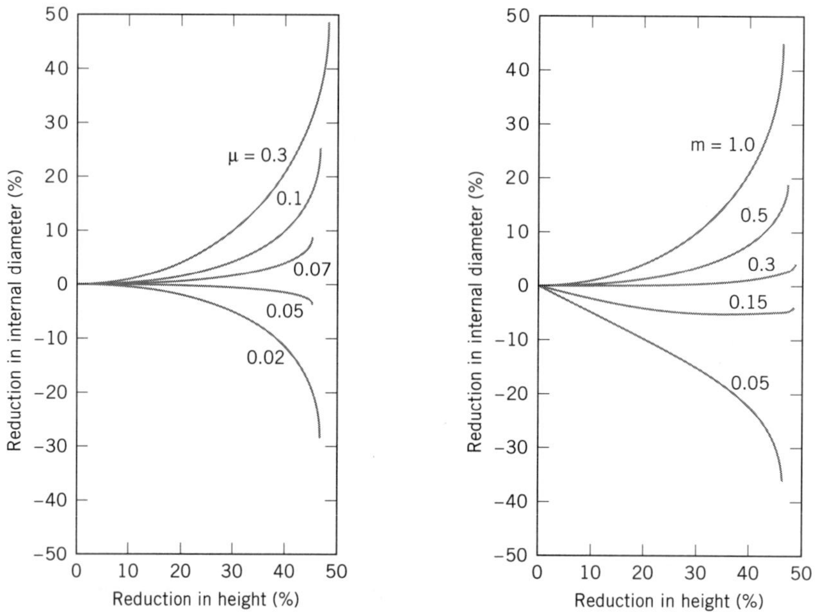

Figure 9.14 Calibration curves used for evaluating the friction coefficient from ring compression test.[11]

The disadvantages are:

1. Because of possible anisotropy introduced by machining or pre-processing, the specimen can acquire an oval shape, in which case an average diameter has to be used.
2. This test is not applicable for operations involving substantial relative motion between the workpiece and the dies under high contact pressures, such as an extrusion operation.

Combined Forward and Backward Flow Extrusion Geometry[15]

This test was developed to study friction conditions for extrusion operations. The geometry of the test is shown in Figure 9.15. The test is used to quantitatively estimate a global friction factor for the test system, by measuring the length of the extruded rod. This length is compared with theoretically computed lengths for a given punch travel to estimate the friction factor. The process was analyzed using an upper bound approach, and the calibration curves for different friction factors were determined by analyzing the flow characteristics of the extruded material. Figure 9.16 shows the results of such comparisons.

---

[15] L. R. Sanchez, K. Jeinmann, and J. M. Story, *Procs. 13th North American Manufacturing Research Conference*, Society of Manufacturing Engineers, Dearborn, Michigan), (May 1985): 110-117.

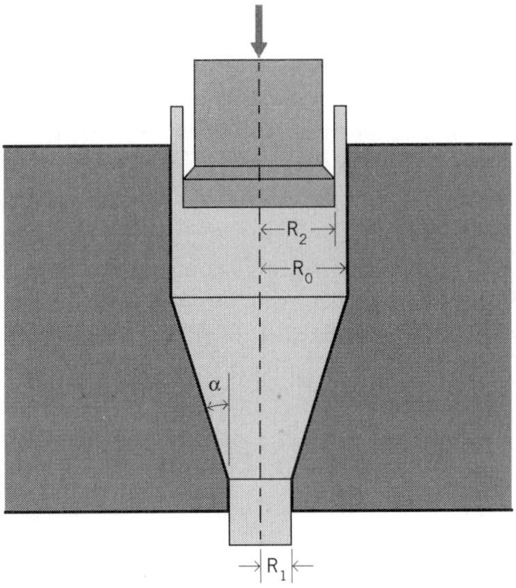

Figure 9.15 Geometry of the backward-and-forward extrusion test.[15]

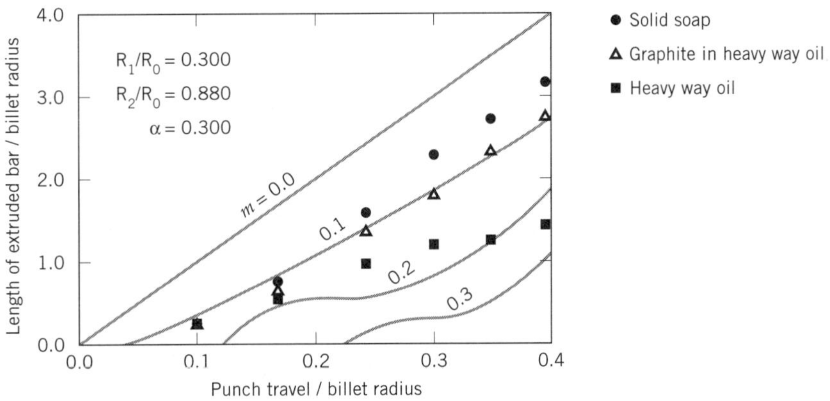

Figure 9.16 Comparison of experimental results with computations for the backward

The advantages of this test are:

1. It closely simulates a production extrusion operation.
2. No load measurements or strain measurements are required to evaluate the friction coefficient.

The disadvantages are:

1. Accuracy of the comparisons depends on the simplifying assumptions used in the upper bound analysis.
2. This test is not very sensitive to small changes in lubricating conditions, and thus is not very useful to rank lubricants.

Double-Backward-Extrusion Geometry for Friction Evaluation [16]

The geometry of this test is shown in Figure 9.17. A billet is forced against a stationary lower punch by a moving upper punch in a stationary die to produce two cups as shown. This test was optimized using the finite element method to design a test geometry (billet diameter, billet height, punch diameter) such that maximum sensitivity in test results was obtained for small changes in lubricating conditions. The ratio of the two cup heights was used as a measure of the friction existing in the system. The difference in the cup heights arises from the difference in relative motion between the two punches and the workpiece and is a measure of lubrication. Figure 9.18 shows the plots of predicted ratios as a function of the punch height for some friction conditions.

To find the calibration curves, an FEM analysis using the program DEFORM[17,18] was carried out. The experimental value of the cup height ratio can then be compared to the predicted ratios to yield the friction coefficient.

The advantages of this test are:

1. The geometry is very sensitive to small changes in lubricating conditions, thereby making the test a good tool to compare lubricant quality.
2. The test is simulative and does not require the use of loads and strains to calculate the friction coefficient.

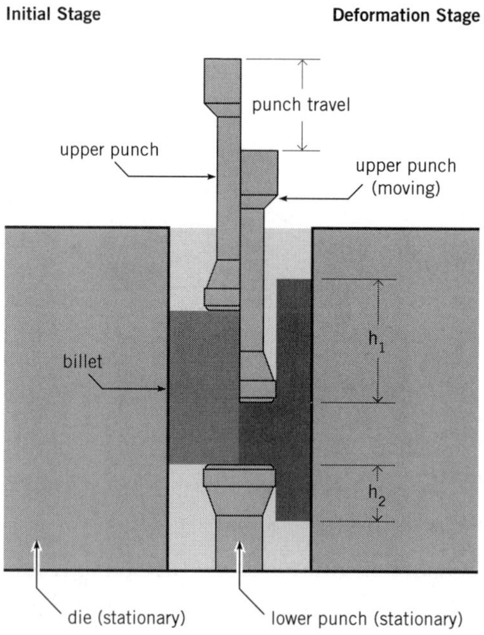

Figure 9.17 Double-backward extrusion geometry.[16]

---

[16] A. Buschhausen, K. J. Weinmann, Y. Lee, and T. Altan, *J. Mat. Proc. Technol.*, 33, 1992: 95-108.

[17] S. Kobayashi, S.-I. Oh, and T. Altan, *Metalforming and the Finite Element Method*, (New York and Oxford, Oxford University Press, 1989).

[18] S.-I. Oh, W. Wu, J. Tang, and A. Vedhanayagam, *J. Mater. Process. Technol.*, 27, 1991: 25-42.

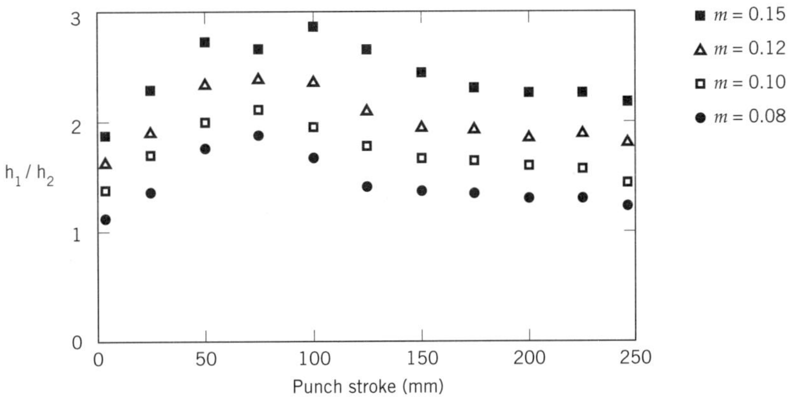

Figure 9.18 Predicted ratios of cup heights versus punch strokes for various values of the friction factor m.[16]

CHAPTER 9 - PROBLEMS

A. Proficiency Problems

1. For the geometry shown in Figure 9.1*b*, find the relationship between the friction coefficient and the ramp angle at which sliding begins.

2. Use the rope formula to obtain the unknown forces in each of the following cases, assuming a friction coefficient of m = 0.2.

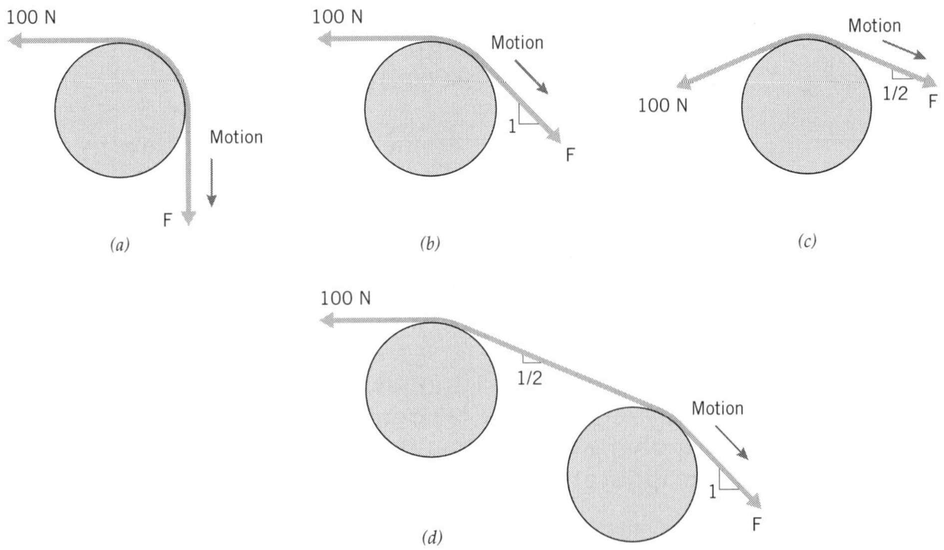

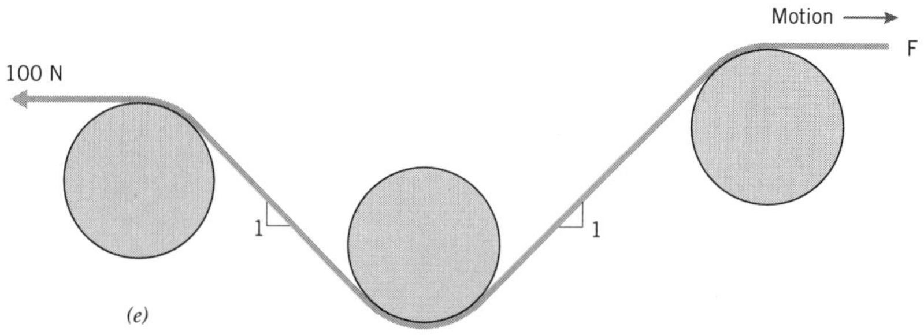

(e)

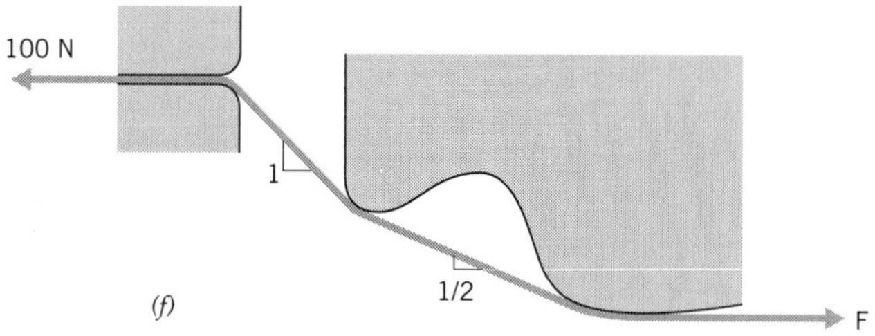

(f)

3. Use the geometries presented in Problem 2 and the following information to find the true *strains* in the region where *F* is applied:

$$A_o = \text{original cross-sectional area of sheet} = 1 \text{ mm}^2$$

$$\bar{\sigma} = 200\,\bar{\varepsilon}^{0.2} \quad \text{(MPa)}$$

Which geometry cannot be deformed as shown?

4. a. A ring compression test is carried out using a standard forging press. The original and final ring dimensions are shown below. Find the value of m or μ from these data using figure 9.14.

Height (original) = 100 mm	Height (final) = 60 mm
Inside radius (original) = 200 mm	Inside radius (final) = 160 mm

b. Given the purpose and use of the ring compression test, which limiting form of friction (Coulomb or sticking) is likely to be the more accurate?

5. What are the friction factors determined for solid soap, graphite in heavy way oil, and way oil from Figure 9.16? What difficulty do you see in applying this test to differentiate lubricant properties for production applications?

6. a. For the double-backward extrusion test, the test is stopped at two punch strokes and the cups removed. Given the following experimental results and using Figure 9.18, determine the friction coefficients and friction factors for this case.

   Punch stroke = 100 mm, h_1 = 80 mm, h_2 = 30 mm

   Punch stroke = 200 mm, h_1 = 120 mm, h_2 = 90 mm

   b. What can you say about the role of sliding distance in determining friction for this configuration, lubricant, and material combination?

B. Depth Problems

7. Derive an equation or equations that could be solved numerically to find ε_1 or ε_2l as a function of H for the test shown in the figure. Assume Coulomb friction, Hollomon hardening ($\sigma = k\ \varepsilon^n$), and uniform tensile strain (and tensile stress) in each leg. For simplicity, assume that the pins have negligible diameter, so that the geometry can be computer simply.

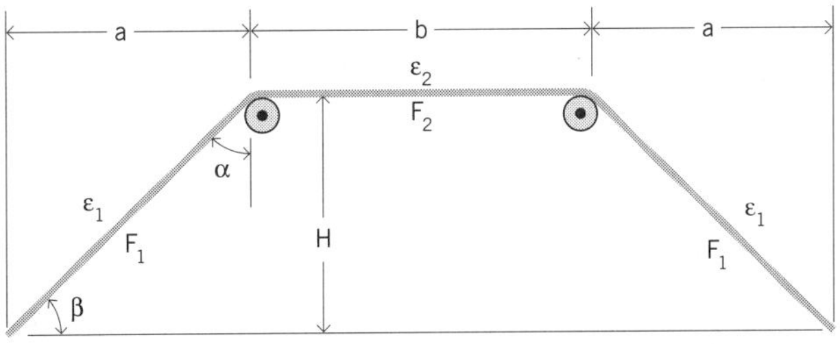

8. Consider now a real test like the one shown in Problem 7.

   a. Why would you expect to see a difference between e_1 and e_2 even for very narrow strips (to insure uniaxial tension) and with nearly perfect lubrication (by using rollers, Teflon, and oil, for example)? How would this effect depend on bend radius and sheet thickness?

   b. Given results from the test described in part a, how would you find the true friction coefficient from a similar test with fixed pins, normal lubrication, and the same material?

9. Use the rope formula and the geometry shown below to determine apparent friction coefficient from F_1 and F_2 for the modified OSU Friction Test (see Fig. 9.11b and dimensions below).

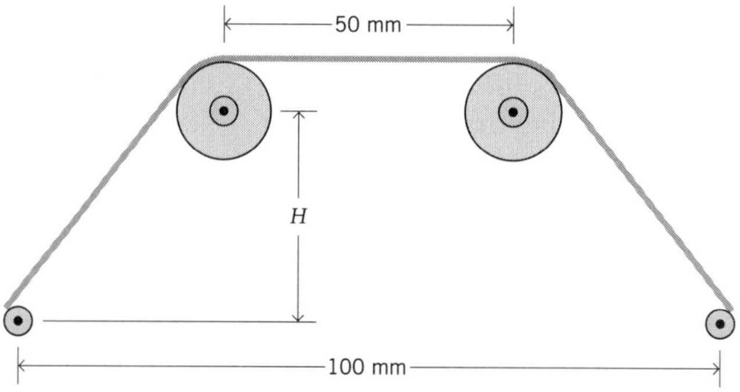

10. a. Explain qualitatively why and how the friction factor or friction coefficient affects the shape change in the ring compression test.

    b. Could you design a plane-strain compression test similar to the ring compression test simply by putting a long square rod between flat platens?

    c. How would you carry out a slab analysis of the ring compression test to generate curves like those shown in Figure 9.14 ?

11. Given the curves shown in Figure 9.14 in terms of m and μ, could you determine a relationship between the normal stress and effective stress for the material used in generating these curves?

C. Numerical Problems

12. Using the formulas derived in Problem 7, plot the relationship between stroke (H) and e_1 or e_2 in the test for m = 0.0, 0.1, 0.2, and 0.3, and n = 0.20 and 0.40.

13. a. Starting from the result and specifications in Problem 4, find the strain rates ε_1 and ε_2 as a function of stroke, assuming that stroke rate is a constant 10 mm/sec.

    b. Find the apparent friction coefficient for a frictionless case where a strain-rate-sensitive material is tested, e.g. m = 0.020, m = 0.10, or m = 0.30. That is, assume that a rate-sensitive constitutive equation is used (from Problem 4) to find μ from measured strains, but that in fact those strains were established by a material exhibiting strain-rate sensitivity.

CHAPTER

10

Classical Forming Analysis

It is instructive and useful to consider a few examples of older, simpler methods of analysis. Until the advent of large-scale digital computation, only these methods were suitable. Today, they are useful for visualizing the physical picture in forming, and are still used for a variety of purposes where computation speed and simplicity are more important than the ability to handle very complex geometries accurately. For example, most real-time control is based on such closed-form solutions.

The examples in this chapter are chosen to be illustrative and informative of the more advanced numerical procedures available today. In particular, a great deal of time is spent on slab analysis of extrusion and drawing, which is very similar in concept to finite element analysis. The results are compared with the simplest computations of ideal work in order to show the additional accuracy and insight provided by even one-dimensional analysis. We include a very brief introduction to the method of slip line fields because this method is occasionally used for plane-strain problems where the expense and time of large computations are not desired. We leave introductions to upper bound and lower bound methods to Chapter 8. We ignore some classical analysis, such as that for deep drawing of a cup, which has little generality and which relies on many unrealistic assumptions in any closed-form treatment.

10.1 IDEAL WORK METHOD

The **ideal work method** is based on calculating the ideal plastic work required to change the shape of a part during a given forming operation, then using an efficiency factor gained by experience to modify this result to include nonideal contributions to the work. As such, it is partially predictive and partially empirical in nature. The method can be used to estimate the power and force requirements for a given forming operation when similar operations have been studied previously.

Ideal Work in Simple Tension

To illustrate the method, let's calculate the ideal work done during the tensile deformation of a rod. For the rod shown in Figure 10.1, we can compute the **total plastic work**[1] of deformation either internally or externally, as follows:

$$\Delta w = \int_{l=l_0}^{l=l} F dl = \pi\left(\frac{d}{2}\right)^2 l \int_{\bar{\varepsilon}=0}^{\bar{\varepsilon}=\bar{\varepsilon}} \bar{\sigma} d\bar{\varepsilon} \equiv \pi\left(\frac{d}{2}\right)^2 \int_{\bar{\varepsilon}_{ij}=0}^{\varepsilon_{ij}=\varepsilon_{ij}} \sigma_{ij} d\varepsilon_{ij} = \text{vol}\int_{\varepsilon=0}^{\varepsilon=\varepsilon} \sigma d\varepsilon \quad (10.1)$$

external work	internal plastic work	internal plastic work	internal tensile work

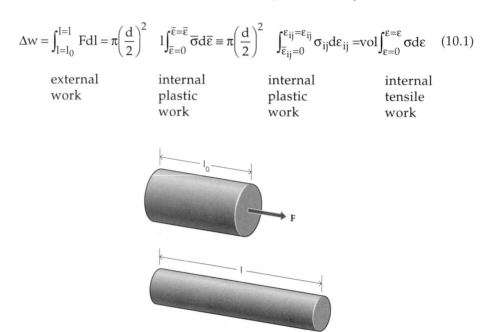

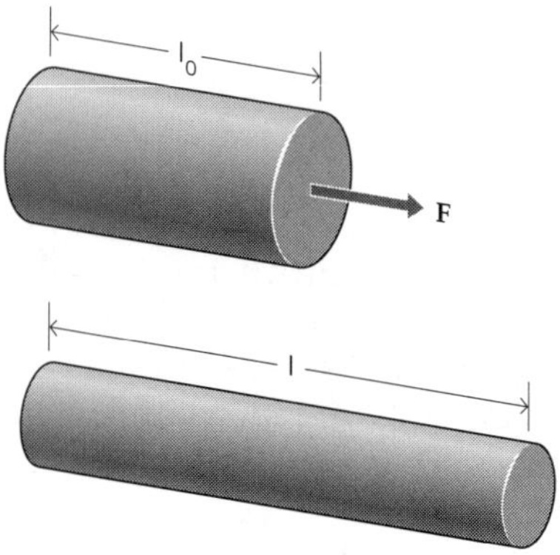

Figure 10.1 Extension of a rod under uniaxial tension.

Mechanical principles (see Chapter 5) tell us that the internal and external work are equal, while the definitions of effective stress and strain satisfy the second equality in Eq. 10.1. The third equality simply recognizes that only the axial stress in an ideal tensile test is non-zero. The connection between internal work and effective stress and strain requires a volume correction because stress and strain are normalized quantities. The total work, however, depends on specimen size. Thus the volume enters the formulation as follows:

$$\Delta w = F dl \text{ (definition of work)}$$

but

$$F = \sigma a \text{ (a = current area) and } dl = l d\varepsilon$$

then

$$\Delta w = F dl = \sigma a \cdot l d\varepsilon = V \cdot \sigma d\varepsilon \quad (10.20)$$

[1] We will not make any distinctions between plastic work and plastic plus elastic work using this approximate method. For all practical forming operations involving large plastic strains, the error in neglecting the elastic component of strain and work will be negligible.

To compute the **ideal work** for the tensile deformation shown in Figure 10.1 from the material behavior, we must quantify the material response. Assuming a Hollomon-type law ($\sigma = k\varepsilon^n$), we can find the ideal work:

$$\Delta w_i = \pi\left(\frac{d}{2}\right)^2 \; l\int_o^{\varepsilon} k\varepsilon^n d\varepsilon = \frac{K\bar{\varepsilon}^{n+1}}{n+1}, \quad \text{where } \bar{\varepsilon} = \ln\frac{l}{l_o} \tag{10.3}$$

Thus the ideal work for a tensile test is readily computed. Note that the ideal work depends only on the starting and final configurations, whereas the real work depends on how the forces were applied (*Fdl* in Eq. 10.2). The external work includes ideal work, **redundant work** (i.e., work done by deformation that is unnecessary to achieve the final shape), and **frictional work** (which is dissipated at the surface of the workpiece and tools and does not contribute to deformation).

Exercise 10.1 **Find the ideal work, redundant work, and actual work done in a two-step forming operation of a square rod 10 mm × 10 mm × 100 mm:**

1. **Uniaxial tension to a length of 110 mm, followed by**
2. **Uniaxial compression to a length of 105 mm**

Assume that there is no friction work and that the material obeys Hollomon's equation:

$$\bar{\sigma} = 500\,\bar{\varepsilon}^{\,0.25} \text{ (MPa)}$$

First, let's compute the actual work done by considering the operation in two steps. The change in effective strain up to the end of the first step is

$$\Delta\bar{\varepsilon}_1 = |\Delta\varepsilon_{axial}| = \ln\frac{110}{100} = 0.095 \tag{10.1-1}$$

because the $\Delta\bar{\varepsilon}$ is defined by a proportional, tensile path. The change for the second step is also proportional and compressive (negative tensile), so we get

$$\bar{\varepsilon}_2 = |\Delta\varepsilon_{axial}| = \ln\frac{105}{100} = 0.049 \tag{10.1-2}$$

where we note that it is impossible to do negative work or have a negative change in $\Delta\bar{\varepsilon}$, so in this case the signs of $\Delta\bar{\varepsilon}$ and $\Delta\varepsilon_{axial}$ are opposite. (Remember, the work done in each case is positive because the compression work requires negative forces and negative length changes.) Then the actual work done (there is no sliding, so the frictional work is zero) over the entire deformation is computed from the total change in effective strain:

$$\Delta\bar{\varepsilon}_{total} = \Delta\varepsilon_1 + \Delta\varepsilon_2 = 0.095 + 0.049 = 0.144 \tag{10.1-3}$$

$$\frac{\Delta w_{actual}}{\text{volume}} = \int_0^{0.144} 500\bar{\varepsilon}^{0.25}\, d\bar{\varepsilon} = \frac{500}{1.25}(0.144)^{1.25} = 35.5 \text{ MPa} \tag{10.1-4}$$

$$w_{actual} = 35.5 \text{ MPa} \times 10 \text{ mm} \times 10 \text{ mm} \times 100 \text{ mm} = 355 \text{ N} \cdot \text{m} \tag{10.1-5}$$

The ideal work calculation for the entire operation ignores what happens during the cycle and is based only on original and final shapes. The ideal change of $\Delta\bar{\varepsilon}$ expected on this basis is simply:

$$\Delta\bar{\varepsilon}_{overall} = \ln\frac{105}{100} = 0.049 \tag{10.1-6}$$

and the overall ideal work is

$$\frac{\Delta w_{ideal}}{\text{volume}} = \int_0^{0.049} 500\,\bar{\varepsilon}^{0.25}\, d\bar{\varepsilon} = \frac{500}{1.25}(0.049)^{1.25} = 9.25 \text{ MPa} \tag{10.1-7}$$

$$\Delta w_{ideal} = 9.2 \text{ MPa} \times 10\text{mm} \times 10\text{mm} \times 100\text{mm} = 92 \text{ N} \cdot \text{m} \tag{10.1-8}$$

The redundant work in the operation is simply the actual work less the ideal work:

$$\Delta w_{redundant} = \Delta w_{actual} - \Delta w_{ideal} = 263 \text{ N} \cdot \text{m} \tag{10.1-9}$$

because no friction work was done. Note that if this forming operation had a third step added thatbrought the length of the rod back to 100 mm, then *no* ideal work would have been done and instead all the work would have been redundant.

Wire Drawing or Round Extrusion

The ideal work done during a uniform tensile test is, in fact, the entire work done, at least while the deformation is uniform. Because there is no contact or friction, there is no surface work done and the assumed uniform, proportional strain is correct.

For real forming operations (such as wire drawing or extrusion, for example—Figure 10.2), contact, friction, and **redundant deformation** are always present, thus affecting the **actual work** of deformation. Although the division is a bit arbitrary, we can picture the actual work as composed of the three components shown in Figure 10.3.

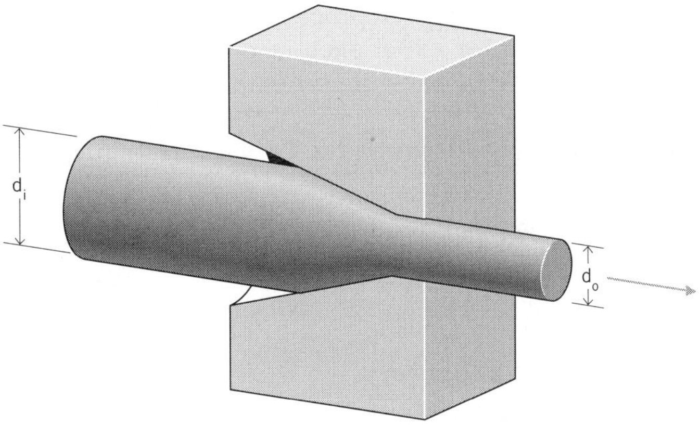
d_i
d_o

Figure 10.2 Schematic of a wire-drawing or extrusion operation.

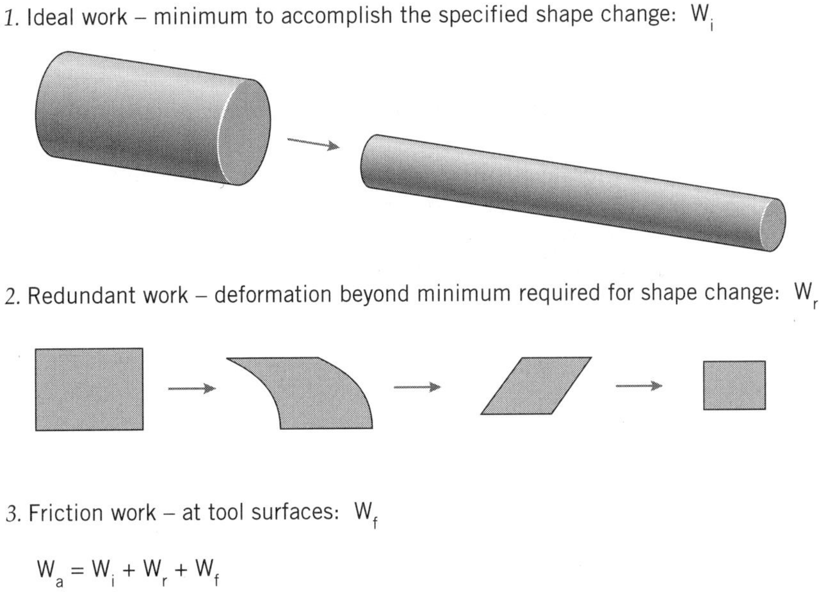
1. Ideal work – minimum to accomplish the specified shape change: W_i
2. Redundant work – deformation beyond minimum required for shape change: W_r
3. Friction work – at tool surfaces: W_f
$W_a = W_i + W_r + W_f$

Figure 10.3 Classification of mechanical work dissipated during forming.

Furthermore, we define a useful quantity, the **efficiency** of the process, as follows:

$$\text{efficiency} = \eta = \frac{\text{ideal work}}{\text{actual work}} \tag{10.4}$$

The efficiency of a given process, die material, machine, and so forth, must be known from prior experience in order to apply the ideal work method. Typical values of efficiency are shown in Table 10.1.

TABLE 10.1 Typical Efficiencies for Various Metal-forming Operations

Uniaxial tension	~ 1.0
Forging[2]	0.20–0.95
Flat rolling	0.80–0.90
Deep drawing	0.75–0.80
Wire drawing	0.55–0.70
Extrusion	0.5–0.65

Note: *It is interesting to consider what happens to the deformation work done during a forming operation. In fact, 90%-95% of the plastic work done is dissipated as heat to the surroundings during cold forming and close to 100% during hot forming. In cold forming, the remaining 5%-10% of the deformation work goes to change the metallurgical structure of the material in the form of more dislocations, grain boundary area, cell boundaries, and so on. During annealing, this stored energy can be recovered in the form of heat when the structure reverts to a lower-energy form.*

To illustrate the application of the ideal work method to a nonideal process such as wire drawing or extrusion (Figure 10.2), we ask what are the power and force requirements for a drawing machine ("draw bench") or extrusion machine, given the following information:

$$d_i = 0.25'' \qquad \sigma = k\varepsilon^n$$
$$d_o = 0.15'' \qquad n = 0.25$$
$$\eta = 0.75 \qquad k = 50{,}000 \text{ psi.}$$

We consider a volume element of material, $\pi(d_i/2)^2 dl_i$, on the inlet side of the dies, corresponding to a volume element $\pi(d_o/2)^2 dl_o$, on the outlet side of the dies. Since the volume does not change during plastic deformation, $dl_o\, d_o^2 = dl_i\, d_i^2$. For this volume element of material, we can calculate the ideal and actual work done:

$$\bar{\varepsilon} = \ln \frac{dl_o}{dl_i} = 2 \ln \frac{d_i}{d_o} = 2 \ln (1.67) = 1.02$$

[2] The very low efficiency for forging occurs in thin parts, where the dissipation is by friction over the large surfaces. Simple upsetting, on the other hand, approaches uniaxial tension except for small friction at the contact areas.

per unit volume of material:

$$\Delta w_i = \frac{k\bar{\varepsilon}^{n+1}}{n+1} \frac{50{,}000 \text{ psi } (1.02)^{1.25}}{1.25} = 41{,}100 \frac{\text{in.} - \text{lb}}{\text{in.}^3}$$

$$\Delta w_a = \frac{\Delta w_i}{\eta} = \frac{41{,}100}{0.75} = 54{,}800 \frac{\text{in.} - \text{lb}}{\text{in.}^3}$$

$$\text{work per volume element: } \quad 54{,}800 \cdot \pi \left(\frac{d_i}{2}\right)^2 dl_i = 54{,}800 \cdot \pi \left(\frac{d_o}{2}\right)^2 dl_o \qquad (10.5)$$

In order to calculate the external load required, we equate the external work with the calculated internal work for a given material volume. For the case of **extrusion**, a pressure is applied on the inlet side of the material. The pressure acts on an area $\pi(d_i/2)^2$ through a distance of dl_i, such that the external work done is $P_{extrusion}\ \pi(d_i/2)^2 dl_i$. We equate this work to the internal work for this volume and find that the extrusion pressure is simply equal to the actual work per unit volume ($P_{extrusion} = W_a$):

$$P_{extrusion}\ \pi \left(\frac{d_i}{2}\right)^2 dl_i = 54{,}800 \cdot \pi \left(\frac{d_i}{2}\right) dl_i \ (\text{in} - \text{lb})$$

(external work) (internal work)

$$P_{extrusion} = 54{,}800 \text{ psi} \qquad (10.6)$$

The method consists of finding the ideal work based on original and final shapes, applying an efficiency factor to find the actual work, then finding external quantities such as force or pressure by equating material work to external work for a given volume element.

A similar result is obtained for the **drawing** case, where a force or **traction** (external force per unit area) is applied on the outlet side of the workpiece in order to draw the material through the dies. In this case, the same volume element is considered, but in terms of the material after deformation:

$$\sigma_{draw}\ \pi \left(\frac{d_o}{2}\right)^2 \cdot dl_o = f_{draw} \cdot dl_o = 54{,}800 \cdot \pi \left(\frac{d_o}{2}\right) dl_o \ (\text{in.} - \text{lb})$$

(external work) (internal work)

$$\sigma_{draw} = W_a = 54{,}800 \text{ psi}, \quad \text{or} \quad f_{draw} = \sigma_{draw} \cdot A_o = 54{,}800 \cdot \pi \left(\frac{d_o}{2}\right)^2 \text{in.}^2 \qquad (10.7)$$

Plane-Strain Extrusion or Drawing

The wire-drawing or extrusion case has one simplification because of the similarity to a tensile test, namely, that the ideal effective strain is equal to the axial strain. For other geometries, we must find the effective strain based on a given yield surface for the ideal deformation in order to compute the ideal work.

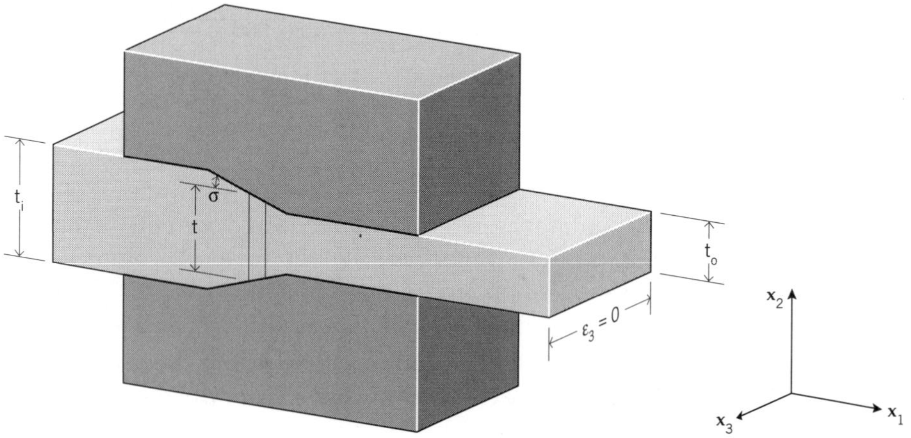

Figure 10.4 Schematic of plane-strain drawing operation.

Consider a plane-strain extrusion or drawing operation, as shown in Figure 10.4. The only difference between this geometry and the wire-draw or extrusion one is the material constraint; that is, there is no deformation in the $\mathbf{x}_3$ direction. Then, because the volume is constant, $\varepsilon_1 = -\varepsilon_2$, we can calculate the three principal strains and the effective strain (as always, assuming a proportional path to obtain the ideal work, and a von Mises material):

$$\varepsilon_1 = -\varepsilon_2 = \ln\left(\frac{t_i}{t_o}\right)$$

$$\bar{\varepsilon} = \left\{\frac{2}{3}\left[\ln\left(\frac{t_i}{t_o}\right)^2 + \ln\left(\frac{t_i}{t_o}\right)^2 + 0\right]\right\}^{1/2} = \left[\frac{4}{3}\ln\left(\frac{t_i}{t_o}\right)^2\right]^{1/2} = \frac{2}{\sqrt{3}}\ln\left(\frac{t_i}{t_o}\right) \tag{10.8}$$

Using the same parameters as for the wire-drawing/extrusion problem posed above, we obtain a different answer:

$t_i = 0.25''$	$\sigma = k\varepsilon^n$
$t_o = 0.15''$	$n = 0.25$
$\eta = 0.75$	$k = 50{,}000$ psi

$$\bar{\varepsilon} = \frac{2}{\sqrt{3}} \ln\left(\frac{0.25}{0.15}\right) = 0.59$$

$$\Delta w = \frac{K\bar{\varepsilon}^{n+1}}{n+1} = \frac{50{,}000 \text{ psi } (0.59)^{1.25}}{1.25} = 20{,}700 \frac{\text{in.} - \text{lb}}{\text{in.}^3} \tag{10.9}$$

However, the equivalence of internal and external work is the same, such that $P_{extrusion} = W_a$ and $\sigma_{draw} = W_a$.

Limiting Draw Reduction

One of the most classical calculations prior to large-scale computation involves the limiting draw reduction for wire drawing or other shape drawing. The idea behind it is very simple, although the calculation based on the ideal work method is not conceptually consistent.

Consider, for example, the wire-drawing operation in Figure 10.2. We can see that the operation will be impossible if the stress required to draw material through the dies (the drawing stress, σ_d) is greater than the stress to deform or break[3] the material (σ_{def}) already through the dies. By estimating the flow stress of the material outside of the dies using the hardening law, and by estimating the drawing stress using the ideal work method, we can find the critical draw reduction, when the two forces are just equal:

$$\sigma_{def} = \sigma_{draw} \text{ (limiting draw reduction)} \tag{10.10}$$

$$\sigma_{def} = \bar{\sigma} \approx K\bar{\varepsilon}^n = K\left[2\ln\left(\frac{d_i}{d_o}\right)\right]^n \tag{10.11}$$

$$\sigma_{draw} = W_a = \frac{k\bar{\varepsilon}^{n+1}}{\eta(n+1)} \tag{10.12}$$

so,

$$K\bar{\varepsilon}^{*n} = \frac{\bar{\varepsilon}^{*n+1}}{\eta(n+1)}, \quad \text{or} \quad \bar{\varepsilon}^* = \eta(n+1) \tag{10.13a}$$

$$\left(\frac{d_i}{d_o}\right)^* = \exp\left[\frac{\eta(n+1)}{2}\right] \tag{10.13b}$$

---

[3] Strictly speaking, it might be possible to deform the material already through the dies in tension by a certain amount and thus harden it, raise the load-carrying capability of the rod, and then continue to pull material through the dies. In fact, this is undesirable in practice because it would lead to a nonuniform wire, and it usually is not possible because the deformation through the dies generally is beyond the uniform tensile strain, such that further deformation produces a continuously dropping force. Thus, practically, the material coming out of the die cannot deform further in any successful forming operation.

where the * indicates the maximum draw ratio or ideal effective strain attainable.

Equation 10.13 is interesting because it shows that the largest reduction depends on the efficiency of the process and the work hardening of the material. However, there are some conceptual difficulties with this classical result. The main problem lies in the approximate equality shown in Eq. 10.11. The result is approximate because the ideal strain is used to estimate the material flow stress outside of the die, although we know that redundant deformation (related to η) will produce additional hardening. In Eq. 10.13 the drawing stress is corrected for friction and redundant deformation by use of the efficiency factor, but the flow stress is uncorrected. Ignoring this inconsistency induces only a small error because the hardening is not large near the limit strains in any case. See Exercise 10.2 and Table 10.2 for verification of this statement.

Exercise 10.2 **Derive a more realistic expression for the limiting draw ratio by assuming that the redundant work is equal to the frictional work.**

With the improved assumption, the redundant work may be computed in terms of the ideal work:

$$W_a = \frac{W_i}{\eta} = W_i + W_r + W_f, \text{ but } W_r = W_f \tag{10.2-1}$$

so,

$$\frac{W_i}{\eta} = W_i + 2\,W_r. \tag{10.2-2}$$

$$W_r = \left(\frac{1-\eta}{2\eta}\right) W_i \tag{10.2-3}$$

The part of the work that contributes to deformation and hardening is $W_i + W_r$, because frictional work is dissipated as heat at the surface. The deformation work is

$$W_i + W_r = W_{def} = W_i + \left(\frac{1-\eta}{2\eta}\right) W_i = \left(\frac{1+\eta}{2\eta}\right) W_i$$

We now find the final effective strain of the material by equating the deformation work as computed by the efficiency factor to that determined by integration of the hardening curve (using Eq. 10.3):

$$W_{def}\,(\text{from known } \eta) = \frac{1+\eta}{2\eta} W_i = \left(\frac{1+\eta}{2\eta}\right)\left(\frac{K}{n+1}\right)\bar{\varepsilon}_i^{\,n+1} \tag{10.2-4}$$

$$W_{def}\,(\text{in terms of } \bar{\varepsilon}_a) = \int_0^{\bar{\varepsilon}_a} \bar{\sigma}\, d\bar{\varepsilon} = \left(\frac{K}{n+1}\right)\bar{\varepsilon}_a^{\,n+1} \tag{10.2-5}$$

which obtains:

$$\bar{\varepsilon}_a = \left(\frac{1+\eta}{2\eta}\right)^{\frac{1}{n+1}} \bar{\varepsilon}_i \tag{10.2-6}$$

$$\bar{\sigma}_a = K\bar{\varepsilon}_a^n = K\left(\frac{1+\eta}{2\eta}\right)^{\frac{n}{n+1}} \bar{\varepsilon}_i^n \tag{10.2-7}$$

The limiting draw reduction is then determined as before, by comparing the strength of the material leaving the die, $\bar{\sigma}_a$, with the draw stress required to pull the material through the die, σ_d:

$$\bar{\sigma}_a = K\left(\frac{1+\eta}{2\eta}\right)^{\frac{n}{n+1}} \bar{\varepsilon}_i{}^n = \frac{1}{\eta}\left(\frac{K}{n+1}\right)^{\bar{\varepsilon}_i{}^{n+1}} = \sigma_d \tag{10.2-8}$$

such that

$$\bar{\varepsilon}_i{}^* = \eta\,(n+1)\left(\frac{1+\eta}{2\eta}\right)^{\frac{n}{n+1}} \tag{10.2-9}$$

or, written in terms of the geometry of the drawing operation:

$$\left(\frac{d_i}{d_o}\right)^* = \exp\left[\frac{\eta\,(n+1)}{2}\left(\frac{1+\eta}{2\eta}\right)^{\frac{n}{n+1}}\right] \tag{10.2-10}$$

where the * indicates the maximum draw reduction or maximum ideal effective strain attainable.

Table 10.2 compares computed limiting draw reductions from the standard expression (Eq. 10.13) and more realistic one (Eqs. 10.2–10). The table shows that both the strain hardening exponent (n) and efficiency (η) have large influences on the draw reduction that is possible, but the choice of expressions is not very significant. In either case, we can expect to achieve draw reductions in the range of 1.4 to 1.8 under normal conditions.

TABLE 10.2 Computed Limiting Draw Reductions

Efficiency, *material*	η	n	$\left(\frac{d_i}{d_o}\right)^*_{std}$ *(Eq. 10.13)*	$\left(\frac{d_i}{d_o}\right)^*_{improved}$ *(Eqs. 10.2–10)*	*Difference* *(%)*
Ideal, brass	1.00	0.50	2.12	2.12	0
Ideal, steel	1.00	0.25	1.87	1.87	0
Medium, steel	0.70	0.25	1.55	1.58	2
Low, steel	0.50	0.25	1.37	1.40	2
Low, aluminum	0.50	0.15	1.33	1.41	6
Very. low, zinc	0.25	0.00	1.15	1.17	2

10.2 SLAB CALCULATIONS (1-D VARIATIONS)

The ideal work method is often unsatisfactory because it relies on a completely empirical value of the efficiency, which takes into account friction, redundant deformation, and any other errors in computation. In essence, the ideal work method resembles a scaling factor that depends on the volume of the part and on the overall shape change, which enter through the uniform strain ratios.

To extend this analysis in the simplest possible way, we consider a one-dimensional variation in stress and strain, while assuming a homogeneous state within a thin slice, or **slab**, normal to this direction. In fact, we have already performed such a calculation[4] in Chapter 1, in order to analyze the tensile test in 1-D. In general, we make the following inherent assumptions in setting up a slab analysis:

- 1-D variation in field quantities only (assume the $\mathbf{x}_1$ direction)
- principal directions are constant, known, and include $\mathbf{x}_1$

[4] The calculation for a tensile test is particularly simple because no friction enters the analysis. However, the basic assumptions and approach are the same.

In addition, other assumptions can be used to simplify the numerical result, but are not required by the basic method:

- fully plastic material (yield criterion is met throughout the part)
- constant flow stress (no strain hardening)
- geometric simplifications (vanishingly small angles, for example)
- certain material and friction models
- simplified boundary conditions

The basic steps in applying the slab method are as follows:

1. Identify the direction with the most significant variation in stress and strain.
2. Consider the equilibrium of a slab of material normal to this direction, including stresses caused by contact and friction.
3. Derive a differential equation relating for the 1-D variation of stress.
4. Use plasticity to reduce to one unknown function.
5. Apply initial and boundary conditions.
6. Solve the single ordinary differential equation to obtain $\sigma_1(x_1)$.

Plane-Strain Drawing: Closed-Form Solutions

Let's solve a typical problem using the slab method. A plane-strain drawing operation (for example, the drawing of a thin sheet) with straight, funnel-shaped dies is shown in Figure 10.4. Also shown is a general slab of material within the die region. The direction of most significant variation is along the direction of material motion, along which we expect a variation of stress from zero (inlet side) to σ_d, the drawing stress, on the outlet side. Because there is no strain or variation of any kind in the $\mathbf{x}_3$ direction, we will take this dimension to be unity in order to simplify the appearance of our force balance and areas.

We will make the following assumptions in order to find a simple, closed-form solution:

- the angle is very small relative to unity
- the material deforms throughout at a constant flow stress, $\bar{\sigma}$
- always-slipping Coulomb friction is present at the boundaries ($\tau_f = \mu P$)

Figure 10.5 illustrates the core of the problem statement. Figure 10.5*a* shows the general slab and the forces acting on it: drawing stress and its variation across the slab, the contact pressure, and the die friction. Figure 10.5*b* shows the differential geometry of the contact region for resolution of forces in the $\mathbf{x}_2$ direction, while Figures 10.5*c* and 10.5*d* are force resolution diagrams used to find equilibrium relationships in the $\mathbf{x}_1$ and $\mathbf{x}_2$ directions. Starting with $\mathbf{x}_2$, we obtain:

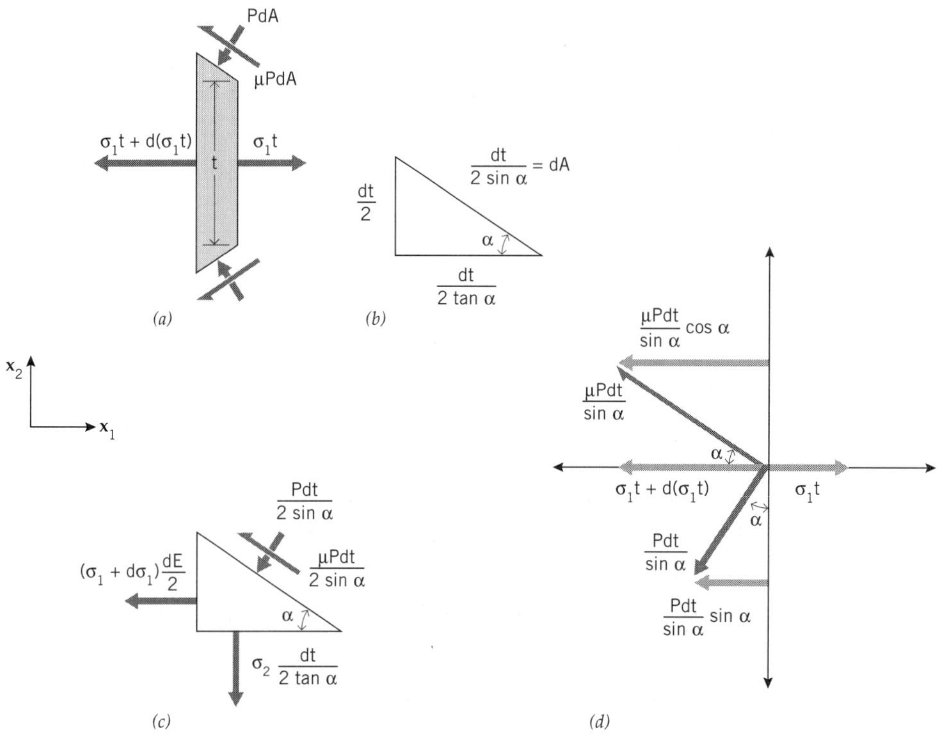

Figure 10.5 Differential slab geometry and force balance for plane-strain drawing.

$$\Sigma F x_2 = 0 = \mu P \frac{dt}{2} - \frac{Pdt}{2\tan\alpha} - \sigma_2 \frac{dt}{2\tan\alpha} \tag{10.14}$$

so,

$$\sigma_2 = P(\mu \tan\alpha - 1) \approx -P \quad (\text{small}\,\alpha) \tag{10.15}$$

In order to set up the governing differential equation, we consider the equilibrium of the slab in the direction of variation, $\mathbf{x}_1$, again using Figure 10.5, in particular the force resolution diagram for the entire slab in this direction, Figure 10.5d:

$$\Sigma F x_1 = 0 = \sigma_1 t - \sigma_1 t - d(\sigma_1 t) - \frac{\mu Pdt}{\tan\alpha} - Pdt \tag{10.16}$$

Substituting for P ($\approx \sigma_2$, small α approximation), combining terms, and noting that $d(\sigma_1 t) = \sigma_1 dt + t d\sigma_1$:

$$\sigma_1 dt + t d\sigma_1 = \sigma_2 (1 + \mu \cot\alpha)\, dt \tag{10.17}$$

or

$$d\sigma_1 = \left[\sigma_2\,(1+B) - \sigma_1\right]\frac{dt}{t}, \tag{10.18}$$

where $B \equiv \mu \cot \alpha$.

Note, however, that Equation 10.18 involves one differential equation and two unknown functions, $\sigma_2(t)$ and $\sigma_1(t)$. But σ_2 and σ_1 cannot be independent because the stress state must lie on the yield surface for continued plastic deformation, and the strain ratios in the slab must satisfy the assumed symmetry of the problem, namely plane strain. We must therefore use knowledge of the material response to find $\sigma_2(\sigma_1)$, so that there remains one differential equation and one unknown.

For plane strain (and proportional straining) as shown in Figure 10.4, $\varepsilon_1 = -\varepsilon_2$, and $\varepsilon_3 = 0$. Assuming an effective stress of $\bar{\sigma}$, and applying the normality condition for a von Mises material, we can solve for σ_2 in terms of σ_1:

$$d\varepsilon_3 = 0 = d\lambda\,(2\sigma_3 - \sigma_1 - \sigma_2) \quad \text{(normality condition)} \tag{10.19}$$

so,

$$\sigma_3 = \frac{\sigma_1 + \sigma_2}{2} \tag{10.20}$$

$$\bar{\sigma} = \frac{1}{\sqrt{2}}\left[(\sigma_1 - \sigma_2)^2 + (\sigma_1 - \sigma_3)^2 + (\sigma_2 - \sigma_3)^2\right]^{1/2}$$

$$\text{(definition of von Mises effective stress)} \tag{10.21}$$

Substituting Eq. 10.20 into Eq. 10.21 obtains

$$\bar{\sigma} = \frac{\sqrt{3}}{2}\left|\sigma_1 - \sigma_2\right| = \frac{\sqrt{3}}{2}(\sigma_1 - \sigma_2) \quad (\text{because } \sigma_1 > \sigma_2) \tag{10.22}$$

or,

$$\sigma_2 = \sigma_1 - H, \quad \text{where } H \equiv \frac{2}{\sqrt{3}}\bar{\sigma}, \text{ a constant for this problem} \tag{10.23}$$

We can now remove $\sigma_2(t)$ in Eq. 10.18, replacing it by the function $\bar{\sigma}(t)$, which is known from the material constitutive equation, $\bar{\sigma} = f(\bar{\varepsilon})$. In this example, we have assumed that the material does not strain harden, so H and $\bar{\sigma}$ are constants, independent of t. This substitution yields our final differential equation for $\sigma_1(t)$, ready for application of boundary conditions and solution:

$$d\sigma_1 = [(1+B)(\sigma_1 - H) - \sigma_1]\frac{dt}{t} \tag{10.24}$$

or

$$\frac{d\sigma_1}{B\sigma_1 - H(1+B)} = \frac{dt}{t} \tag{10.25}$$

Because B and H are constants (straight dies and no work hardening), we are able to separate the variables as shown in Eq. 10.25 and may therefore solve the problem by direct integration, once the boundary conditions are known. The boundary conditions are simply that when $t = t_i$, $\sigma_1 = 0$ (because there is no force restraining the flow of material on the inlet side of the dies), and at $t = t_0$, $\sigma_1 = \sigma_d$ the unknown draw stress. Applying these conditions, we can solve the definite integral suggested by Eq. 10.25:

$$\int_{\sigma_1=0}^{\sigma_1=\sigma_d} \frac{d\sigma_1}{B\sigma_1 - H(1+B)} = \int_{t=t_i}^{t=t_o} \frac{dt}{t} \tag{10.26}$$

$$\frac{1}{B}\ln\left[B\sigma_1 - H(1+B)\right] \Big/_0^{\sigma_d} = \ln\ t \Big/_{t_i}^{t_o} \tag{10.27}$$

$$\frac{1}{B}\ln\left[\frac{B\sigma_1 - H(1+B)}{-H(1+B)}\right] = \ln\frac{t_o}{t_i} \tag{10.28}$$

By exponentiating each side and rearranging, we obtain the slab result for the plane-strain drawing stress using straight dies, constant hardness material, and Coulomb friction:

$$\sigma_d = \frac{H(1+B)}{B}\left[1 - \left(\frac{t_o}{t_i}\right)^B\right] \tag{10.29}$$

$$\text{where } H = \frac{2}{\sqrt{3}}\bar{\sigma} \text{ and } B = \mu\ \cot\ \alpha \text{ are constant}$$

Thus we have calculated the drawing stress without the need for empirical knowledge about similar operations, as was required for the ideal work method. In addition, we have calculated the variation of σ_1, σ_2, and die pressure (P) throughout the dies as part of the same result:[5]

---

[5] This can be easily seen by integrating Eq. 10.26 from t_i to some general thickness t, during which the stress along x_1 will vary from zero to some value σ_1.

$$\sigma_1(t) = \frac{H(1+B)}{B}\left[1-\left(\frac{t}{t_i}\right)^B\right] \tag{10.30a}$$

$$\sigma_2(t) = \sigma_1(t) - H = \frac{H(1+B)}{B}\left[1-\left(\frac{t}{t_i}\right)^B\right] - H \tag{10.30b}$$

$$P(t) \cong -\sigma_2(t) = \frac{H(1+B)}{B}\left[1-\left(\frac{t}{t_i}\right)^B\right] - H \tag{10.30c}$$

Exercise 10.3 **Derive the axisymmetric-slab-drawing equation comparable to Eq. 10.25, and find the drawing stress for a wire-drawing operation using the slab method. Assume that α can be of any size and that the contact follows sticking friction, such that $\tau_f = \bar{\sigma}/\sqrt{3}$.**

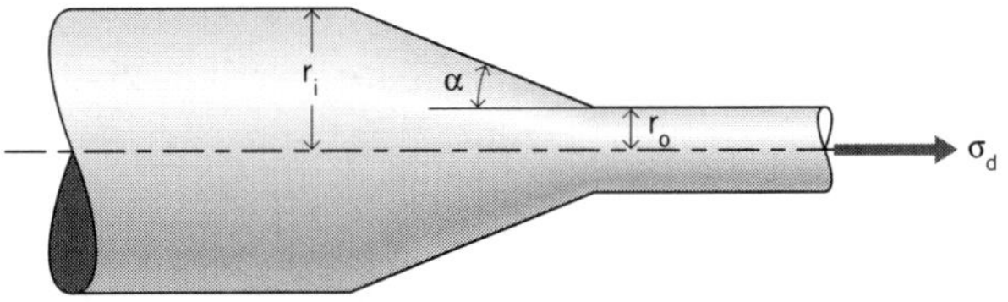

We proceed precisely as before, with the only differences lying in the geometry, the friction law, the small-angle approximation, and the strain state. We proceed in this order to re-derive briefly the governing slab equations.

The basic geometry is defined by the truncated conic section taken from it. The area of the conical face, dA, is equal to $2\pi r dr/\sin\alpha$, where $2\pi r$ is the circumference, and $dr/\sin\alpha$ is the width along the surface corresponding to the section width of dx_1 ($= dr/\tan\alpha$), as shown in the figure below. ($d\theta$ is the differential angle about X_1 corresponding to a small element of contact area.)

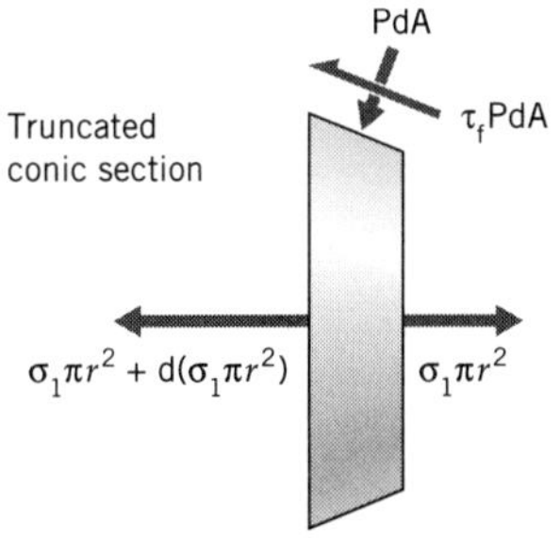

Starting with modification of the geometry, and with reference to this general geometry, we are able to draw the $\mathbf{x}_1$-force-balance diagram, as shown in the following figure:

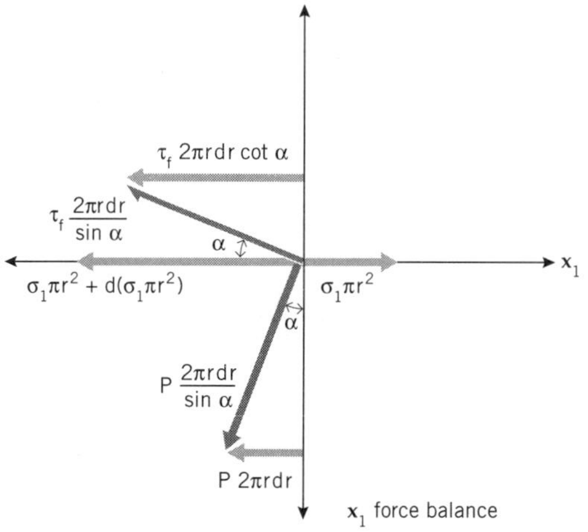

We enforce equilibrium to obtain

$$\Sigma Fx_1 = 0 = \sigma_1\ \pi r^2 - \left[\sigma_1\ \pi r^2 + d(\sigma_1\ \pi r^2)\right] - \tau_f\ 2\pi r dr \cot\ \alpha - P2\ \pi r dr \tag{10.3-1}$$

or,

$$0 = d\,(\sigma_1\ r^2) + 2(\tau_f \cot\ \alpha + P) r dr \tag{10.3-2}$$

or,

$$0 = r d\sigma_1 + 2(\sigma_1 + \tau_f \cot\alpha + P) dr \tag{10.3-3}$$

Note that Eqs. 10.3-1 to 10.3-3 are general, applicable to all slab calculations for axisymmetric drawing or extrusion.

The sticking friction law is now easily incorporated by substituting $\tau_f = \overline{\sigma}/\sqrt{3}$ (see Chapter 9) into Eq. 10.3-3:

$$0 = r d\sigma_1 + 2(\sigma_1 + \frac{\overline{\sigma}}{\sqrt{3}} \cot\alpha + P) dr \tag{10.3-4}$$

In order to find a necessary relationship among τ_f, P, and the internal stresses, we consider equilibrium in the radial direction at a small element of surface area, as shown in the following figure and companion force-balance diagram. With the aid of these diagrams, we obtain

$$\Sigma F_r = 0 = \tau_f\ r\,dr\,d\theta - P\,r\,dr\,d\theta \cot\alpha - \sigma_r\ r\,dr\,d\theta \cot\alpha \tag{10.3-5}$$

such that

$$P = \tau_f \tan\alpha - \sigma_r \quad (= -\delta_r, \text{ small } \alpha) \tag{10.3-6}$$

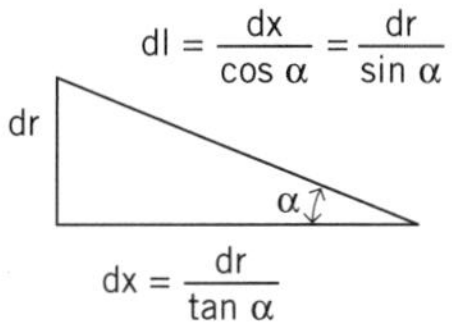

Contact-section geometry

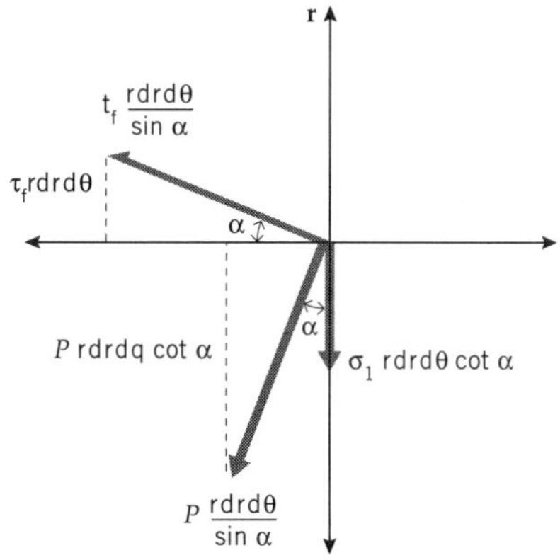

r force balance

While Eqs. 10.3-5 and 10.3-6 are general for axisymmetric cases, we specialize to the sticking friction case by substituting $\bar{\sigma}/\sqrt{3}$ for τ_f:

$$P = \frac{\bar{\sigma}}{\sqrt{3}} \tan\alpha - \sigma_r \tag{10.3-7}$$

We can substitute Eq. 10.3-7 into Eq. 10.3-4 in order to remove P from the governing equation (while introducing the corresponding internal stress, σ_r:

$$0 = r\,d\sigma_1 + 2\left(\sigma_1 + \frac{\bar{\sigma}}{\sqrt{3}}\cot\alpha - \frac{\bar{\sigma}}{\sqrt{3}}\tan\alpha + \sigma_r\right)dr \tag{10.3-8}$$

The last remaining consideration is the assumed plastic yielding of the material, which we use to obtain a relationship among σ_1, σ_r, and $\bar{\sigma}$.

We could use the axisymmetric symmetry in several ways to find the desired relationship. In terms of strain increments, $d\varepsilon_r = d\varepsilon_1 = d\varepsilon_2$. For an assumed isotropic material, $\sigma_r = \sigma_2 = \sigma_3$ is sufficient to substitute in the von Mises yield condition:

$$2\bar{\sigma}^2 = (\sigma_1 - \sigma_r)^2 + (\sigma_1 - \sigma_r)^2 + (\sigma_r - \sigma_r)^2 \tag{10.3-9}$$

to obtain

$$\sigma_r = \sigma_1 - \bar{\sigma} \tag{10.3-10}$$

A more general method, which makes no assumptions about the yield surface, is to enforce the strain symmetry: $\varepsilon_r = \varepsilon_1 = \varepsilon_2$ and to work back through the normality to obtain the stress condition. (For von Mises yield, the same result is obtained.)

We then substitute Eq. 10.3-10 into Eq. 10.3-8 to obtain the final governing slab equation for the case of axisymmetric draw of a von Mises material with sticking friction:

$$0 = r d\sigma_1 + 2\left[2\sigma_1 + \frac{\bar{\sigma}}{\sqrt{3}}(\cot\alpha - \tan\alpha - \sqrt{3})\right]dr \tag{10.3-11}$$

or

$$\frac{d\sigma_1}{4\sigma_1 + B\bar{\sigma}} = \frac{dr}{r} \qquad \text{where } B \equiv \sqrt{3}\,(\cot\alpha - \tan\alpha - \sqrt{3}) \tag{10.3-12}$$

Once the governing differential equation has been obtained, Eq. 10.3-12, we simply apply boundary conditions and integrate to obtain the wire-drawing stress following the procedure presented for plane-strain drawing:

$$\int_{\sigma_1=0}^{\sigma_1=\sigma_d} \frac{d\bar{\sigma}_1}{4\sigma_1 + B\bar{\sigma}} = \int_{r_i}^{r_o} \frac{dr}{r} \tag{10.3-13}$$

$$\frac{1}{4}\lim \frac{4\sigma_d + B\bar{\sigma}}{B\bar{\sigma}} = \ln\frac{r_o}{r_i}, \text{ or} \tag{10.3-14}$$

$$\sigma_d = \frac{B}{4}\bar{\sigma}\left[\left(\frac{r_o}{r_i}\right)^4 - 1\right] \tag{10.3-15}$$

Plane-Strain Compression

Let's use the slab method to investigate another forming operation, plane-strain compression, or **upsetting**, of a long bar. Much of the construction will be similar to the plane-strain drawing or extrusion problem, with the exception that this is not a **steady state** problem. (This can easily be seen because a fixed quantity of material must be considered, one that changes shape throughout the entire operation. This is quite different from extrusion, where the "action" is only in a small region of the die contact even though material flows continuously through the region.)

What do we hope to accomplish by a slab analysis of this operation? We should be able to estimate the overall loads required to forge the bar, and to estimate the pressure distribution acting on the die faces. Such information is necessary in the design of sufficiently strong dies and in choosing a press of sufficient force and power to complete the forging operation without failure.

Figure 10.6 illustrates what happens in the real upsetting of a long bar, along with the idealization assumed in the slab method. In the real operation (Figure 10.6a), the unconstrained sides bulge out because friction constrains the ends in contact with the upper and lower dies. As better lubrication is applied, this

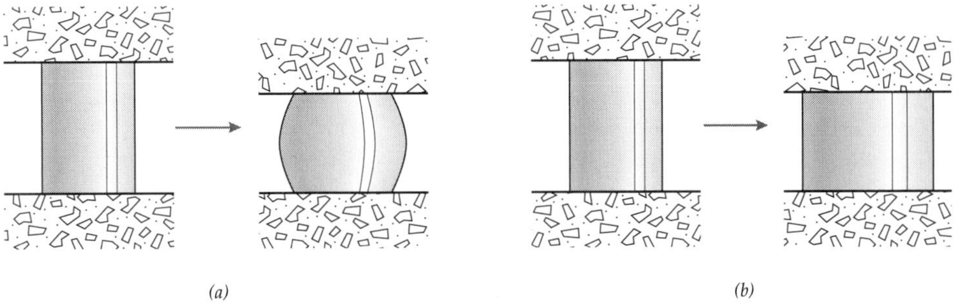

Figure 10.6 Real (*a*) and slab (*b*) visualization of upsetting of a long bar.

bulging is reduced, but it can never be eliminated unless frictional forces are eliminated. However, in the slab approach we assume that the principal directions are constant and known, with the result that we do not allow the characteristic bulging (also called **barreling**). Instead, we assume that the intermediate workpiece shapes are those for the zero-friction case (Figure 10.6*b*).

> ***Note:** We made the same implicit assumption in the extrusion case. In fact, straight vertical lines in the original material are curved toward the inlet side in the final part because of friction acting on the surfaces of the material in contact with the dies. Rolling, which we will consider next, has similar frictional effects that are integral to the operation itself: without friction, no rolling could take place.*

With our idealized deformation geometry, we can easily choose the important direction for analysis, $\mathbf{x}_1$. The frictional forces will build up as we move along $\mathbf{x}_1$ toward the center of the bar, because progressively more surface area is subjected to the friction restraint opposing outward motion. With the most important direction identified for our 1-D slab analysis, we proceed as for the extrusion/drawing case. We will assume sticking friction, with the frictional stress equal to $\bar{\sigma}/\sqrt{3}$ (see Chapter 9).

Figure 10.7 shows the relevant slab geometry, which differs from the extrusion/drawing case in detail only. For example, the die half-angle (α) is zero, and we choose a coordinate system at the symmetry line of the bar in order to define clearly the friction force direction operating on the bar surface, which is always toward the center (i.e., resisting the outward sliding direction).

Figure 10.7*c* defines the equilibrium equation for this case:

$$\Sigma F_{x_1} = 0 = (\sigma_1 + d\sigma_1)h - \sigma_1 h - \frac{2\bar{\sigma}}{\sqrt{3}} dx_1 \tag{10.31}$$

Using the plane-strain relationship for von Mises flow (Eq. 10.23), we can substitute

$$P = -\sigma_2 = -\sigma_1 + \frac{2}{\sqrt{3}}\bar{\sigma} \tag{10.32}$$

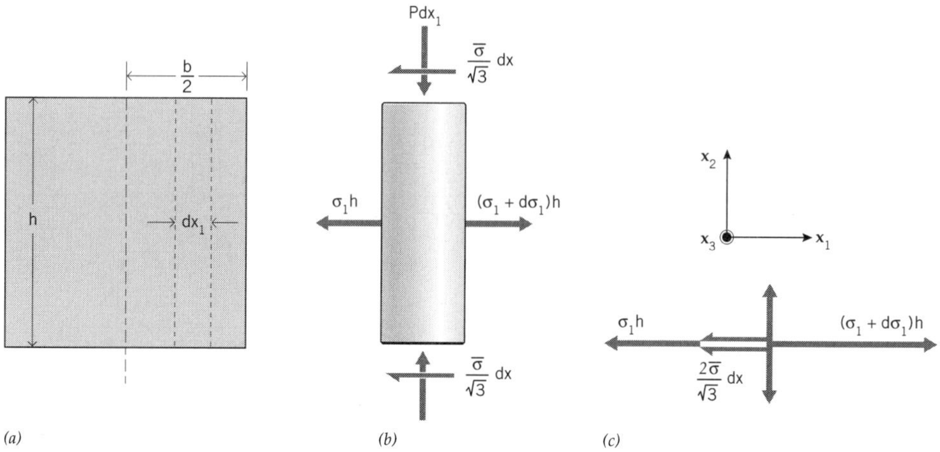

Figure 10.7 Slab geometry and force balance for plane-strain compression.

For the constant $\bar{\sigma}$ case (non-work-hardening), $dP = -d\sigma_1$, and Eq. 10.31 reduces to

$$dP = -\frac{2\bar{\sigma}}{h\sqrt{3}} dx_1 \tag{10.33}$$

In order to find the real pressure distribution acting on the die and workpiece, we must integrate Eq. 10.33 subject to known boundary conditions. In this case, we know that $\sigma_1 = 0$ at the outer surface of the workpiece (at $x = b/2$) because this is a "free" surface—that is, there are no forces transmitted through it. Equation 10.32 gives the pressure acting on the die at the outside edge:

$$\text{at} \quad x = \frac{b}{2}, \quad P = \frac{2}{\sqrt{3}}\bar{\sigma} \tag{10.34}$$

We can now perform the definite integration of Eq. 10.33 to find the pressure distribution:

$$\int_{P=\frac{2\bar{\sigma}}{\sqrt{3}}}^{P=P} dP = \frac{-2\bar{\sigma}}{h\sqrt{3}} \int_{x_1=b/2}^{x_1=x_1} dx_1 \tag{10.35}$$

or

$$P = \frac{2\bar{\sigma}}{n\sqrt{3}}\left(1 + \frac{b}{2h} - \frac{x_1}{h}\right) \tag{10.36}$$

Eq. 10.36 is plotted in Figure 10.8, and the characteristic pressure distribution is known as the **friction hill** because of its origin in the die friction. (Under frictionless conditions, the hill is flat, as demonstrated in Exercise 10.4.)

As shown in Figure 10.8a, the maximum pressure is easily calculated using Eq. 10.36 and by noting that the maximum always occurs at the centerline. In order to find the total load needed to forge the bar, we simply integrate the die pressure over the contact area:

$$\frac{\text{load}}{\text{length}} = 2\frac{2\bar{\sigma}}{\sqrt{3}}\int_{x_1=0}^{x_1=b/2}\left(1+\frac{b}{2h}-\frac{x_1}{h}\right)dx_1 = \frac{2\bar{\sigma}}{\sqrt{3}}\,b\left(1+\frac{b}{4h}\right) \tag{10.37}$$

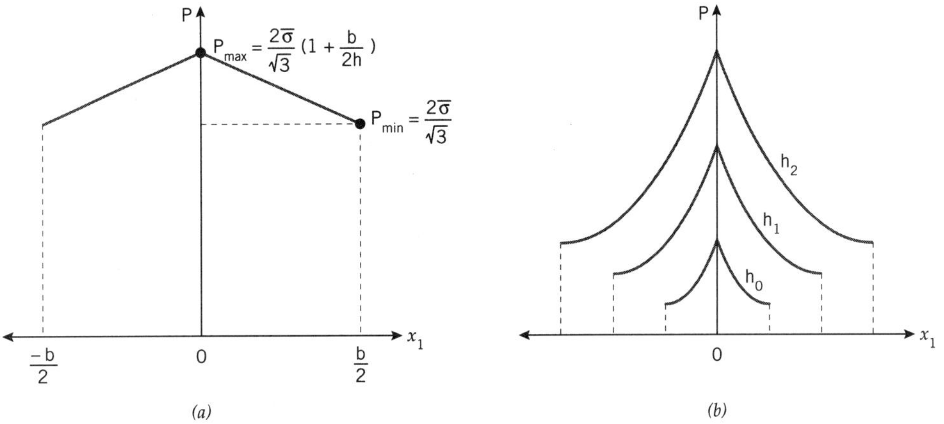

Figure 10.8 "Friction hill" for the case of plane-strain compression.

Exercise 10.4 **Reproduce the key equations and results for plane-strain forging for the case of Coulomb friction. Show that the frictionless case can be obtained from either Coulomb- or sticking-friction calculations.**

The only difference between Figure 10.6*a* and *b* is the form of the frictional forces acting on the ends of the slab, as shown, which leads to an equilibrium equation of

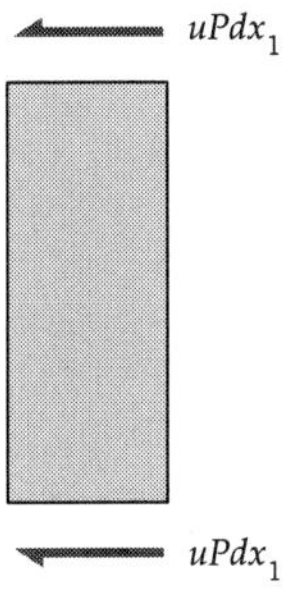

$$\Sigma Fx_1 = 0 = (\sigma_1 + d\sigma_1)h - \sigma_1 h - 2\mu P dx_1 \tag{10.4-1}$$

or

$$dP = \frac{-2\mu P}{h} dx_1 \tag{10.4-2}$$

where the same substitutions have been made for P in terms of σ_1 (because plane strain is still assured). The boundary conditions are the same, so integrating yields

$$P = \frac{2\bar{\sigma}}{\sqrt{3}} \exp\left[\frac{2\mu}{h}(\frac{b}{2} - x)\right] \tag{10.4-3}$$

on the RHS of the forging, with the LHS following by symmetry.

The maximum and minimum pressures and the total load are obtained as before:

$$P(x=0) = \ P_{max} = \frac{2\bar{\sigma}}{\sqrt{3}} \exp\left(\frac{\mu b}{h}\right) \tag{10.4-4}$$

$$P\left(x = \frac{b}{2}\right) = P_{min} = \frac{2\bar{\sigma}}{\sqrt{3}} \tag{10.4-5}$$

$$\frac{\text{load}}{\text{length}} = \frac{2\bar{\sigma}}{\sqrt{3}} \frac{h}{\mu}\left[\exp\left(\frac{\mu b}{h}\right) - 1\right] \tag{10.4-6}$$

and the friction hill is illustrated in the figure below.

The frictionless pressure distribution can be obtained by direct substitution in either the sticking-friction case ($\bar{\sigma}$ = 0 in the frictional component,[6] Eq. 10.31) or the Coulomb-friction case (m = 0 in Eq. 10.4-3), with the following result:

$$P = \frac{2}{3}\bar{\sigma} \tag{10.4-7}$$

---

[6] Note that the frictionless case does not imply a strength-free material, so a nonzero integration limit of $2\bar{\sigma}/\sqrt{3}$ remains in Eq. 10.3-5.

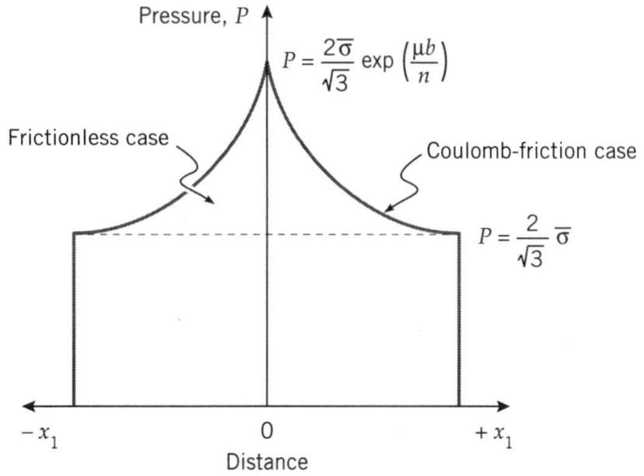

Plane-Strain Compression: Work-Hardening Case

We have assumed so far that the material does not work harden—that $\overline{\sigma}$ = constant. In fact, $\overline{\sigma}$ will change as the bar is forged, but at any instant we maintain the assumption that it will remain uniform for all x_1. Thus, the only adjustment needed to our derivations is for the change of $\overline{\sigma}$ as the forging height changes from some initial h_o to some general height h.

Equations 10.31 to 10.36 are unchanged because they express the equilibrium condition among the slabs at one time, when $\overline{\sigma}$ is constant. However, it remains necessary to express all variables in terms of one, let's say the die closure height, h. Clearly, b changes as h does, and we can find the relationship by noting that the volume of the bar does not change during plastic deformation:

$$h_o b_o l_o = hbl \tag{10.38}$$

so

$$b = \frac{h_o b_o}{h} \tag{10.39}$$

because $l = l_o$ for plane strain.

If we know the tensile-test stress-strain curve for the material, we can express $\overline{\sigma}$ in terms of h. Given:

$$\overline{\sigma} = f(\overline{\varepsilon}) \tag{10.40}$$

we need only to relate $\overline{\varepsilon}$ to h. For plane-strain deformation (proportional at all times), we have $\varepsilon_2 = -\varepsilon_1 = \varepsilon_h$ and ε_{23}(along l) = 0. Thus, for this proportional path we have

$$\overline{\varepsilon} = \left[\frac{2}{3}\left(\varepsilon_h^2 + \varepsilon_h^2 + 0^2\right)\right]^{1/2} = \frac{2}{\sqrt{3}}|\varepsilon_h| \tag{10.41}$$

and
$$|\varepsilon_h| = \ln\frac{h_o}{h} \tag{10.42}$$

$$\bar{\varepsilon} = \frac{2}{\sqrt{3}}\ln\frac{h_o}{h} \tag{10.43}$$

$$\bar{\sigma} = f\left(\frac{2}{\sqrt{3}}\ln\frac{h_o}{h}\right) \tag{10.44}$$

With these two substitutions—Eqs. 10.39 and 10.44 into Eqs. 10.36 and 10.4-3—we can obtain the pressure distribution for the plane-strain forging of a bar throughout the forging cycle, as a function of x_1 and h:

$$P = \frac{2}{\sqrt{3}} f\left(\frac{2}{\sqrt{3}}\ln\frac{h_o}{h}\right)\left(1+\frac{h_o b_o}{2h^2}-\frac{x_1}{h}\right)$$
(sticking friction) (10.45)

$$P = \frac{2}{\sqrt{3}} f\left(\frac{2}{\sqrt{3}}\ln\frac{h_o}{h}\right)\exp\left[\frac{2\mu}{h}\left(\frac{b_o h_o}{2h}-x_1\right)\right]$$
(Coulomb friction) (10.46)

A schematic representation of the progression of the friction hill in time, Figure 10.8*b*, illustrates how rapidly the peak pressure increases because of the combination of increasing width b and increasing edge hardness.

Wire Drawing: Numerical Solution

So far, we have chosen conditions in our slab analyses such that a simple integration could be used to solve the differential equations. In more general cases, it may be impossible or impractical to solve the equation in this manner, and a numerical method may be desirable. Furthermore, the numerical solution of simple slab problems illustrates many of the methods that are applied to more complex situations.

As an example, let's consider the wire drawing of a metal which exhibits von Mises yield, power-law hardening ($\sigma = K\bar{\varepsilon}^n$), and Coulomb friction. The wire-drawing derivation presented in Exercise 10.1 can easily be extended to this case by realizing that $\bar{\sigma}$ is no longer constant, and by noting that the friction stress, τ_f, is equal to μP. We make use of Eq. 10.3-3, which is unchanged because the cylindrical cross-sectional geometry is the same, Eq. 10.3-6, which is the same force balance in the r direction with the substitution that $\tau_f = mP$; and Eq. 10.3-10, which is the same because the yield surface is unchanged. Combining these and rearranging obtains

$$0 = r d\sigma_1 + 2\left[(1+B)\sigma_1 - \beta\bar{\sigma}\right] dr$$

where

$$B = \frac{\mu \cos\alpha - \sin\alpha}{1 - \mu \tan\alpha} \tag{10.47}$$

We cannot integrate Eq. 10.47 directly because $\bar{\sigma}$ is a function of $\bar{\varepsilon}$, which is a function of r. For axisymmetric geometry and proportional straining,

$$\varepsilon_1 = -2\varepsilon_2 = -2\varepsilon_3 \tag{10.48}$$

and

$$\bar{\varepsilon} \equiv \left[\frac{2}{3}(\varepsilon_1^2 + \varepsilon_2^2 + \varepsilon_3^2)\right]^{\frac{1}{2}} = \varepsilon_1 = -2\varepsilon_r = 2\ln\frac{r_i}{r} \tag{10.49}$$

Thus we write Eq. 10.47 to emphasize that $\bar{\sigma}$ is a function of r and to relate the two differential variables, σ_1 and r, to one another:

$$d\sigma_1 = 2\left[(1+B)\sigma_1 - B\,\bar{\sigma}\,(r)\right]\frac{dr}{r} \tag{10.50}$$

Note that Eq. 10.50 holds for generally-shaped dies; that is, B may depend on r via $\alpha(r)$. This does not introduce any additional complexity because the term $B\,\bar{\sigma}\,(r)$ is already an arbitrary function of r.

In order to solve Eq. 10.50 to obtain σ_1 as a function of r [and finally to obtain σ_{draw}, which is $\sigma_1\,(r_o)$], we introduce a simple, explicit numerical integration scheme[7] that is easily followed. To illustrate the method, let's rewrite Eq. 10.50 to again emphasize the functional dependencies:

$$d\sigma_1 = \left[F(\sigma_1) + G(r)\right]\frac{dr}{r} \tag{10.51}$$

where $F(\sigma_1) = 2(1+B)\,\sigma_1 \quad G(r) = -B\bar{\sigma}(r)$, an arbitrary hardening law.

[7] See Chapter 1 for similar numerical procedures.

TABLE 10.3 Numerical Integration of Equation 10.50

Step number	r	$F(\sigma_1^{i-1})$	$G(r^i)$	$\frac{r^i - r^{i-1}}{r^i}$	σ^i
0	r^o	-	-	-	0
1	r^1	$F(0) = \alpha_1$	$G(r^1) = \beta_1$	$\frac{r^1 - r^0}{r^1} = \gamma_1$	$\sigma_1^1 = (\alpha_1 + \beta_1)\gamma_1$
2	r^2	$F(\sigma^2) = \alpha_2$	$G(r^2) = \beta_2$	$\frac{r^2 - r^1}{r^2} = \gamma^2$	$\sigma_1^2 = \sigma_1^1 + (\alpha_2 + \beta_2)\gamma_2$
:	:	:	:	:	:
n	r^n	$F(\sigma^{n-1}) = \alpha_n$	$G(r^n) = \beta_n$	$\frac{r^n - r^{n-1}}{r^n} = \gamma_n$	$\sigma^n = \sigma^{n-1} (\alpha_n + (\beta_n))\gamma_n$

Now, we transform the differential equation to a finite-increment form:

$$\Delta \sigma_1 = [F(\sigma_1) + G(r)] \frac{\Delta r}{r} \tag{10.52}$$

and we choose an explicit update method for simplicity:

$$\sigma_1^i = \sigma_1^{i-1} + \left\{F\left[\sigma^{(i-1)}\right] + G(r^i)\right\} \frac{(r^i - r^{i-1})}{r^i} \tag{10.53}$$

Note that we use the current radius, r^i, which is known explicitly for our update procedure, but we use the stress from the previous position (i–1, at r^{i-1}) because we know it explicitly. If we had chosen to use the current stress, which is more consistent mathematically, Eq. 10.53 would need to be solved iteratively at each value of r.

To begin the numerical integration, we apply the initial conditions, or starting boundary conditions at

$$r = r_o, \sigma_1 = 0 \tag{10.54}$$

Next we begin to calculate the updated stress as we move toward reduced r's (Table 10.3).

Note that the procedure would be identical for any kind of functional variation of r and σ_1. So, varying wall angle— for "streamlined" dies, for example—could be handled similarly.

Exercise 10.5 **Find the wire-drawing stress for a work-hardening material that obeys $\bar{\sigma} = 500\ \text{MPa}\ \bar{\varepsilon}^{0.20}$. Assume that the wire-drawing operation reduces the wire radius from 1 mm to 0.6 mm using dies with a constant half-angle of 30° (α = 30°).**

We start from Eq. 10.53 but must make explicit the various terms in the equation.

$$F(\sigma_1) = 2(1+\beta)\sigma_1 = 2\left(1 + \frac{\mu\cos 30^\circ - \sin 30^\circ}{1-\mu\tan 30^\circ}\right)\sigma_1 = 1.42\quad \sigma_1 \tag{10.5-1}$$

and

$$G(r) = -B\,\bar{\sigma}(r) = -\left(\frac{\mu\cos 30^\circ - \sin 30^\circ}{1-\mu\tan 30^\circ}\right)500\,(\bar{\varepsilon})^{0.20} = -145\,(2\ln\frac{r_i}{r})^{0.20}\ (\text{MPa}) \tag{10.5-2}$$

We now carry out the numerical procedure, in four steps as presented in Table 10.3.

Step number	r	$F(\sigma_1^{i-1})$	$G(r^i)$	$\dfrac{r^i - r^{i-1}}{r^i}$	$\sigma^i = \sigma^{i-1} + (F+G)\Delta r$
	(mm)	(MPa)	(MPa)	(mm)	(MPa)
0	1	-	-	-	0
1	0.9	0	–106	–0.111	11.7
2	0.8	17	–124	–0.125	25.1

3	0.7	36	–136	–0.143	39.4
4	0.6	56	-146	-0.167	54.4

Thus, σ_{draw} = 54 MPa.

Flat Rolling: Slab Equations

The slab method may also be applied to the rolling of flat sheet or plate. The set-up of the differential equations, which we will present, is straightforward and follows the previous examples closely. The solution of the equations, even for simple material models (such as no work hardening), has an additional complexity that requires numerical treatment. We will leave such solutions to exercises for the interested reader.

A flat-rolling arrangement is shown in Figure 10.9. In fact, this figure is exaggerated to illustrate the method better because changes in sheet thickness in typical rolling operations are relatively small.

The sheet, which is assumed to be quite wide relative to the length of the contact zone ($R\alpha$), enters the **roll gap** with a thickness of t_i and exits with a thickness of t_o. This geometry requires that the deformation be near plane

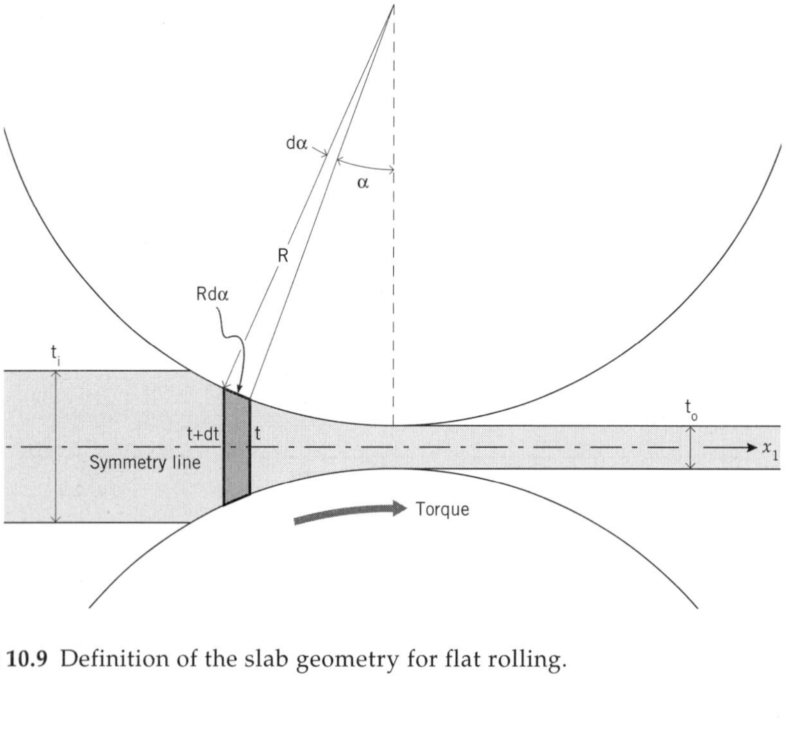

Figure 10.9 Definition of the slab geometry for flat rolling.

strain, which is commonly observed. Note that all the power required for this forming operation is supplied by the rotation of the rolls. No other external work is done.

For the slab calculation, we proceed as before, keeping in mind what we wish to accomplish with a 1-D analysis. By considering a variation in stress along $\mathbf{x}_1$, we can calculate the pressure distribution acting on the rolls. Figure 10.10 illustrates the slab arrangement for rolling, along with the vector face diagram needed to establish the basic equilibrium equation. For this example, we assure Coulomb friction and we choose a unit width normal to the paper in order to eliminate this variable from our equations.

We derive the equations by starting with equilibrium along the principal direction of variation, x_1. It is possible to express the resulting equation in any combination of t, x_1, and a (and their differentials), as shown in the inset to Figure 10.10, but we keep all three variables to obtain a typical form:

$$\Sigma F_{x_1} = 0 = d(\sigma_1 t) + 2\,PR \sin\alpha\, d\alpha - 2\mu\, PR \cos\alpha\, d\alpha \tag{10.55}$$

but

$$d\alpha = \frac{dx_1}{R\cos\alpha}$$

so

$$0 = d(\sigma_1 t) - 2\,P(\mu - \tan\alpha)\,dx_1 \tag{10.56}$$

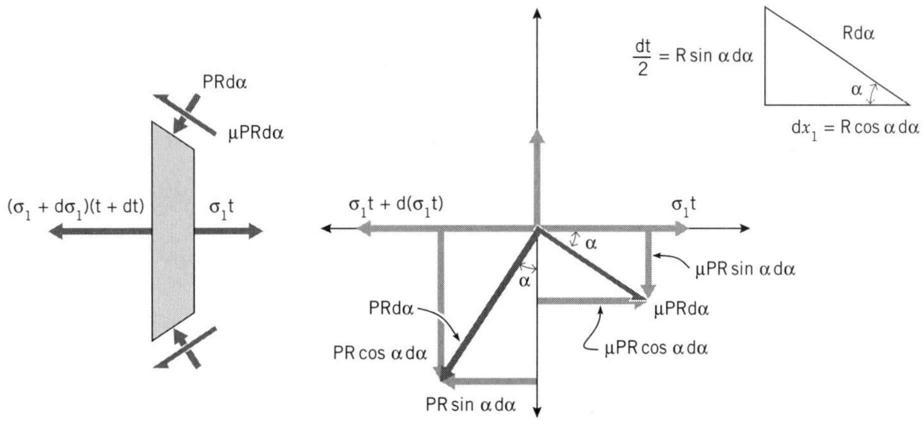

Figure 10.10 Slab equilibrium in flat rolling.

Note that Eq. 10.56 was derived by *assuming* a direction of the friction force with a positive $\mathbf{x}_1$ component. This is clearly the case as the sheet enters the rolls because the rolls pull it in. However, a simple analysis of the velocities of the sheet surface relative to the velocities of the rolls (which are constant) shows that the friction must reverse sign so that the rolls tend to impede the progress of the sheet as it leaves the rolls. (The velocity of the sheet surface varies because the sheet elongates in the roll contact zone and the velocity thus increases.) Figure 10.11 shows schematically the variation of the relative velocity of a roll (i.e., v_{roll}-v_{sheet}) at the contact surface. At the point marked N, the roll velocity just matches the sheet velocity, hence this is called the *neutral point*.

With the addition of this physical insight, Eq. 10.56 splits into two equations, depending on which side of the neutral point is considered:

$$0 = d(\sigma_1 t) + 2P(\tan\alpha - \mu)dx_1 \quad \text{(before N)} \tag{10.57a}$$

$$0 = d(\sigma_1 t) + 2P(\tan\alpha + \mu)dx_1 \quad \text{(after N)} \tag{10.57b}$$

Equations 10.57 are forms of the well-known ***von Karman* equations of rolling**.

In general, Eqs. 10.57 require considerable approximation and manipulation to solve. Examples are deferred to the Problems for this chapter, but a brief discussion may be helpful.

The first step in any analysis requires finding the position of the neutral point, N. However, in the case of Coulomb friction, whose friction force depends on contact pressure, the position of N depends on P(*t*) or P(*x*), which in turn depends on N. To approach the problem in a general way, we can write the governing equations as follows:

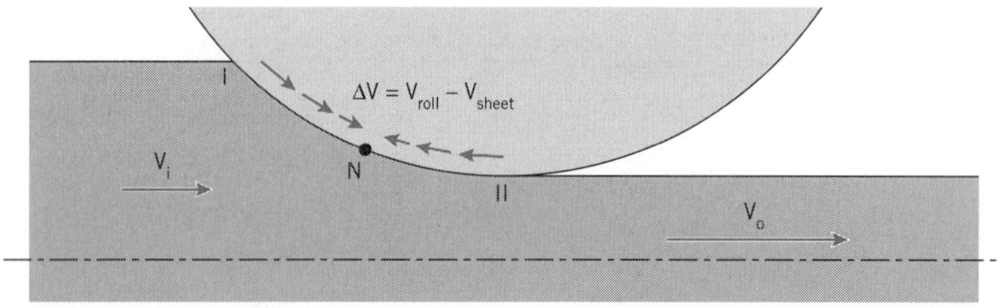

Figure 10.11 Definition of the neutral point.

$$d\sigma_1 = F(\sigma_1, x_1)\,dx_1 \quad I < x_1 < N \tag{10.58a}$$

$$d\sigma_1 = G(\sigma_1, x_1)\,dx_1 \quad N < x_1 < II \tag{10.58b}$$

The points N, I, and II are shown on Figure 10.11. In order to find N, we note that the boundary conditions are as follows:

$$\sigma_1 = 0 \quad \text{at } x_1 = I, \qquad \sigma_1 = 0 \quad \text{at } x_1 = II$$

because the only forces applied to the sheet are by the rollers, which are not in contact outside the region $I < X_1 < II$. Furthermore, the stress σ_1 at the neutral point must be single-valued, so the following equality must hold:

$$\sigma_1\,(N)^{\text{left}} = \sigma_1\,(N)^{\text{right}} \tag{10.59}$$

$$\int_{x_1=I}^{N} F(\sigma_1, t)\,dx_1 = \int_{x_2=II}^{N} G(\sigma_1, t)\,dx_1 \tag{10.60}$$

Eq. 10.60 shows that the maximum value of s_1 occurs at the neutral point for cases where F and G are monotonic functions of x_1 or t. To solve Eq. 10.60 numerically, the two sides may be integrated numerically for all possible values of N, and the point may be located that satisfies the equality.

10.3 THE SLIP-LINE FIELD METHOD FOR PLANE-STRAIN ANALYSIS

Slip-line field theory can cope with rather complicated two-dimensional geometries, but cannot be extended easily to elastoplastic or viscoplastic analysis, or to complicated boundary conditions. The complete theory of slip-line fields will not be described in detail here; for a comprehensive mathematical presentation the reader is referred to Hill[8], or to Hosford and Caddell[9]. The basic idea is to use the Tresca yield criterion for rigid plastic materials, and to express the equilibrium equations in a reference system where it takes a much simpler form. For plane-strain rigid plastic deformation, the basic equations were outlined in Chapter 8. The stress equations are:

[8] R. Hill, *The Mathematical Theory of Plasticity*, (Oxford: Clarendon Press, 1989; 1950).

[9] W. F. Hosford and R. M. Caddell, *Metal forming; Mechanics and Metallurgy*: (Englewood Cliffs: Prentice-Hall, Inc., 1983).

equilibrium equations, when inertia and gravity are neglected:

$$\frac{\partial \sigma_{xx}}{\partial x} + \frac{\partial \sigma_{xy}}{\partial y} = 0 \quad (10.61a)$$

$$\frac{\partial \sigma_{xy}}{\partial x} + \frac{\partial \sigma_{yy}}{\partial y} = 0 \quad (10.61b)$$

von Mises yield criterion:

$$(\sigma_{xx} - \sigma_{yy})^2 + 4\sigma_{xy}^2 = \frac{4\sigma_0^2}{3} = 4\,k^2 \quad (10.62)$$

the plane strain condition on stress for plastic zone:

$$P = -\frac{1}{2}(\sigma_{xx} + \sigma_{yy}) = -\sigma_{zz} \quad (10.63)$$

where P is the hydrostatic pressure.

These equations are transformed according to a change of coordinates and of unknown stress functions. At any point M where plastic deformation takes place, we denote by MI the first principal axis, corresponding to the maximum principal stress σ_I; and by MII the second principal axis, corresponding to σ_{II}. MI is referred by its angle θ with the **x** axis. A new reference system (M, **a**, **b**) is defined locally: its unitary vectors **a** and **b** make, respectively, angles of –45° and 45° with MI (Figure 10.12).

Two families of curves can be defined: an α curve is tangential to **a**, and a β curve is tangential to **b** in M. The transformed equilibrium equations can be more easily solved on α and β lines, called **slip lines**, in terms of the hydrostatic

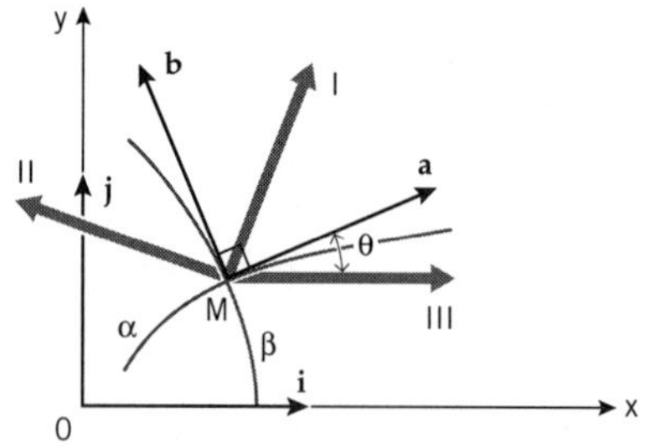

Figure 10.12 Representation of the local reference system for the slip lines α and β.

pressure P and the angle θ. To prove this statement, Eqs. 10.61 to 10.63 will be rewritten in terms of P and θ, on the α and β lines. First it is easy to verify that in the (MI, MII) reference system, the unit vectors **i** and **j**, on 0x and 0y respectively, have components so that

$$[i]=\begin{bmatrix}\cos\theta\\-\sin\theta\end{bmatrix}\quad\text{and}\quad [j]=\begin{bmatrix}\sin\theta\\ \cos\theta\end{bmatrix} \tag{10.64}$$

The stress components in the initial reference system can now be expressed in terms of the principal stresses σ_I and σ_{II}, so we have:

$$\begin{bmatrix}\sigma_{xx}\ \sigma_{xy}\\ \sigma_{xy}\ \sigma_{yy}\end{bmatrix}=\begin{bmatrix}\cos\theta-\sin\theta\\ \sin\theta\quad\cos\theta\end{bmatrix}\begin{bmatrix}\sigma_I & 0\\ 0 & \sigma_{II}\end{bmatrix}\begin{bmatrix}\cos\theta & \sin\theta\\ -\sin\theta & \cos\theta\end{bmatrix}$$
$$=\begin{bmatrix}\sigma_I\cos^2\theta+\sigma_{II}\sin^2\theta & 2(\sigma_I-\sigma_{II})\sin\theta\cos\theta\\ 2(\sigma_I-\sigma_{II})\sin\theta\ \cos\theta & \sigma_I\sin^2\theta+\sigma_{II}\cos^2\theta\end{bmatrix} \tag{10.65}$$

The plastic yield criterion is written for the principal stresses, taking into account that $\sigma_I > \sigma_{II}$, and we obtain for the plastic region:

$$(\sigma_I-\sigma_{II})^2=4\,k^2 \Leftrightarrow \sigma_I-\sigma_{II}=2\,k \tag{10.66}$$

We are now able to make the change of unknown functions: Eqs. 10.65 to 10.67 take the simpler form:

$$\sigma_{xx}=\frac{\sigma_I+\sigma_{II}}{2}+\frac{\sigma_I-\sigma_{II}}{2}\cos 2\theta=-P+k\cos 2\theta \tag{10.67}$$

$$\sigma_{yy}=\frac{\sigma_I+\sigma_{II}}{2}+\frac{\sigma_{II}-\sigma_I}{2}\cos 2\theta=-P-k\cos 2\theta \tag{10.68}$$

$$\sigma_{xy}=(\sigma_I-\sigma_{II})\sin 2\theta=k\sin 2\theta \tag{10.69}$$

Using Eqs. 10.67 to 10.69 to rewrite the equilibrium equations (Eqs. 10.61a and 10.61b), we get:

$$-\frac{\partial P}{\partial x}-2k\sin 2\theta\frac{\partial\theta}{\partial x}+2k\cos 2\theta\frac{\partial\theta}{\partial y}=0 \tag{10.70a}$$

$$-\frac{\partial P}{\partial y} + 2k \cos 2\theta \frac{\partial \theta}{\partial x} + 2k \sin 2\theta \frac{\partial \theta}{\partial y} = 0 \qquad (10.70b)$$

We still have to rewrite these two equations in the SLF (local) reference system (α) and (β): To do this transformation, a change of orthonormal basis must be done (see Section 2.5). The **R** matrix is defined in the usual way (see Section 2.5):

$$[R] = \begin{bmatrix} \cos\left(\theta - \frac{\pi}{4}\right) & \sin\left(\theta - \frac{\pi}{4}\right) \\ -\sin\left(\theta - \frac{\pi}{4}\right) & \cos\left(\theta - \frac{\pi}{4}\right) \end{bmatrix} \qquad (10.71)$$

which allows expression of the (infinitesimal) components:

$$\begin{bmatrix} dx \\ dy \end{bmatrix} = [R]^T \begin{bmatrix} d\alpha \\ d\beta \end{bmatrix} = \begin{bmatrix} \frac{\partial x}{\partial \alpha} & \frac{\partial x}{\partial \beta} \\ \frac{\partial y}{\partial \alpha} & \frac{\partial y}{\partial \beta} \end{bmatrix} \begin{bmatrix} d\alpha \\ d\beta \end{bmatrix} \qquad (10.72)$$

If Eqs. 10.11 and 10.12 are combined, we observe that

$$[R] = \begin{bmatrix} \frac{\partial x}{\partial \alpha} & \frac{\partial y}{\partial \alpha} \\ \frac{\partial x}{\partial \beta} & \frac{\partial y}{\partial \beta} \end{bmatrix} = \begin{bmatrix} \cos\left(\theta - \frac{\pi}{4}\right) & \sin\left(\theta - \frac{\pi}{4}\right) \\ -\sin\left(\theta - \frac{\pi}{4}\right) & \cos\left(\theta - \frac{\pi}{4}\right) \end{bmatrix} \qquad (10.73)$$

It is now possible to consider Eqs. 10.70a and 10.70b as a column vector and multiply it by the [R] matrix; the first component of the resulting vector will be

$$\begin{aligned} -\left(\frac{\partial P}{\partial x}\frac{\partial x}{\partial \alpha} + \frac{\partial P}{\partial y}\frac{\partial y}{\partial \alpha}\right) - 2\,k\left(-\sin 2\theta \cos\left(\theta - \frac{\pi}{4}\right) + \cos 2\theta \sin\left(\theta - \frac{\pi}{4}\right)\right)\frac{\partial \theta}{\partial x} \\ +2\,k\left(\cos 2\theta \cos\left(\theta - \frac{\pi}{4}\right) + \sin 2\theta \sin\left(\theta - \frac{\pi}{4}\right)\right)\frac{\partial \theta}{\partial y} = 0 \end{aligned} \qquad (10.74)$$

After some handling of trigonometric functions, Eq. 10.74 becomes

$$\frac{\partial P}{\partial \alpha} + 2\,k\cos\left(\theta - \frac{\pi}{4}\right)\frac{\partial \theta}{\partial x} + 2\,k\sin\left(\theta - \frac{\pi}{4}\right)\frac{\partial \theta}{\partial y} = 0 \tag{10.75}$$

Using Eq. 10.73 again allows us to further transform Eq. 10.75 into

$$\frac{\partial P}{\partial \alpha} + 2\,k\frac{\partial x}{\partial \alpha}\frac{\partial \theta}{\partial x} + 2\,k\frac{\partial y}{\partial \alpha}\frac{\partial \theta}{\partial y} = \frac{\partial P}{\partial \alpha} + 2\,k\frac{\partial \theta}{\partial \alpha} = 0 \tag{10.76}$$

The second component of the equilibrium equations is calculated in a similar way so that

$$\frac{\partial P}{\partial \beta} - 2\,k\frac{\partial \theta}{\partial \beta} = 0 \tag{10.77}$$

Eqs. 10.76 and 10.77 are generally written as

$$P + 2\,k\,\theta = \text{constant on an } (\alpha) \text{ line} \tag{10.78a}$$

$$P - 2\,k\,\theta = \text{constant on a } (\beta) \text{ line} \tag{10.78b}$$

and are called the **Hencky equations**.

The same approach can be followed for the components v_α and v_β of the velocity field in the local SLF reference system. The resulting **Geringer equations** are

$$dv_\alpha - v_\beta\,d\theta = 0 \quad \text{on an } (\alpha) \text{ line} \tag{10.79a}$$

$$dv_\beta + v_\alpha\,d\theta = 0 \quad \text{on a } (\beta) \text{ line} \tag{10.79b}$$

The method of practical resolution will be outlined only when P and θ are known on a curve Γ. As pictured in Figure 10.13, a small part AA´ of an (α) line is to be calculated together with a small part BA´ of a (β) line.

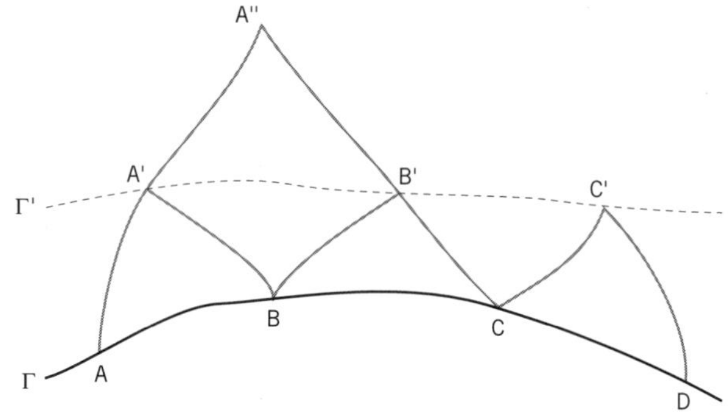

Figure 10.13 Construction of the SLF from a curve (Γ).

The Hencky equations on the (α) line AA´ and the (β) line BA´ give, respectively,

$$P_A + 2\,k\,\theta_A = P_{A'} + 2\,k\,\theta_{A'} \tag{10.80a}$$

$$P_B - 2\,k\,\theta_B = P_{A'} - 2\,k\,\theta_{A'} \tag{10.80b}$$

This is a linear system where the unknowns are $P_{A'}$ and $\theta_{A'}$, the solution of which allows the determination of the position of the geometrical point A´ if a hypothesis is made on the approximation of the slip lines: linear segments, arcs of circle, and so forth. In the same manner, from B and C, point B´ can be determined; then from C and D point C´ is calculated; and so on. A new discretized curve (Γ′) is obtained (A´B´C´...), which is used to compute the next series of geometrical points.

The analysis can be extended to axisymmetrical problems, provided suitable hypotheses are selected.

CHAPTER 10 - PROBLEMS

A. PROFICIENCY PROBLEMS

1. Calculate the ideal work to strain a unit volume of material under uniaxial tension from $(\bar{\varepsilon} = 0)$ to $(\bar{\varepsilon} = \varepsilon)$ for each of the following hardening laws:

   a. $\bar{\sigma} = K\,(\bar{\varepsilon} + \varepsilon_o)^n$ (Ludwik law)

   b. $\bar{\sigma} = \sigma_o + K\,(\bar{\varepsilon} + \varepsilon_o)^n$ (Swift law)

c. $\bar{\sigma} = \sigma_o(1 - A\,e^{-\beta\bar{\varepsilon}})$ (Voce law)

d. $\bar{\sigma} = K$ (ideal plastic)

e. $\bar{\sigma} = \sigma_o + K\bar{\varepsilon}$ (linear hardening)

2. Calculate the drawing stress and force using the ideal work method for a wire-drawing operation from 2 mm to 1 mm diameter of a Voce material where σ_o = 500 MPa, A = 0.5, β = 0.2 and where the efficiency factor is assumed to be 0.5.

3. Calculate the extrusion force for the same conditions as Problem 2.

4. Calculate the plane-strain-drawing stress and force from 2 mm to 1 mm thickness for the material and eefficiency defined in Problem 2. How does the drawing stress compare to the wire drawing stress computed in Problem 2?

5. Repeat Problem 4 for plane-strain extension.

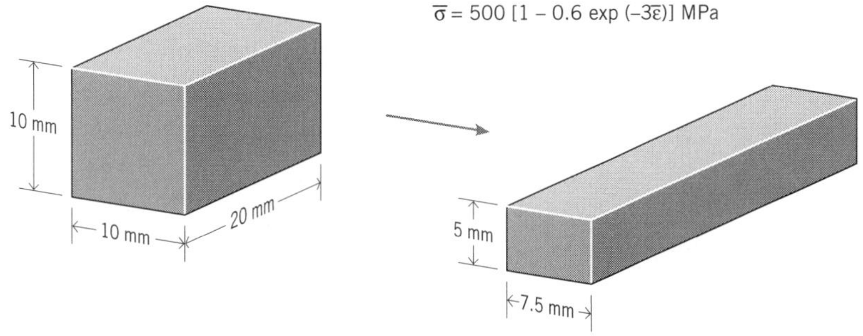

6. Calculate the total ideal work done for the operation diagrammed below, given the hardening law shown. Assume the material obeys von Mises yield.

7. A steel deforms at high temperature at a constant stress of 100 MPa. For a given forming operation, the strain path may be approximated by two proportional paths, the first from ($\varepsilon_1 = 0$, $\varepsilon_2 = 0$) to ($\varepsilon_1 = 0.5$, $\varepsilon_2 = 0.25$), and the second from ($\varepsilon_1 = 0.5$, $\varepsilon_2 = 0.25$) to ($\varepsilon_1 = 0.6$, $\varepsilon_2 = 0.5$).

   a. What is the ideal work per volume of material?

   b. What is the tensile strain equivalent to this forming deformation?

   c. Assume that a single proportional path was followed from the start to finish. How would the answers to a and b change?

8. Repeat Problem 7 for a different forming operation for which the initial, intermediate, and final geometric strains are as follows:

initial:	$(\varepsilon_1 = 0, \varepsilon_2 = 0)$
intermediate:	$(\varepsilon_1 = 0.5, \varepsilon_2 = 0.25)$
final:	$(\varepsilon_1 = 0, \varepsilon_2 = 0)$

9. a. Use L'Hôspital's rule to find the plane-strain-drawing stress for a frictionless case starting from Eq. 10.29. Note that as $\mu \to 0$, $\beta \to 0$. L'Hôspital's rule states that the limit of a composite function

$$\left[\lim_{x \to a} \frac{f(x)}{g(x)}\right] \text{ in cases where } \left[\lim_{x \to a} f(x) \to 0, \quad \lim_{x \to a} g(x) \to 0\right] \text{ or}$$

$$\left[\lim_{x \to a} f(x) \to \infty, \quad \lim_{x \to a} g(x) \to \infty\right] \text{ may be found by the ratio of the}$$

derivatives: $\displaystyle \lim_{x \to a} \frac{f(x)}{g(x)} = \frac{f'(x)}{g'(x)}$

if these limits exist.

b. Compare the frictionless result obtained in part a with the drawing stress obtained from the ideal work method.

10. Following the procedure of Exercise 10.3, derive an expression of the wire-drawing stress for the Coulomb-friction case. Show all of your steps clearly.

11. Following derivations in the text and in Problems 9 and 10, complete the following table assuming constant material hardness, $\bar{\sigma}$:

Calculation type	*Wire draw* σ_d	*Round extrusion* P_{ext}	*Sheet draw* σ_d	*Sheet extrusion* P_{ext}
Slab (general α, μ)				
Slab (any μ, small α)				
Slab ($\mu = 0$)				
Ideal Work				

12. For each forming operation in Problem 11, show how much the external stress or pressure changes by using the small-angle approximation.
    a. Use "typical values": r_{in}, t_{in} = 30 mm; r_{out}, t_{out}, t = 20 mm; $\alpha = 20°$, $\mu = 0.25$
    b. Use "extreme values": r_{in}, t_{in} = 40 mm; r_{out}, t_{out}, t = 20 mm; $\alpha = 45°$, $\mu = 0.5$

13. Perform a slab calculation for plane-strain compression of a Hollomon material $\bar{\sigma} = 600 \text{ MPa } \bar{\varepsilon}^{0.25}$ subject to Coulomb friction ($\mu = 0.3$). Assuming an initial geometry of $h_0 = 100$ mm and $b_o = 50$ mm, plot the friction hill at several values of h/h_0 (0.75, 0.5, 0.25) and find P_{max} and $P_{average}$ as a function of h/h_0.

14. Consider the forging operation shown here. If the die material can withstand a contact pressure of 1500 MPa and the workpiece exhibits a constant effective stress of 500 MPa, what is the minimum height to which the billet can be forged? Assume two cases: (a) $\mu = 0.25$, and (b) sticking friction?

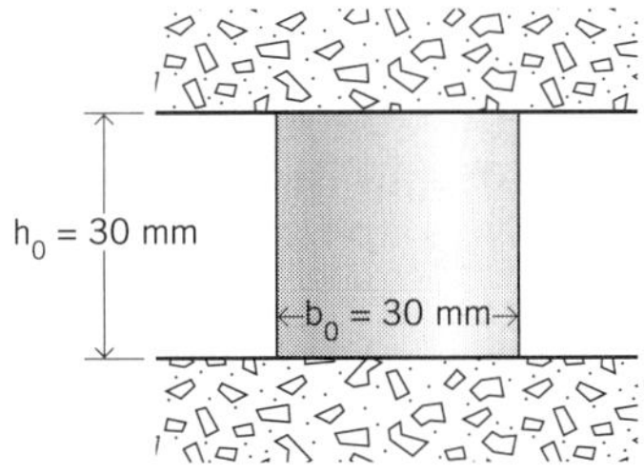

15. Consider the drawing of a sheet that reduces both of its dimensions in the cross sections as shown.

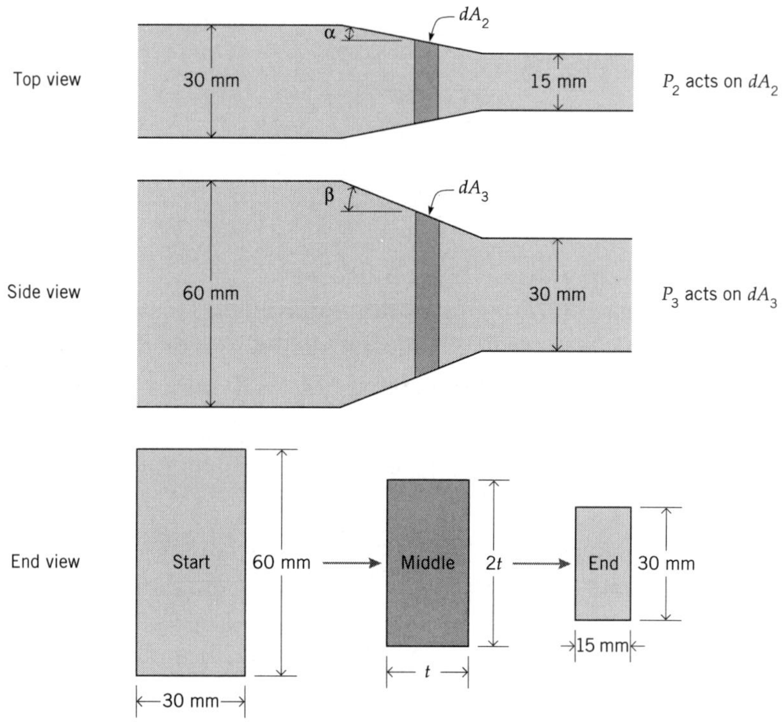

a. Set up and derive the differential equation that governs the drawing operation. Show all of your work and leave the resulting equation in the form of the following variables:

P_2, P_3, dA_2, dA_3, α, β, μ, t, σ_1

b. Considering the symmetry of the operation and, assuming a von Mises material, find the ratio of the stresses $\sigma_1/\sigma_2/\sigma_3$.

B. DEPTH PROBLEMS

16. Find the efficiency of a wire-drawing operation, as estimated by a slab calculation, assuming the following parameters:

$\bar{\sigma} = 1000$ MPa, $\alpha = 10°$, $r_{in} = 50$mm, $r_{out} = 40$mm, $\mu = 0.25$

17. For the conditions presented in Problem 16, what is the maximum $\frac{r_{in}}{r_{out}}$ that can be performed?

18. a. Repeat Exercise 10.2 for the most general assumption that the redundant work is a fixed fraction of the friction work:

$$\alpha = \frac{w_r}{w_f}, \quad 0 \le \alpha \le 1.$$

b. Show that your general expression reduces to the standard expression (Eq. 10.13) when $\alpha = 0$.

c. Show that your general expression reduces to the improved expression (Eq. 10.2-10) when $\alpha = 1$.

d. Compute $(d_i/d_0)^*$ for the cases in Table 10.2 using your general expression with $\alpha \to \infty$ (i.e., all nonideal work is redundant work). What are the differences compared with standard expression for these cases?

e. Use your result and your knowledge of scaling of W_f and W_r with respect to size to predict whether $(d_i/d_0)^*$ would go up or down as the size of the drawing operation increases. (Hint: How does α depend on size?)

19. a. Derive equations similar to Eqs. 10.57 for the case of sticking friction.

b. Assuming plane-strain conditions, eliminate P from your equations.

c. By making appropriate geometric substitutions, write your equations in terms of α and $d\alpha$ alone, and t and dt alone.

20. Real rolling operations exhibit several characteristics that are not easily seen in simple slab analysis. What do you think is the origin of the effects?

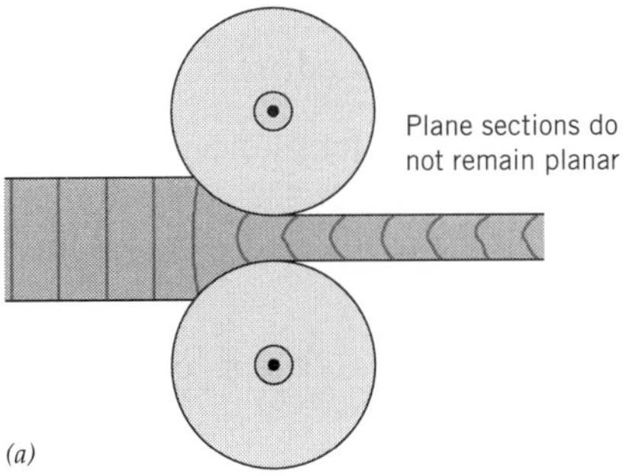

(a)

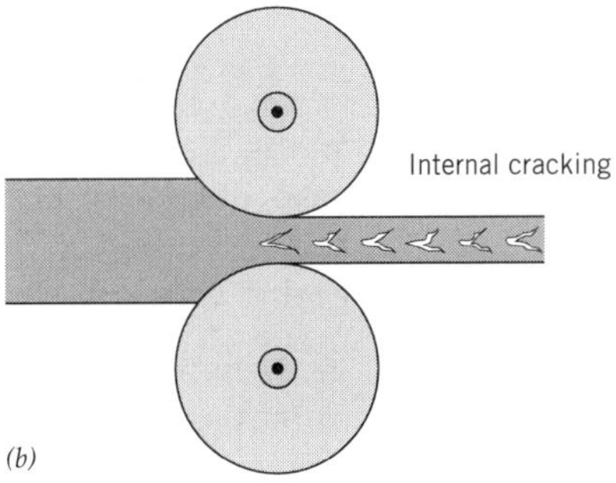

(b)

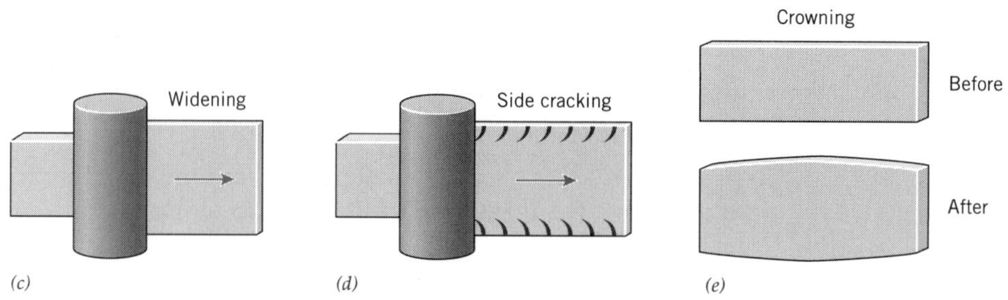

21. Consider a sheet-drawing operation (plane strain) that uses streamlined dies, as shown.

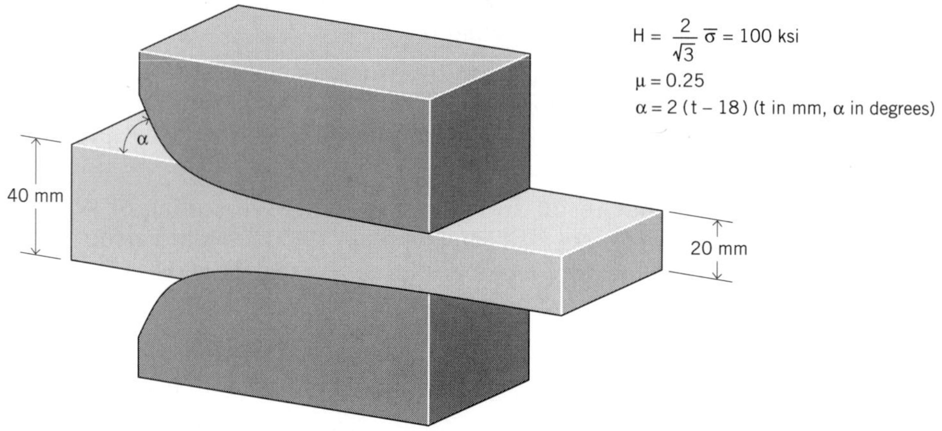

a. Use a simple five-step numerical procedure based on the small-angle formula to find σ_1 at t = 28 mm, 26 mm, 24 mm, 22 mm, and 20 mm.

b. What is the limiting $\frac{t_{in}}{t_{out}}$ ratio that can be attained with these dies, for t_i = 40mm, based on your numerical procedure?

22. Using a slab analysis similar to the one for the plane-strain-compression case, derive an expression for pressure as a function of radial position for the compression of a cylinder. Consider cases with (a) Coulomb friction and (b) sticking friction.

23. How would the sheet-drawing stresses depend on normal plastic anisotropy (r) based on Hill's theory?

C. NUMERICAL AND COMPUTATIONAL PROBLEMS

23. Write a program to carry out the numerical integration of general slab equations for drawing and extrusion as outlined in this chapter. Check your program for the special cases derived in closed form to make certain that the correct answers are obtained. How many steps are required in order to obtain a result within 0.1% for the cases presented in Problem 12? Use at least this many steps in subsequent simulations.

24. a. Using the straight-sided geometry of Problem 12 and the program you wrote in Problem 23, find the drawing (extrusion) stresses (pressures) for a power-law hardening material and compare these with the closed-form solutions using $\bar{\sigma}$ = constant. (Find the appropriate constant by using the flow stress of the hardening material strained to a strain intermediate between the original state, $\bar{\varepsilon} = 0$, and final state $\bar{\varepsilon}^f$.

$$\bar{\sigma} = 500 \text{ MPa}\, \bar{\varepsilon}^{0.25}$$

b. Compare your results for the typical values of Problem 12, but using a range of die half-angles: α = 10°, 20°, 30°, 40°, 50°, 60°, 70°, and 80°. What effect does α have on the result and on the accuracy of the constant $\bar{\sigma}$ assumption?

c. Repeat the simulation of part a for materials with several hardening laws. In each case, compare the results with the equivalent closed-form solution for constant $\bar{\sigma}$. Material laws are:

$\bar{\sigma} = 500\,(\bar{\varepsilon} + 0.05)^{0.25}$ (Ludwik)

$\bar{\sigma} = 100 + 500\,(\bar{\varepsilon} + 0.05)^{0.25}$ (Swift)

$\bar{\sigma} = 500\,(1 - 0.6\, e^{-3\bar{\varepsilon}})$ (Voce)

$\bar{\sigma} = 250 + 350\,\bar{\varepsilon}$ (linear)

How does the choice of hardening law affect the numerical result relative to the closed-form one?

d. For parts b and c, plot the die contact pressure as a function of x_1,and comment on which designs are likely to be best for die wear.

e. For parts a, b, and c, compute the efficiency of the process and comment on how it relates to material law and friction.

25. Perform a slab analysis for the die geometries shown. Use various combinations of friction and geometry as shown in the following diagram, and consider which kind of die shape is best in terms of drawing/extrusions stress and die pressure.

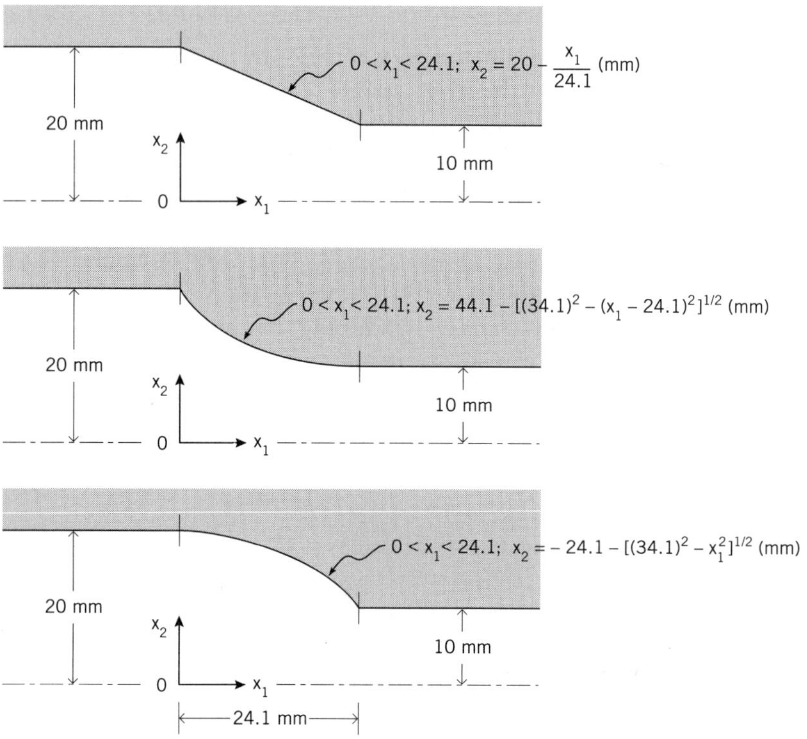

friction:	(a) m = 0.3; (b) sticking
material:	choices in Problem 24
geometry:	plane-strain or wire in die shapes as shown

26. For the various conditions presented in Problem 24, find the limiting draw ratio for each case by comparing $\bar{\sigma}$ and σ_1 as the material moves through various positions in the dies. (Modify your program, written in Problem 23, to make this comparison and return the limiting draw reduction.) How does this ratio depend on die angle?

27. For the various material laws presented in Problem 24, derive a closed form equation similar to Eq. 10.13b. How do these results compare with those obtained in Problem 26?

28. Modify your program, written in Problem 23, to use an implicit update scheme. That is, write $F(\sigma_1)$ in terms of the current stress rather than the stress obtained at the end of the last step. It will then be necessary to solve the equation at each time step iteratively, using a Newton-Raphson procedure.

See Chapter 1 for guidance. How many time steps are required to obtain the accuracy demanded of your explicit program? What is the relative computation time for the two approaches to obtain similar accuracy?

29. Write a computer program to integrate and solve the von Karman rolling equations, Eqs. 10.57. You may wish to start with the program written for Problem 23 and modify it accordingly.

30. Using the rolling program written in Problem 29,

    a. Plot how the position of the neutral point varies with μ (Coulomb coefficient) and n (work-hardening exponent).
    b. Plot the pressure profiles as a function of μ and n.

31. Use your rolling program written in Problem 29 to analyze the second die geometry (plane-strain only), now treated as a rolling operation. How do the die pressures and efficiencies compare for the two processes?

Index